Performance-Based Gear Metrology
Kinematic-Transmission-Error Computation and Diagnosis

Performance-Based Gear Metrology
Kinematic-Transmission-Error Computation and Diagnosis

William D. Mark, Ph.D.

Drivetrain Technology Center,
Applied Research Laboratory
and
Graduate Program in Acoustics
The Pennsylvania State University, USA

A John Wiley & Sons, Ltd., Publication

Registered office
John Wiley & Sons Ltd, The Atrium, Southern Gate, Chichester, West Sussex, PO19 8SQ, United Kingdom

For details of our global editorial offices, for customer services and for information about how to apply for permission to reuse the copyright material in this book please see our website at www.wiley.com.

Library of Congress Cataloging-in-Publication Data

Mark, William D.
   Performance-based gear metrology: kinematic-transmission-error computation
and diagnosis / William D. Mark.
        p. cm.
   Includes bibliographical references and index.
   ISBN 978-1-119-96169-7 (cloth)
  1. Motor vehicles–Transmission devices.   2. Motor vehicles–Vibration–Measurement.
3. Gearing–Vibration–Measurement.   4. Gearing–Vibration–Mathematical models.   I. Title.
   TL262.M37 2013
   621.8'330287–dc23

                                                                        2012023995

A catalogue record for this book is available from the British Library.

Print ISBN: 9781119961697

Typeset in 10/12.5 Palatino by Laserwords Private Limited, Chennai, India
Printed and bound in Malaysia by Vivar Printing Sdn Bhd

This book is dedicated to my wife Nancy.

Contents

Preface

A fictitious pair of meshing parallel-axis spur or helical gears with rigid equi-spaced perfect involute tooth-working-surfaces would transmit an *exactly* constant speed ratio. Real gears are not rigid, and their working surfaces contain inten-tional geometric modifications and manufacturing errors which together comprise the geometric working-surface deviations of the individual teeth. The combined elastic deformations and geometric working-surface deviations cause a time-varying perturbation superimposed on the otherwise transmission of an exactly constant speed ratio. At high operating speeds, gear-system inertial properties will affect this time-varying perturbation. The resultant time-varying perturbation superimposed on the transmission of a constant speed ratio is called the "transmission error".

For sufficiently slow operating speeds, inertial effects become negligible. If a gear-pair is operating at such a slow speed and transmitting a constant load or torque, the above-described time-varying perturbation is called the "static trans-mission error", which is the principal source of vibratory excitation arising from a meshing-gear-pair. The static transmission error has two components, one arising from loading-dependent elastic deformations and the other arising from geomet-ric deviations of the working surfaces from equispaced perfect involute surfaces. This latter component, which is inertia and loading independent, is the kinematic transmission-error contribution. It is caused by geometric deviations of the working surfaces of the individual teeth on a gear from equispaced perfect involute surfaces. The purpose of gear metrology equipment is to measure such deviations or their contributions to the transmission error.

The subject of gear metrology (Scoles and Kirk, 1969; Goch, 2003) generally is subdivided into two different measurement disciplines: elemental measurements such as tooth spacing, profile, lead (alignment), and so on, and functional mea-surements such as single-flank or double-flank measurements. The introduction of computer numerically controlled (CNC) dedicated gear measurement equipment now allows a gear-measurement machine to be programmed to measure a gear in great detail, with no operator interaction, permitting the operator to carry out other tasks while the gear is being measured. Although the machine measurements made on each tooth, for example, profile or lead, are elemental measurements, if they are made in sufficient detail they describe the working-surface deviations of all of the teeth on the measured gear. Hence, such a detailed measurement set contains all of the tooth-deviation information that is exercised in a functional test, such as a single-flank test.

It therefore is possible, in principle, to compute the working-surface-deviation contributions, provided by a single measured gear, to the contribution that gear would provide if mated with another gear and run in a single-flank test. The method to compute this contribution, obtained from CNC gear metrology measurements, to yield the contribution of the measured gear to the static transmission error obtainable from a single-flank test, is the subject of this book. This contribution is the *kinematic* (force and mass independent) contribution of the measured gear to its transmission error contribution (Merritt, 1971, p. 84). The analysis is carried out for parallel-axis helical and spur gears with nominally involute working surfaces.

Useful diagnostic methods are also included: specifically, methods to compute working-surface deviations that are the *cause* of any user-identified transmission-error rotational harmonic, for example, "sideband" harmonics and "ghost-tone" harmonics. This capability should enable a user to identify the manufacturing source and working-surface amplitude causing any such identified harmonic. It also should enable the reader and user to *understand* the causes of such harmonics.

Linear-system methods have been used to carry out the analysis and to organize the book. Such analysis methods generally involve an "input" to a linear system which operates on (modifies) the input, yielding an output – that is, system response. The relationship between input and output, when described in the "frequency domain" is a product, and the system characterization in the frequency domain is called a "transfer function". A brief description of this approach to general systems can be found in Lanczos (1956, pp. 248–259). A more comprehensive treatment is found in Gaskill (1978) Chapters 1–9.

In the present application, the "input" is the geometric deviations from equispaced perfect involute surfaces of the working surfaces of all teeth on the subject gear within the same rectangular contact region on the working surfaces. Involute gear geometry is described in Baxter (1962), Buckingham (1949) Chapters 4 and 8; Lynwander (1983) Chapter 2; and comprehensively in Colbourne (1987). Chapter 2 of this book summarizes properties of involute helical gears required in the remainder of the book.

A method to mathematically represent these "system input" working-surface geometric deviations is required and described in Chapter 3. For a number of reasons described in Chapter 1, normalized two-dimensional Legendre polynomials have been chosen to represent the working-surface deviations. Accessible treatments of Legendre polynomials can be found in Jackson (1941) and Bell (1968).

Because sufficiently accurate currently available CNC gear-measurement machines are capable of making only line-scanning radial (profile) measurements and/or line-scanning axial (lead, that is, alignment) measurements, a method to effectively "interpolate" between these line-scanning measurements is required. The method of "Gaussian quadrature" (Lanczos, 1956, pp. 396–400; Cheney, 1982, pp. 110–111; Hildebrand, 1974, pp. 387–392) is used to evaluate the Legendre polynomial expansion coefficients, which also interpolates the line-scanning measurements.

The basis for Gaussian quadrature is the Lagrange interpolation formula (Lanczos, 1961, pp. 5, 6).

Because a gear is circular, one normally would expect Fourier series (Lanczos, 1956, p. 254; Gaskill, 1978, p. 107) to be the appropriate frequency-domain representation of the above-described working-surface-deviation "system input" functions. But as shown heuristically at the beginning of Chapter 4, and *rigorously required* as explained in Chapter 1, because the working-surface deviations at any single fixed location on the N tooth-working-surfaces on a gear are a set of N equispaced discrete working-surface-deviation samples, the Fourier series representation of this discrete sequence of N samples reduces to the discrete Fourier transform (DFT) of this sequence of samples. Cooley, Lewis, and Welch (1969, 1972) call, what is generally called the DFT, the "Finite Fourier Transform", and treat it *not* as an approximation to the continuous Fourier transform, but rather, as the exact counterpart to Fourier series, but defined on a discrete set of N (equispaced) samples, which is periodic with period N. Their treatment is the exact mathematical tool required for description of the "system input" tooth-working-surface geometric deviations eventually leading to computation of the kinematic contribution to transmission-error vibratory excitations. The inverse DFT provides the necessary formulation for computation of the working-surface deviations that are the cause of any specific rotational-harmonic contribution to the kinematic transmission error. Beginning with Chapter 4, use of complex variables is required. An excellent summary of their properties can be found in Hildebrand (1976) Chapter 10.

The analysis required to describe the "system input" through Chapter 4 is relatively straightforward, but the "exact" analysis required to compute the kinematic transmission-error contributions is significantly more involved. Yet, conceptually, the computed effect on the working-surface deviations of the meshing action with a mating gear is very simple: it is an averaging action on the working-surface deviations. How this averaging action takes place is described in Chapter 5, but without the detailed analysis required to compute it, which is put off until Chapter 7. Each two-dimensional normalized Legendre-polynomial representation term of the working-surface deviations has its own system "transfer function", which describes the above-mentioned averaging action on the working-surface deviations for that particular two-dimensional Legendre term. Hence, we have called these individual "transfer functions" "mesh-attenuation functions".

Analytical approximations to these individual "mesh-attenuation functions" are provided in Chapter 6, which enable the reader to understand the physical sources of transmission-error low-order rotational harmonics, so-called "sideband" harmonics of the tooth-meshing harmonics, and "ghost tones".

The detailed derivation of the "mesh-attenuation functions" and reduction of all results to real (not complex) quantities is provided in Chapter 7, including formulas for representing kinematic transmission-error contributions in the "frequency" domain as a function of rotational harmonic number, and in the "time" domain as a function of gear rotational position. At one step in the derivations the Jacobian (Hildebrand, 1976, p. 353) of a coordinate transformation is required.

Because a major purpose of the book is to provide the means to measure a gear and compute the kinematic-transmission-error contributions from such measurements, the measurement requirements, and formulas for computing the kinematic transmission error in "time" and "frequency" domains, are summarized in Chapter 8, along with useful metrics of transmission-error contributions.

General Background References

Gear Metrology

Goch, G. (2003) Gear metrology. *CIRP Annals–Manufacturing Technology*, **52** (2), 659–695.
Scoles, C.A. and Kirk, R. (1969) *Gear Metrology*, Macdonald & Co., London.

The book by Scoles and Kirk is useful for general gear metrology terminology and practice; the article by Goch reviews the state of the art as of 2003. The American Gear Manufacturers Association (AGMA) has a number of publications pertaining to gear metrology.

Involute Gears

Baxter, M.L. Jr., (1962) Basic theory of gear-tooth action and generation, in *Gear Handbook*, Chapter 1, 1st edn (ed. D.W. Dudley), McGraw-Hill, New York, pp. **1-1**, 1-21.
Buckingham, E. (1949) *Analytical Mechanics of Gears*, McGraw-Hill, New York. Republished by Dover, New York.
Colbourne, J.R. (1987) *The Geometry of Involute Gears*, Springer-Verlag, New York.
Lynwander, P. (1983) *Gear Drive Systems*, Marcel Dekker, New York.
Merritt, H.E. (1971) *Gear Engineering*, John Wiley & Sons, Inc., New York.

Baxter, Buckingham, and Lynwander describe properties of involute gear teeth. Colbourne's treatment is comprehensive. Merritt discusses "Kinematic Error". In Chapter 2, I have attempted to describe all involute properties required in the remainder of the book.

Linear Systems

Gaskill, J.D. (1978) *Linear Systems, Fourier Transforms, and Optics*, John Wiley & Sons, Inc., New York.
Lanczos, C. (1956) *Applied Analysis*, Prentice-Hall, Englewood Cliffs, NJ. Republished by Dover, New York.

Lanczos provides brief treatments of complex Fourier series, Fourier integral transform, and linear-system input-output relations. Gaskill provides a more comprehensive treatment of these topics, including the convolution operation and two-dimensional systems.

Legendre Polynomials

Bell, W.W. (1968) *Special Functions for Scientists and Engineers*, D. Van Nostrand, London. Republished by Dover, Mineola, NY.
Jackson, D. (1941) *Fourier Series and Orthogonal Polynomials*, Mathematical Association of America, Buffalo, NY. Republished by Dover, Mineola, NY.

Jackson and Bell both discuss expansions of arbitrary functions in series of Legendre polynomials. Jackson proves convergence and the important unweighted least-squares property of expansions in Legendre polynomials.

Gaussian Quadrature

Cheney, E.W. (1982) *Introduction to Approximation Theory*, 2nd edn, Chelsea, New York.

Hildebrand, F.B. (1974) *Introduction to Numerical Analysis*, 2nd edn, McGraw-Hill, New York. Republished by Dover, New York.

Lanczos, C. (1956) *Applied Analysis*, Prentice-Hall, Englewood Cliffs, N.J. Republished by Dover, New York.

Lanczos, C. (1961) *Linear Differential Operators*, D. Van Nostrand, London.

Lanczos (1956) provides an exceptionally clear treatment and explanation of the remarkable properties of Gaussian quadrature. The basis for this treatment is the Lagrangian interpolation formula which he derives with exceptional clarity in Lanczos (1961). Cheney provides a clear statement of the accuracy capability of Gaussian quadrature and provides a convergence proof. Hildebrand (1974) provides convenient formulas for the weights required in evaluation of Gaussian quadrature integrations.

Discrete Fourier Transform

Cooley, J.W., Lewis, P.A.W., and Welch, P.D. (1969) The finite fourier transform. *IEEE Transactions on Audio and Electroacoustics*, **AU-17**, 77–85. Reprinted in Rabiner, L.R. and Rader, C.M. (eds) (1972) *Digital Signal Processing*, IEEE Press, New York, pp. 251–259.

This paper develops Fourier analysis for a periodic function defined on N (equispaced) integers $j = 0,1,2,\ldots, N-1$. It is the *exact* Fourier theory required for representation of deviations of the working surfaces of gear teeth from equispaced perfect involute surfaces. It is absolutely fundamental to the developments contained in this book.

Complex Variables

Hildebrand, F.B. (1976) *Advanced Calculus for Applications*, 2nd edn, Prentice-Hall, Englewood Cliffs, NJ.

Chapter 10 of Hildebrand provides a convenient summary of properties of complex variables. Section 7.4 on pp. 352–353 also includes a treatment of Jacobians, which are required for a coordinate transformation carried out in Chapter 7.

Further Reading

Drago, R.J. (1988) *Fundamentals of Gear Design*, Butterworths, Boston.

Dudley, D.W. (1984) *Handbook of Practical Gear Design*, McGraw-Hill, New York.

Smith, J.D. (2003) *Gear Noise and Vibration*, 2nd edn, Marcel Dekker, New York.

Townsend, D.P. (ed.) (1991) *Dudley's Gear Handbook*, McGraw-Hill, New York.

Acknowledgments

I first wish to acknowledge the superb software skills of Dr. Cameron P. Reagor <cameron@intensityconsulting.com>, who wrote all of the software used in the transmission-error and working-surface computations displayed in this book, and who carried out these computations.

Preparation of this book has been partially supported by The AGMA Foundation through the Gear Research Institute located at Penn State University's Applied Research Laboratory. This support is most gratefully acknowledged. Refinements to analyses carried out earlier and preparation of software utilized in the transmission-error and diagnostic computations displayed herein were supported by the Department of Commerce National Institute of Standards and Technology (NIST) Advanced Technology Program Motor Vehicle Manufacturing Competition through M&M Precision Systems Corporation. I am indebted to Al Lemanski for helping to facilitate this support. Additional refinements, documentation, and user-friendly software preparation were supported by the United States Council for Automotive Research (USCAR). All of this support is gratefully acknowledged.

I also wish to acknowledge the collaborative work carried out with General Motors. The acoustical measurements and spectrum analysis of these measurements, yielding the continuous spectrum measurements shown in Figure 1.5, were carried out by GM, thereby enabling the correlation shown there between the kinematic transmission-error computation and the acoustic spectrum. GM also carried out on their Gleason M&M metrology equipment the repeated measurements and kinematic transmission-error computations described in Section 6.6, using the same gear as in our computations of Figures 6.7 and 6.8, and yielding virtually identical "ghost-tone" transmission-error amplitude, thereby verifying that such very small working-surface deviations causing ghost tones can be measured and their ghost-tone transmission-error amplitudes computed yielding virtually identical results to the results we had obtained using different Gleason M&M metrology equipment. The interest and efforts in this work of Neil Anderson (formally with GM) and Arvo Siismets of GM are gratefully acknowledged.

The above-described remarkable consistency in computation of ghost-tone transmission-error amplitudes, resulting from gear measurements made on different Gleason M&M measurement machines, is a tribute to the very high quality of these machines. I also wish to acknowledge the many helpful discussions with Mark Cowan of Gleason M&M pertaining to the repeatability and accuracy of their machines.

Modeling by my former students Matthew Alulis, Edward Jankowich, and William Welker showed that it is possible to accurately compute the tooth-stiffness characterization used in the analytical developments of Chapter 7 (but *not* required for computation of the kinematic transmission error of a measured gear).

Earlier preliminary analytical developments and software implementations were carried out by the author at the firm of Bolt, Beranek, and Newman, Inc. The confidence in these developments shown by Sig Leimonas and George Nagorny enabled this effort to move forward. I also wish to gratefully acknowledge early software implementations carried out at BBN by Raya Stern, Bob Fabrizio, Ray Fischer, and especially Jeanne Hladky. Early discussions with Dr. Fred Kern also were helpful. I most sincerely wish to express my appreciation to Professor Stephen H. Crandall for making all of this possible.

The office and administrative assistance provided by The Pennsylvania State University Applied Research Laboratory Drivetrain Technology Center, and the encouragement by Dr. Suren Rao, have made the task of manuscript preparation much easier than it would have been without these accommodations. I am especially indebted to Linda L. Jones for word processing of the manuscript and to Lawrence C. Miller for final preparation of the figures.

1

Introduction

As described in the Preface, the subject of this book is how to efficiently measure a gear and compute from these measurements the kinematic (force and mass independent) contribution of the measured gear to its transmission error, how to compute the working-surface-deviations that are the cause of any user-identified transmission-error rotational harmonic, and how to understand the relationship between such working-surface-deviations and the resultant rotational-harmonic contributions caused by these deviations.

Using computer numerically controlled (CNC) dedicated gear metrology equipment, measurements on a helical gear in sufficient detail to accurately carry out the above-described computations can take from a few to several hours. Hence, this methodology is not generally suitable for continuous production checking, but it is suitable, and is being used, for intermittent checking. Because the manufacturing errors generated by each individual manufacturing machine generally are consistent from one manufactured gear to another in a manufacturing run, it is sensible to regard the methods described herein as suitable for assessing the quality of a particular manufacturing machine, or process. Furthermore, because transmission-error contributions from working-surface errors are caused by the *collective working-surface error pattern of all teeth* on a gear, the methods described herein also may be suitable for establishing performance-based gear-accuracy standards.

The analytical relationships derived herein between tooth-working-surface-deviations and resulting transmission-error frequency spectra allow the reader to understand the causes of certain transmission-error tones, such as "sideband tones" and "ghost tones." Moreover, because tooth damage, such as surface damage and bending-fatigue damage, cause transmission-error contributions in the same manner as manufacturing deviations, the analysis contained herein is applicable to gear-health monitoring considerations.

Because the means by which a gear must be measured to enable computation of its kinematic transmission-error contributions, and the method of kinematic transmission-error computation, both involve considerable detail, a general orientation to the content of the book is described in the remainder of this Introduction. It is hoped this orientation might give the reader a broad perspective before he or she begins working through the details of the analysis.

1.1 Transmission Error

The transmission error of a meshing-gear-pair describes the deviation from the transmission of an *exactly constant* speed ratio. It can be displayed in the time domain as a function of the rotational position of one of the two meshing gears, or in the frequency domain. The transmission error is the principal source of vibration excitation caused by meshing-gear-pairs, for example, Mark (1992b), Smith (2003), and Houser (2007). The subject of this book is transmission-error contributions caused by geometric deviations from equispaced perfect involute surfaces of the tooth-working-surfaces of parallel-axis helical or spur gears. Because an idealized pair of parallel-axis helical gears, each with equispaced rigid perfect involute teeth would transmit an *exactly constant* speed ratio, the transmission-error contribution from each gear of a meshing-pair of nominally involute gears is the instantaneous deviation of the position of that gear from the position of its rigid perfect involute counterpart. Therefore, as described earlier in Chapter 3 by Equation (3.2), and in more detail in Chapters 5 and 7, the transmission-error contributions from each of two meshing gears are additive to yield the transmission error of the gear-pair. Consequently, especially with regard to the geometric deviations of the working-surfaces from equispaced perfect involute surfaces, it is rigorously meaningful to define and compute the transmission-error contribution arising from the geometric deviations of the working-surfaces of a single gear.

Frequency Spectrum

Vibration analyses and measurements are very often carried out in the frequency domain. It generally is impossible to directly measure the source of vibration; normally what is measured is the structural response, or in the case of noise, the acoustic response. Because the structural path between source of vibration and response measurement location is usually well modeled as a linear time-invariant system, source vibration *tones* retain their identity between source and receiver, although their amplitudes and phases are affected by the transmission media. But such media normally cause substantial changes in temporal vibration signals.

Because gears are circular and normally have equispaced teeth, dealing with their vibration signatures in the frequency domain is especially useful. Thus, dealing only with a single gear of a meshing-pair, as described above, because the gear is circular, its fundamental rotational frequency has a period of one gear rotation, and all of its harmonics are integer multiples of the fundamental frequency associated with one gear rotation.

Figure 1.1 is a sketch of the stronger harmonics that typically are observed from a single gear. The harmonic labeled 1 near the origin is the fundamental frequency associated with the rotational period of the gear. All harmonics shown are integer multiples of this fundamental rotational harmonic. If the gear has N teeth, then the

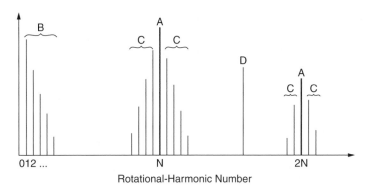

Figure 1.1 Sketch of dominant rotational harmonics caused by a single gear of a meshing-pair operating at constant speed and transmitting constant loading. Abscissa labels rotational-harmonic numbers n = 1, 2, ... The period of rotational harmonic n = 1 is the rotation period of the gear. All harmonics are integer multiples of n = 1. Rotational harmonic n = N is the tooth-meshing fundamental harmonic with period equal to the gear rotation period divided by the number of teeth, N. Low-order rotational harmonics B and ''sideband'' harmonics C typically are strong. ''Ghost tone,'' when present, is labeled D (Adapted from Mark (1991))

Nth rotational harmonic is the *tooth-meshing fundamental* harmonic. The period of the tooth-meshing fundamental is the gear rotation period divided by the number N of teeth. The first heavy harmonic, rotational harmonic N in Figure 1.1, is the tooth-meshing fundamental. Also shown heavy is another harmonic at rotational harmonic $2N$, which is twice the frequency of the tooth-meshing fundamental harmonic. These two harmonics are labeled A. In the neighborhoods of these tooth-meshing-harmonics are so-called "sideband rotational harmonics" labeled C. Also shown are low-order rotational harmonics 1, 2, ... labeled B, and a "ghost-tone" rotational harmonic labeled D. The physical sources of all such harmonics are explained in this book. Although not shown, unless a gear is geometrically perfect, there normally will be weak contributions to all integer multiples of the rotational fundamental harmonic.

When a frequency analysis of the structural response to the vibration excitation caused by a pair of meshing gears operating at constant speed and loading is carried out, a superposition of two such spectra as that shown in Figure 1.1 is obtained. Because of the manner in which the teeth of the two gears mesh, the locations of the tooth-meshing-harmonics A from each of the two meshing gears will coincide, but generally, the locations of all other rotational harmonics will differ, because the rotation periods of the two individual gears generally will differ (unless they both have the same number of teeth).

Physical Sources of Harmonic Contributions

It is important to understand the physical sources of the various harmonics shown in Figure 1.1. If the subject gear is geometrically *perfect*, such that every tooth is geometrically identical with no spacing errors and modified (e.g., with tip and end relief) exactly the same, then the only harmonics that would be present in Figure 1.1 would be the tooth-meshing-harmonics labeled A (assuming exactly constant rotational speed and loading). These tooth-meshing-harmonics A are caused by deviations of the elastically deformed teeth from perfect involute surfaces. Such deviations are the combined superposition of intentional (and unintentional) geometric modifications of the teeth and tooth/gearbody elastic deformations. (This fact is easily understood from the observation that if all elastically deformed tooth-working-surfaces are identical with no spacing errors, the only lack of smooth transmission is that associated with a period equal to the tooth-meshing period.) Hence, the generation of all rotational harmonics B, C, and D is caused by geometric variations of the individual tooth-working-surfaces from the mean (average) modification (intentional or otherwise) of the working-surfaces, assuming constant speed and constant loading. These deviations causing rotational harmonics B, C, and D thus are tooth-to-tooth geometric variations of the working-surfaces, including tooth-spacing errors, from the mean working-surface. Computation of the transmission-error amplitudes of these rotational harmonics, and diagnosing their working-surface sources, is the principal subject of this book. The gear measurements required to successfully accomplish this, in any specific application, also yield a very accurate three-dimensional determination of the working-surface modification, averaged over all teeth, enabling this achieved modification to be compared with that specified by the design engineer.

1.2 Mathematical Model

The mathematical analysis contained herein, leading to a systematic method for measuring a helical (or spur) gear, and computing from those measurements the locations and amplitudes of any transmission-error rotational harmonics, such as B, C, and D of Figure 1.1, has been possible because of the elegant relative simplicity of the meshing action of involute helical gears, described in more detail in Chapter 2. How this analysis has been possible can be partially understood with the aid of Figure 1.2, reproduced again as Figure 2.6.

Contact between the teeth of perfect involute helical gears takes place in a plane surface, the plane of contact shown in Figure 1.2, also called the plane of action. Real teeth have intentional modifications and manufacturing errors, however small, and they elastically deform. The (lineal) transmission error contribution from either gear in Figure 1.2 is, simply, the error in the instantaneous position of that gear, "measured" in the direction of the plane of contact of the lower figure, relative to the instantaneous position of its rigid perfect involute counterpart. The (lineal) transmission error of the gear-pair then is the superposition (algebraic sum) of the contributions from each of the two gears. Recall that rigid perfect involute gears transmit an *exactly* constant speed ratio.

It can be seen from the lower part of Figure 1.2 that, in any plane cut normal to the gear axes, tooth-pair contact takes place at a single point on each tooth. That point is in the plane of contact. When projected axially on tooth-working-surfaces,

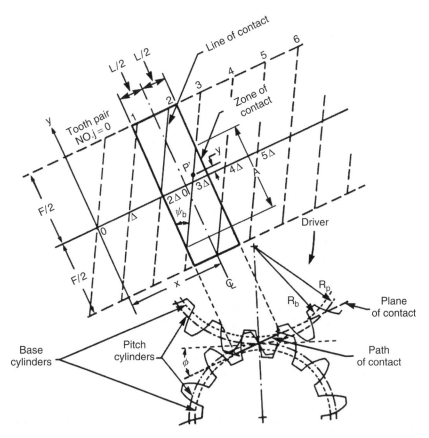

Figure 1.2 Lines of tooth-pair contact and zone of contact in plane of contact of a meshing-pair of geometrically perfect involute helical gears (Adapted from Mark (1978))

these points become lines of contact as illustrated in the upper part of Figure 1.2, which is the plane of contact of the lower part of the figure. As the gears rotate, these lines of contact move across the tooth-working-surfaces, and through the zone of contact shown in the upper part of the figure. (In reality, these "lines" of contact have finite width due to the local "Hertizian deformation" of the tooth-working-surfaces.)

It can be seen from the upper portion of Figure 1.2 that the transmission-error amount by which one gear approaches its mating gear, in the plane of contact, is a function of the deviations of the tooth-working-surfaces of *all* teeth in contact, on the lines of contact, at that particular instant of gear-pair rotation. As the gear-pair rotates, each line of contact sweeps across a tooth-working-surface. Thus, the mathematical problem is to describe, in a useful way, the simultaneous deviations of all teeth in contact, along the above-described lines of contact on the individual teeth, as a function of the rotational positions of the gears. The relative simplicity of involute helical gear geometry has made this possible. "Linear-system analysis" methods have been utilized. The details are carried out in Chapter 7. But the book has been organized to show how gears must be measured, and transmission-error computations carried out and understood, without perhaps full comprehension of every detail of Chapter 7.

Role of Discrete Fourier Transform (DFT)

In carrying out the original analysis (Mark, 1978) that is the foundation analysis for this book, the expression for the transmission-error Fourier series coefficients, Equations (112) and (111) of that reference, could be made useful (and understandable) only by mathematically representing the tooth-working-surface-deviations as a linear superposition of "elementary errors," that is, by Equation (131) of the above-cited reference. This representation led to Equations (134) and (135) of the above-cited reference for the Fourier series coefficients of the transmission error contributions, Equation (134), expressed as a function of the discrete Fourier transform (DFT), Equation (135), of the expansion coefficients of the above-described superposition of elementary errors. (The counterparts to these equations in Chapter 7 are, respectively, Equations (7.49), (7.48), (7.50), (7.60), and (7.59).)

   This original analysis, and its counterpart in Chapter 7, illustrates that the DFT (Cooley, Lewis, and Welch, 1969, 1972) is the *exact* mathematical tool required to compute, understand, and diagnose transmission-error rotational harmonic contributions, such as those illustrated by the B, C, and D contributions in Figure 1.1. As one might guess, use of the DFT arises because the working-surface geometric deviations at any fixed location on each of the tooth-working-surfaces of a gear with N teeth constitute a discrete equispaced sequence of N deviations, which because a gear is circular, is periodic with period N. The DFT is the exact mathematical tooth *required* to describe the frequency content of such phenomena. Cooley, Lewis, and Welch (1969, 1972) refer to the DFT as the "Finite Fourier Transform." Our use of its definition and properties in Chapter 4, and beyond, is consistent with that of Cooley, Lewis, and Welch.

1.3 Measurable Mathematical Representation of Working-Surface-Deviations

The above-described requirement to mathematically represent the tooth-working-surface-deviations as a linear superposition of "elementary errors" has suggested this representation method as a sensible starting point in the overall developments to be carried out. Requirements for a satisfactory representation method of tooth-working-surface-deviations are: it be capable of representing *any* deviations, that is, mathematically "complete," appropriately normalized so that expansion coefficients can be interpreted, measureable by dedicated CNC gear metrology equipment, "efficient" in some sense, and amendable to Fourier integral transformation of simple form. This last requirement is a consequence of Equation (125) of Mark (1978), which is Equation (7.62) of Chapter 7.

Among known methods of representation, two-dimensional normalized Legendre polynomials meet all of the above-mentioned requirements. If enough terms are used they can accurately represent any deviation surface, that is, they are complete, for example, Jackson (1941, pp. 63–68) and Bell (1968, p. 57). When appropriately normalized, Equations (3.13) and (3.14), their expansion coefficients can be directly interpreted, as in Equation (3.23). As a consequence of their important (unweighted) least-squares property (Jackson, 1941, pp. 215, 216) they are generally efficient. Moreover, the lowest-order two-dimensional Legendre term is a constant, representing exactly a tooth-spacing error, and the next-order linear terms represent straight-line errors commonly observed in lead (alignment) measurements and profile measurements. The Fourier integral transform of a generic Legendre polynomial is a spherical Bessel function of the first kind (Bateman, 1954, p. 122; Antosiewicz, 1964, p. 437). This very important property allows the final form of the analytical results to be represented as simply as is possible considering the complexity of the physical problem being analyzed.

Measurement Compatibility

Present-day dedicated CNC gear metrology equipment can carry out line-scanning (lead) measurements in an axial direction and line-scanning (profile) measurements in a radial direction on tooth-working-surfaces. Consequently, to obtain a representation of the working-surface-deviations over the entire rectangular working-surfaces, some sort of interpolation procedure is required to provide an (approximate) determination of the working-surface-deviations between the line-scanning measurements.

CNC gear measurement machines can be programmed to carry out such scanning measurements at any location, that is, locations of the lead-scanning measurements at any radial locations, and locations of the profile-scanning measurements at any axial locations. Consider, for example, obtaining an approximation to the working-surface-deviations utilizing only working-surface measurements provided by scanning lead measurements. Suppose scanning lead measurements are made at n different radial locations on a tooth. Then, by using the Lagrange interpolation formula (Lanczos, 1961, pp. 5, 6; Lanczos, 1956, pp. 397, 398) it is known that at any axial location, a polynomial of degree $n-1$ can be constructed to interpolate radially across the n scanning lead measurements to provide an approximation to the profile deviations at that axial location, which will agree exactly with the n scanning lead measurements at that axial location. Then, by providing such an interpolation at each axial location, an approximation to the working-surface could be obtained everywhere, which would agree at all points on the n scanning lead measurements. But, it is possible to do much better.

To generate the Legendre expansion coefficients of the working-surface-deviations, integrations over the working-surfaces, Equation (3.20), are required. Continuing to consider lead-scanning measurements, these integrations over the working-surfaces are to be treated as iterated integrals, first along the line-scanning measurements, as in Equation (3.30), then across the line-scanning measurements, as in Equation (3.31). The accuracy of the integrals, Equation (3.30), *along* the line-scanning measurements, is limited only by the density of sample points along the scanning lines (and the accuracy of the measurements). The second integrals, Equation (3.31), *across* the line-scanning measurements, are integrations involving the expansion coefficients obtained *along* the line-scanning measurements. We generally can expect these expansion coefficients to vary more smoothly in the direction *across* the line-scanning measurements than the raw measurements, especially in the case of the lower-order expansion coefficients.

As suggested above, these second integrations, Equation (3.31), could be carried out by utilizing the Lagrange interpolation formula to interpolate the integrands, yielding a polynomial representation of each integrand of degree $n - 1$, assuming there have been n scanning lead measurements. But a CNC gear-measurement machine can be programmed to locate the n scanning lead measurements at any radial locations. If these radial locations are chosen to be at the n zero locations of an appropriately normalized Legendre polynomial of degree n, then the accuracy achievable in the integrations is comparable to what would normally be achieved by $2n$ scanning lead measurements, for *any* radial locations of these *additional n* scanning lead measurements (Lanczos, 1956, pp. 396–400). The resulting integration procedure is called Gaussian quadrature. The above reference by Lanczos provides a very clear proof and explanation of this remarkable result. A comparable mathematically exact statement can be found in Cheney (1982, p. 110). The radial coordinate used in our analysis is "roll distance," Equation (3.3).

The case where profile scanning measurements are used instead of lead scanning measurements is completely analogous to that described above; in this case, say m scanning profile measurements would be located axially at the zeros of a normalized Legendre polynomial of degree m.

As mentioned in Hildebrand (1974, p. 467) and shown explicitly in Mark (1983), use of Gaussian quadrature to evaluate the Legendre expansion coefficients yields Legendre polynomial expansions that agree, exactly, with the data values at the n Legendre polynomial zeros used to evaluate the expansion coefficients. That is, the resultant Legendre polynomial expansions interpolate, exactly, the data values. However, it is shown below that the Legendre polynomial expansion interpretation is the preferred interpretation.

It is known that as $n \rightarrow \infty$, the errors in Gaussian quadrature converge to zero for any continuous function (Cheney, 1982, p. 111). This behavior is in contrast to equispaced polynomial interpolation which can exhibit non-convergent oscillatory behavior (Lanczos, 1961, pp. 12, 13; Lanczos, 1956, p. 348). Consequently, all results contained in this book assume that line scanning lead and/or profile measurements are made at the zeros of appropriately normalized Legendre polynomials.

An essential requirement of a tooth-working-surface representation and measurement method is that it be capable of accurately representing sinusoidal working-surface-deviations responsible (as shown herein) for causing "ghost tones," which are strong rotational-harmonic transmission-error tones, as illustrated by D in Figure 1.1. The lower curve in Figure 1.3 (reproduced again as Figure 3.B.2), shows a sinusoid that has been sampled by 32 samples, indicated by the small circles. The abscissa

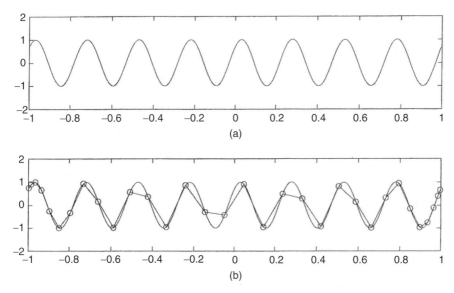

Figure 1.3 Upper curve (a) is Legendre polynomial reconstruction of the sinusoid, where the Legendre polynomial expansion coefficients were evaluated by Gaussian quadrature using only the 32 sample values of the sinusoid from the lower curve (b). Lower curve (b) shows 32 discrete samples of a sinusoid with 8 full cycles. Abscissa sample locations are the zero locations of a Legendre polynomial of degree 32 (From Mark and Reagor (2001), reproduced by permission of the American Gear Manufacturers Association)

locations of these samples are the zero locations of a Legendre polynomial of degree 32. The upper curve in Figure 1.3 is a Legendre polynomial reconstruction, where the Legendre polynomial expansion coefficients were evaluated by Gaussian quadrature using *only* the 32 sample values of the sinusoid of the lower curve. The upper curve is a virtually identical reconstruction of the lower curve, which illustrates the accuracy obtainable by utilizing Gaussian quadrature to evaluate the expansion coefficients of the Legendre polynomial reconstruction of the upper curve.

The reconstruction of the upper curve in Figure 1.3 also illustrates why the least-squares property of Legendre polynomial representations is the interpretation preferred over the above-mentioned interpolation property. Although the upper curve clearly agrees with every one of the 32 sample points of the lower curve, if given only those 32 sample points, one would be hard pressed to provide a smooth curve through them. Yet, Gaussian quadrature combined with the least-squares property of Legendre polynomial expansions provides the remarkable reconstruction shown by the upper curve. Rules are provided in Appendix 3.B for choosing the minimum number of line-scanning measurements required to accurately represent working-surface-deviations causing any ghost tone of user-specified rotational-harmonic number.

1.4 Final Form of Kinematic-Transmission-Error Predictions

The fact that meshing-gear-pairs generate tones suggests immediately the use of frequency analysis methods. Moreover, because a gear is circular, implying a fundamental rotational frequency with period equal to the gear rotational period, the frequency analysis method to be used is Fourier series. However, because the tooth-working-surface-deviations on all N teeth on a gear at any fixed location y,z on the tooth-working-surfaces, Figure 1.4, constitute an equispaced sequence of N discrete values, which are periodic with period N, the fundamental mathematical tool required to describe these deviations in the frequency domain is the discrete Fourier transform (DFT).

In Chapter 4, this DFT representation is shown by Equation (4.22) to be

$$\hat{\eta}_C\left(n;y,z\right) = \sum_{k=0}^{\infty}\sum_{\ell=0}^{\infty} B_{k\ell}\left(n\right)\psi_{yk}\left(y\right)\psi_{z\ell}\left(z\right)\,, \quad n=0,\pm1,\pm2,\cdots \tag{1.1}$$

where $B_{k\ell}(n)$ is the DFT, Equation (4.21), of the expansion coefficients of the two-dimensional normalized Legendre polynomials $\psi_{yk}\left(y\right)\psi_{z\ell}(z)$ used to represent the working-surface-deviations. The above equation expresses the collective deviations (e.g., errors) of all N teeth, at each working-surface location y,z, in the frequency domain, where n is rotational-harmonic number. (The terms "deviation" and "error" are used here and everywhere in the book to describe deviations of tooth-working-surfaces from equispaced perfect involute surfaces. These deviations can include intentional modifications.) The mean-square error spectrum then is shown by Equation (4.30) to be

$$G_{\eta}(n) = 2\sum_{k=0}^{\infty}\sum_{\ell=0}^{\infty}\left|B_{k\ell}(n)\right|^2\,, \quad n=1,2,3,\cdots \tag{1.2}$$

which describes, in the frequency domain as a function of rotational-harmonic number n, the collective mean-square contributions of the tooth-working-surface-deviations at all locations y,z, but not including any of the attenuating effects that would be attributable to the meshing action with a mating gear. That is, $G_{\eta}(n)$

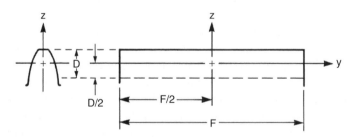

Figure 1.4 Tooth-working-surface coordinate system. The location of a generic point on the tooth-working-surfaces is described by Cartesian coordinates y, z

describes in the frequency domain only properties arising from working-surface-deviations from equispaced perfect involute surfaces.

The (complex) Fourier series coefficients of the kinematic transmission-error contributions arising from the working-surface-deviations of a single gear are given by Equation (5.16) as

$$\alpha_n = \sum_{k=0}^{\infty}\sum_{\ell=0}^{\infty} B_{k\ell}(n)\,\hat{\phi}_{k\ell}\left(\frac{n}{N}\right), \quad n = 0, \pm 1, \pm 2, \cdots. \tag{1.3}$$

Comparison of Equations (1.1) and (1.3) permits a simple interpretation of the "mesh-attenuation functions" $\hat{\phi}_{k\ell}(n/N)$. Each pair of normalized Legendre polynomial terms $\psi_{yk}(y)\psi_{z\ell}(z)$ in Equation (1.1) represents a unique two-dimensional "error pattern" on the tooth-working-surface illustrated in Figure 1.4. The amplitude of this error pattern is described in the frequency domain by $B_{k\ell}(n)$, which of course will differ for every different measured gear. By a direct comparison of Equations (1.1) and (1.3) one can conclude (correctly) that the function $\hat{\phi}_{k\ell}(n/N)$, for each Legendre $k\ell$ pair, describes the computed attenuation, in the frequency domain, of the specific error pattern characterized by the pair of normalized Legendre terms $\psi_{yk}(y)\psi_{z\ell}(z)$. This is the attenuation of working-surface-deviations that a mating gear would provide, if the measured gear were meshed with a mating gear and the transmission-error contributions arising from the measured gear were somehow measured, for example, in a single-flank test. The mean-square spectrum of the computed kinematic transmission-error contributions is given by Equation (5.18),

$$G_\zeta(n) = 2\,|\alpha_n|^2, \quad n = 1, 2, 3, \cdots \tag{1.4}$$

where α_n is given by Equation (1.3).

Especially in the case of the higher-order rotational harmonics n, the mesh-attenuation functions $\hat{\phi}_{k\ell}(n/N)$ in Equation (1.3) usually provide significant attenuation to the harmonics of the spectra $B_{k\ell}(n)$ which are caused by the unattenuated working-surface-deviations. This reduction in amplitude can be observed for each rotational harmonic n by direct comparison of the after-attenuation spectrum, Equation (1.4), with the before-attenuation spectrum, Equation (1.2). (In example computations, we compare the *rms* values of these two spectra.) However, in the case of "ghost tones," very little, if any, attenuation is observed. It is for this reason that "ghost tones" create unwanted noise problems.

The kinematic transmission-error contributions from the measured gear are computed in the "time" domain, that is, as a function of "roll distance" x, by Equation (7.177),

$$\zeta(x) = \alpha_{0e} + 2\sum_{n=1}^{\infty}\left[\alpha_{ne}\cos\left(2\pi nx/N\Delta\right) + \alpha_{no}\sin\left(2\pi nx/N\Delta\right)\right]. \tag{1.5}$$

The algorithms by which all computations can be carried out are summarized in Chapter 8.

Fundamental Assumptions and Parametric Dependence of Mesh-Attenuation Functions $\hat{\phi}_{k\ell}(n/N)$

To carry out the kinematic transmission-error computations it is assumed that every tooth on a gear is measured over a rectangular region, as illustrated in Figure 1.4, defined by axial facewidth F, and radial distance D determined from tip and root roll angle values by Equation (3.7). Consequently, a fundamental assumption is that the tooth-pair contact region on the subject gear is over the same rectangular region $(-F/2) < y < (F/2)$, $(-D/2) < z < (D/2)$ of every tooth on the gear. This rectangular region generally will be smaller than the full region illustrated in Figure 1.4, and is chosen by the engineer responsible for the gear measurements. Therefore, it is assumed that tooth contact with a mating gear would take place over this full rectangular region used in the gear measurements. The locations of the normalized Legendre polynomial zeros used to specify scanning lead and/or scanning profile measurement locations are then normalized to D in the case of the lead measurements and F in the case of the profile measurements, as described in Appendices 3.A and 3.B.

For any parallel-axis helical (or spur) gear of nominal involute design, this rectangular region together with the gear nominal parametric descriptions determines both the axial Q_a and transverse Q_t contact ratios, as can be seen from Equations (6.5) to (6.8). These contact ratios thus effectively describe, for any helical or spur gear, the rectangular contact region on the tooth-working-surfaces. The computed mesh-attenuation functions $\hat{\phi}_{k\ell}(n/N)$ are parametrically dependent on both Q_a and Q_t. In addition, their rotational-harmonic dependence is a function only of the ratio n/N, where N is the number of teeth. The tooth-meshing fundamental rotational harmonic is $n = N$.

Finally, it is shown in Chapter 5 by Equation (5.13), and more rigorously in Chapter 7, that if the tooth-pair stiffness per unit length of line of contact is assumed to be constant, the mesh-attenuation functions $\hat{\phi}_{k\ell}(n/N)$ are independent of tooth-pair stiffness. Consequently, in this practically important case, apart from the tooth-meshing-harmonic contributions at $n/N = 1, 2, 3, \ldots$, the transmission-error contributions are dependent only on the geometric working-surface-deviations from equispaced perfect involute tooth surfaces, and for any rotational-harmonic values $n = 1, 2, 3 \ldots, (n/N) \neq 1, 2, 3, \ldots$ only on the axial Q_a and transverse Q_t contact ratios. These loading and inertia independent transmission-error contributions are the *kinematic* contributions to the transmission error (Merritt, 1971, p. 84).

1.5 Diagnosing Transmission-Error Contributions

Tooth-Meshing-Harmonic Contributions

The tooth-meshing-harmonic contributions A in Figure 1.1 to the transmission-error spectrum of a meshing-gear-pair operating at constant speed and constant torque are the harmonics at rotational harmonics $n = N, 2N, 3N, \ldots$, where n is rotational-harmonic number of one of the meshing gears and N is the number of teeth on that gear. These harmonics are caused by the additive contributions of the tooth-pair elastic deformations and the combined geometric deviations of the mean working-surface from a perfect involute surface of each of the two gears. The gear measurement, interpolation, and analysis methods developed in this book provide very accurate determination of the three-dimensional mean-working-surface-deviation of a measured gear, limited only by the accuracy of the CNC gear measurements. But tooth elastic deformations are not dealt with herein, and therefore, tooth-meshing-harmonic contributions to the transmission error are not computed.

Kinematic-Transmission-Error Contributions

Once any rotational harmonic n, in either a measured noise spectrum or structure-borne transducer spectrum, is identified to be of interest, the method described in Section 4.5 can be used to compute the three-dimensional working-surface error-pattern on some or all of the teeth that is the *cause* of the particular identified rotational harmonic n, that is, by Equation (4.42). To accomplish this, the working-surfaces of all teeth on the subject gear must be measured as described in Chapter 3. This computation does not require the more involved computation of the kinematic transmission-error spectrum.

However, if the rotational-harmonic n of interest is identified in the computed transmission-error spectrum, Equation (1.4), then the same computation described by Equation (4.42) is to be utilized to generate the working-surface error pattern causing the identified rotational harmonic. The resultant error pattern, computed by Equation (4.42), provides the information required to diagnose the manufacturing source of this error pattern, which is the cause of the identified rotational-harmonic tone.

Analytical approximations to the mesh-attenuation functions are described in Chapter 6 for various classes of errors, that is, accumulated tooth-spacing (index) errors, higher-order polynomial errors, and undulation errors. The behavior of these mesh-attenuation-function approximations allows one to diagnose, with some confidence, the manufacturing-error sources of low-order rotational harmonics $n = 1, 2, \ldots, B$ in Figure 1.1, so-called "sideband harmonics," C, and "ghost tones," D.

1.6 Application to Gear-Health Monitoring

Because tooth-working-surface damage affects the transmission-error spectrum in the same way that working-surface-deviations affect it, the general results and insights provided by the analysis in this book are applicable to gear-health monitoring considerations. In particular, because damage on one tooth or a few teeth will cause only a small change to the mean-working-surface-deviation, detection methods utilizing *changes* in rotational-harmonic amplitudes, $n \neq N, 2N, \ldots$, should enable the earliest detections, for example, Mark *et al.* (2010). Furthermore, because the non-tooth-meshing rotational-harmonic amplitudes of the transmission error, $n \neq N, 2N, \ldots$, are almost entirely independent of tooth stiffness, these harmonic amplitudes normally are virtually unaffected by modest changes in gear loading.

Legendre polynomials provide a very efficient system for representing the working-surface-deviations caused by tooth-bending-fatigue damage (Mark, Reagor, and McPherson, 2007, Figure 5); therefore, the methods described herein can be used to compute the changes in transmission error caused by tooth-bending-fatigue damage (Mark and Reagor, 2007).

Although application to gear metrology of the overall methodology described herein utilizes normalized two-dimensional Legendre polynomials to mathematically represent tooth-working-surface-deviations, the general method of computing transmission-error contributions described in Section 7.3 initially uses a completely generic method of mathematically representing working-surface-deviations, Equation (7.50), which later is specialized to the use of Legendre polynomials, beginning with Equation (7.75). This more general representation method has been motivated, in part, because of the potential use of other working-surface-deviation representations for application in gear-health monitoring. A wide range of machinery monitoring methods can be found in Randall (2011).

1.7 Verification of Kinematic Transmission Error as a Source of Vibration Excitation and Noise

After publication of Mark and Reagor (2001) illustrating computer implementation of the gear measurement and computational methods described herein, we were asked to exercise these methods on a helical gear that we were told was responsible for causing an unwanted "ghost tone." Because we knew the manufacturing errors causing this problem were likely to be very small, we decided to carry out two independent complete measurements of the gear using line-scanning profile and lead measurements made at the zero locations of normalized Legendre polynomials, as described in Chapter 3 of this book. Computation of the kinematic-transmission-error rotational-harmonic spectrum utilizing measurements from each of the two independent sets of measurements was carried out using the algorithms outlined in Chapter 8. If the two computed spectra were found to be in good agreement, we could be confident in our predictions of the spectra.

The measured helical gear had 51 teeth. The only information provided to us was the physical gear, the nominal design parameters of the gear, and a statement that the ghost-tone harmonic was located modestly above the tooth-meshing fundamental harmonic, $n = N = 51$. Transverse and axial contact ratios were both over 2.0.

In one of the two sets of measurements, 17 scanning profile measurements and seven scanning lead measurements were made on each tooth, and in the other set seven scanning profile measurements and 17 scanning lead measurements were made on each tooth. Because relatively good agreement of the predicted kinematic transmission-error spectrum line amplitudes was obtained from the two sets of measurements, only the prediction obtained from the measurement set using seven profile measurements and 17 lead measurements was provided to the gear owner. The dominant "ghost-tone" harmonic was found to be $n = 72$. Its computed *rms* amplitude was 0.102 μm (4.02 μin.). For the other measurement set, the computed dominant "ghost-tone" harmonic also was $n = 72$ with a computed *rms* amplitude of 0.127 μm (5.00 μin.).

After submission of the written report to the gear owner, the acoustic spectrum, obtained by spectrum analysis of a microphone output, was provided to us by the gear owner. This noise spectrum had been obtained by operating the gear we had measured with another higher quality gear. We had no way to know the transfer function (attenuation characteristic) between the operating gears and the microphone. But we could compare our computed kinematic transmission-error spectra with that obtained from the microphone output by forcing agreement of the amplitudes of rotational harmonic $n = 72$, and comparing the remaining harmonic amplitudes obtained from the acoustic measurements and our two computations. The result of this comparison is shown in Figure 1.5, where our computed amplitudes of the neighboring harmonics in the vicinity of $n = 72$ are shown encircled. Computed and measured amplitudes are linear (not logarithmic) measure.

This example clearly demonstrates that it is possible to measure a gear in detail and compute from those measurements the rotational-harmonic location of a dominant ghost tone, $n = 72$ in this case, even for a transmission-error contribution of exceedingly small amplitude. Moreover, it has demonstrated the kinematic transmission error as a source of vibration excitation and noise. (The very small working-surface errors causing rotational harmonic $n = 72$ are delineated in Chapter 6.) Furthermore, because the subject gear had 51 teeth, rotational harmonic $n = 51$ is the tooth-meshing fundamental harmonic, which exhibits a much lower amplitude in Figure 1.5 than the ghost tone located at $n = 72$, thereby illustrating the relative importance of ghost tones in this particular example of a helical gear-pair transmitting substantial loading.

Figure 1.5 also shows good (but imperfect) correlation between computed and acoustically measured sideband rotational harmonics $n = 68-76$ in the immediate neighborhood of $n = 72$. Considering that the computed *rms* amplitude of the transmission error causing $n = 72$ is about 0.1 μm (4 μin.), it is truly remarkable that it had been possible to compute with some success these much smaller amplitude "sideband" harmonics. An explanation of how such sideband harmonics of ghost tones can be generated is found in Mark (1992a, p. 175 Case III).

1.8 Gear Measurement Capabilities

The discussion and examples provided in Chapter 6 illustrate that "ghost tones" typically are caused by periodic sinusoidal-like manufacturing errors on gear-tooth working-surfaces that are almost entirely unattenuated by the meshing action with a mating gear. In other words, such a sinusoidal (undulation) error of amplitude "a" will result in a kinematic transmission-error contribution of about the same amplitude "a." (A reasonably complete mathematical discussion explaining this can be found in Mark (1992a), which is summarized in Section 6.6.) The above-discussed example, illustrated in Figure 1.5, therefore suggests a requirement to be able to successfully measure an undulation error with *rms* amplitude of about 0.1 μm. Is this requirement reasonable? Houser (2007) shows results for a helical gear with axial and transverse

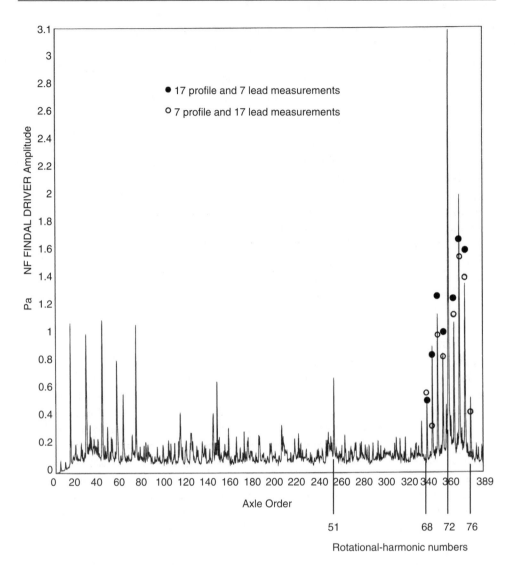

Figure 1.5 High-resolution sound spectrum generated by a 51 tooth helical gear meshing with a higher-quality gear compared with two kinematic-transmission-error spectrum computations. Both kinematic-transmission-error predictions were forced to agree with strongest acoustic-harmonic amplitude at $n=72$, thereby showing agreement of predicted rotational-harmonic amplitudes at $n=68-76$ relative to measured acoustic harmonic amplitudes. Linear-amplitude ordinate. Solid-point amplitudes computed from 17 profile and 7 lead measurements made on every tooth; hollow-point amplitudes computed from 7 profile and 17 lead measurements

contact ratios both modestly over 2.0 yielding a peak-to-peak transmission-error amplitude at design torque of slightly under 0.1 μm (4 μin.). This result implies one-sided transmission-error amplitude of about 0.05 μm (2 μin.), and an even smaller *rms* value of 0.035 μm (1.4 μin.). Hence, a "ghost-tone" transmission-error contribution with *rms* value of 0.1 μm (4 μin.) would exceed the above-mentioned amplitude, indicating that a requirement to be able to satisfactorily measure an undulation-error amplitude of about 0.1 μm (4 μin.), or even smaller, to be entirely reasonable. In repeated cases involving sinusoidal-like undulation errors causing "ghost tones," we have successfully measured the gears, computed the "ghost-tone" transmission-error contributions, and the sinusoidal errors on the working-surfaces causing these ghost tones, with typical *rms* sinusoidal errors and transmission-error amplitudes of about 0.1 μm (4 μin.). Independent gear measurements made on the same gears, using different CNC gear-measurement machines, have yielded almost exactly the same amplitudes. Yet, the generally accepted absolute accuracy of gear metrology equipment is considered to be about 1 μm (39.37 μin.). A partial explanation of how the above-described "ghost-tone" measurements and computations are possible is provided below.

It was described above that the transmission-error contributions of the tooth-meshing-harmonics, labeled A in Figure 1.1, are caused by the *mean* deviation of the elastically deformed working-surfaces from equispaced perfect involute surfaces, which includes the *mean* geometric working-surface-deviations and tooth elastic deformations. The remaining rotational-harmonic contributions labeled B, C, and D are caused by geometric deviations of the individual tooth-working-surfaces from the mean-working-surface – that is, geometric tooth-to-tooth variations in the working-surfaces, including tooth-spacing errors.

Contributions of Linear-Axis Errors

The typical dedicated CNC gear-measurement machine has three linear axes, and a rotary axis on which the gear to be measured is mounted. At any *fixed* location y, z (Figure 1.4), each of the linear axes is in the same position when every tooth is measured. Therefore, a consistent position error in any linear axis will be the same for *every* measured tooth on a gear at each tooth location y, z in Figure 1.4. Consequently, *consistent* linear-axis errors provide errors in the measurement of the geometric deviation of the mean (average) working-surface-deviations, which therefore will contribute only to errors in the computed tooth-meshing-harmonic contributions of the transmission error, but will not contribute errors to the remaining computed rotational-harmonic contributions, B, C, and D in Figure 1.1.

Contributions of Rotary-Axis Errors

Now consider rotary-axis errors. In making scanning profile measurements and scanning lead measurements on a *single* tooth of either a spur or helical gear, the rotary-axis position will vary from a minimum to a maximum rotational position. The rotary-axis error at the average rotary-axis position during these measurements can be regarded as contributing an error to the measured absolute spacing error of that tooth. *Differential* rotary-axis errors from this average rotary-axis position on each tooth will contribute to the above-described errors in the working-surface measurements from the mean (average) working-surface. That is, these differential rotary-axis errors can contribute to the computed transmission-error non-tooth-meshing rotational harmonics. But the maximum rotary-axis motion in measuring any *single* tooth is only a very small fraction of 360°. Consequently, only very short-span rotary-axis differential errors can contribute errors in computation of ghost-tone rotational harmonics. The longer span rotary-axis errors will contribute primarily to errors in (accumulated) tooth-spacing error computations.

Because a rotary scale is circular, it is useful to describe its errors using Fourier series. This decomposes its errors into a superposition of sinusoidal rotational-harmonic contributions. The effects of these rotational-harmonic rotary-axis error contributions on spur-gear transmission-error computations differs from their effects on helical-gear computations.

Consider their effects on spur-gear computations first. It can be seen directly from the involute construction described in Chapter 2, illustrated in Figure 2.3, and from the definition of transmission error given by Equation (3.2), that each rotational-harmonic amplitude of rotary-axis errors will be superimposed on the same rotational harmonic of the computed transmission error of a spur gear (with rotary-axis error amplitude proportional to base-circle radius), but with unknown phase. Hence, for successful transmission-error computations of spur gears, the higher harmonic rotary-axis errors must be minimized.

Fortunately, the effects of such rotary-axis errors are smaller on helical-gear transmission-error computations. This can be seen most easily from the stepped-gear analogy to helical gears illustrated in Figure 2.4. Because the root location of each step on the base cylinder occurs at a different cylinder rotational position, the rotary-axis higher-order rotational-harmonic error components will tend to be averaged out along the lines of tooth contact illustrated in Figures 1.2 and 3.2 in transmission-error computations. This averaging effect is likely at least partially responsible for the capability, shown herein, for successful computation of very small amplitude high order "ghost-tone" rotational-harmonic transmission-error contributions (D in Figure 1.1).

Contributions of Probe Errors

As in the case of linear-axis errors, in the case of probe errors it is convenient to distinguish consistent errors (i.e., bias errors and offset errors) from differential errors (errors in probe difference readings). Because the non-tooth-meshing rotational harmonic contributions to the transmission error are caused by geometric differences of the individual tooth-working-surfaces from the mean (average) working-surface on a gear, consistent probe errors will provide errors only to the mean-working-surface measurements, and therefore, errors only to the tooth-meshing-harmonic contributions A in Figure 1.1. Differential probe errors (errors in measurements of differences in amplitude) will provide errors to computed non-tooth-meshing rotational-harmonic amplitudes B, C, and D in Figure 1.1. Fortunately, very high quality probes are available.

Effects of Statistical Averaging

An enormous number of individual measurement samples (data values) are utilized in the computation of any transmission-error rotational-harmonic amplitude, and in computation of the working-surface-deviations that are the cause of any particular rotational harmonic. Very-short-wavelength surface-roughness characteristics and random measurement errors are averaged out in carrying out the required computations.

Summary of Required Measurement Capabilities

To be able to accurately measure and compute kinematic-transmission-error rotational-harmonic amplitudes, including "ghost-tone" amplitudes, and working-surface-deviations causing individual rotational harmonics, requirements on gear-measurement machines are:

- Linear-axis consistency (repeatability).
- Rotary-axis absolute accuracy (especially short-span high-harmonic accuracy).
- Measurement-probe consistency of bias errors and accuracy in differential measurements (measurement of differences).

Role of Working-Surface-Deviation Representation Method

Two-dimensional normalized Legendre polynomials are used in Chapter 3, and beyond, for representation of tooth-working-surface-deviations. The orthogonal property of this representation method guarantees that constant and long-wavelength measurement errors will provide *no contribution* to measured short-wavelength working-surface-deviations.

How Small is 0.1 μm (4 μin.)?

A tightly packed package of 500 sheets of copier paper is 2 in. thick. One-thousand sheets therefore is 4 in. thick, and one sheet is 4×10^{-3} in. thick. Therefore, 4×10^{-6} in. is 1/1000 of thickness of a sheet of copier paper, which is 4 μin. (0.1 μm). Such manufacturing accuracies are at the high-end of precision grinding (Nakazawa, 1994, p. 12). The wavelength of the center of the visible spectrum of light is about 0.5 μm (20 μin.).

References

Antosiewicz, H.A. (1964) Bessel functions of fractional order, in *Handbook of Mathematical Functions With Formulas, Graphs, and Mathematical Tables*, Chapter 10 (eds M. Abramowitz and I.A. Stegun), U.S. Government Printing Office, Washington, DC. Republished by Dover, Mineola, NY, pp. 435–478.

Bateman, H. (1954) in *Tables of Integral Transforms*, vol. 1 (ed. A. Erdelyi), McGraw-Hill, New York.

Bell, W.W. (1968) *Special Functions for Scientists and Engineers*, D. Van Nostrand, London. Republished by Dover, Mineola, NY.

Cheney, E.W. (1982) *Introduction to Approximation Theory*, 2nd edn, Chelsea Publishing Company, New York.

Cooley, J.W., Lewis, P.A.W., and Welch, P.D. (1969) The finite fourier transform. *IEEE Transactions on Audio and Electroacoustics*, **AU-17**, 77–85. Reprinted in Rabiner, L.R. and Rader, C.M. (eds) (1972) *Digital Signal Processing*, IEEE Press, New York, pp. 251–259.

Hildebrand, F.B. (1974) *Introduction to Numerical Analysis*, 2nd edn, McGraw-Hill, New York. Republished by Dover, New York.

Houser, D.R. (2007) Gear noise and vibration prediction and control methods, in *Handbook of Noise and Vibration Control*, Chapter 69 (ed. M.J. Crocker), John Wiley & Sons, Inc., New York.

Jackson, D. (1941) *Fourier Series and Orthogonal Polynomials*, The Mathematical Association of America, Buffalo, New York. Republished by Dover, Mineola, NY.

Lanczos, C. (1956) *Applied Analysis*, Prentice-Hall, Englewood Cliffs, NJ. Republished by Dover, New York.

Lanczos, C. (1961) *Linear Differential Operators*, D. Van Nostrand Company, London.

Mark, W.D. (1978) Analysis of the vibratory excitation of gear systems: basic theory. *Journal of the Acoustical Society of America*, **63**, 1409–1430.

Mark, W.D. (1983) Analytical reconstruction of the running surfaces of gear teeth using standard profile and lead measurements. *ASME Journal of Mechanisms, Transmissions, and Automation in Design*, **105**, 725–735.

Mark, W.D. (1991) Gear noise, in *Handbook of Acoustical Measurements and Noise Control*, Chapter 36 (ed. C.M. Harris), 3rd edn, McGraw-Hill, New York.

Mark, W.D. (1992a) Contributions to the vibratory excitation of gear systems from periodic undulations on tooth running surfaces. *Journal of the Acoustical Society of America*, **91**, 166–186.

Mark, W.D. (1992b) Elements of gear noise prediction, in *Noise and Vibration Control Engineering: Principles and Applications*, Chapter 21 (eds L.L. Beranek and I.L. Ver), John Wiley & Sons, Inc., New York, pp. 735–770.

Mark, W.D., Lee, H., Patrick, R., and Coker, J.D. (2010) A simple frequency-domain algorithm for early detection of damaged gear teeth. *Mechanical Systems and Signal Processing*, **24**, 2807–2823.

Mark, W.D. and Reagor, C.P. (2001) *Performance-Based Gear-Error Inspection, Specification, and Manufacturing-Source Diagnostics*, AGMA Technical Paper 01FTM6, American Gear Manufacturing Association, Alexandria, Virginia.

Mark, W.D. and Reagor, C.P. (2007) Static-transmission-error vibratory-excitation contributions from plastically deformed gear teeth caused by tooth bending-fatigue damage. *Mechanical Systems and Signal Processing*, **21**, 885–905.

Mark, W.D., Reagor, C.P., and McPherson, D.R. (2007) Assessing the role of plastic deformation in gear-health monitoring by precision measurement of failed gears. *Mechanical Systems and Signal Processing*, **21**, 177–192.

Merritt, H.E. (1971) *Gear Engineering*, John Wiley & Sons, Inc., New York.

Nakazawa, H. (1994) *Principles of Precision Engineering*, Oxford University Press, Oxford.

Randall, R.B. (2011) *Vibration-Based Condition Monitoring*, John Wiley & Sons, Inc., Chichester, West Sussex.

Smith, J.D. (2003) *Gear Noise and Vibration*, 2nd edn, Marcel Dekker, New York.

2

Parallel-Axis Involute Gears

Virtually all present-day parallel-axis gears utilize the involute tooth form. Meshing-gear-pairs with rigid equispaced perfect involute teeth would transmit exactly constant speed ratios resulting in zero transmission error. Because the subject of this book is computation of the kinematic transmission-error contributions arising from geometric deviations of the working surfaces of the teeth from equispaced perfect involute surfaces, it is appropriate to describe in this chapter the geometric properties of the equispaced perfect involute tooth-working-surfaces of parallel-axis helical gears that these geometric deviations can depart from. Spur gears are the special case of helical gears with zero helix angle.

2.1 The Involute Tooth Profile

Figure 2.1 illustrates four common types of parallel-axis gears: spur, single helical, herringbone, and internal spur. A double-helical gear is very similar to a herringbone gear, except in a double-helical gear there is a gap with no teeth located at the apex between the two helices to allow for clearance in manufacturing. Internal gears also can be single helical or herringbone/double helical.

Almost all power-transmission gears are designed to transmit exactly constant rotational speed ratios. Tooth-working-surfaces (contact surfaces, Figure 2.2) that transmit constant speed ratios are called conjugate surfaces. To transmit constant rotational speed ratios, such surfaces must satisfy the "law of gearing" (Drago, 1988, p. 49; Colbourne, 1987, p. 22). Except in a few specialized applications, virtually all present-day parallel-axis power-transmission gear-tooth forms are based on the *involute* tooth profile, which satisfies the law of gearing.

The *transverse plane* is the plane perpendicular to the axes of parallel-axis gears. An involute tooth profile describes the shape of the working-surface of a tooth in the transverse plane. An involute curve is the curve generated by the end of a taut cord as it is unwound from a right circular cylinder (Figure 2.3a). The cylinder from which the cord is unwound is called the *base cylinder*.

Performance-Based Gear Metrology: Kinematic-Transmission-Error Computation and Diagnosis,
First Edition. William D. Mark.
© 2013 John Wiley & Sons, Ltd. Published 2013 by John Wiley & Sons, Ltd.

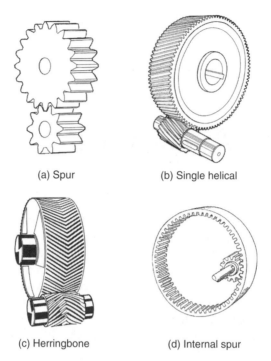

(a) Spur   (b) Single helical

(c) Herringbone   (d) Internal spur

Figure 2.1 Parallel-axis gear pairs. (a) Spur, (b) single helical, (c) herringbone (double-helical), and (d) internal spur (Gear figures courtesy of NASA)

Figure 2.2   Tooth-working-surfaces

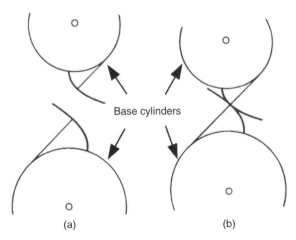

Figure 2.3 (a) Endpoints of taut cords unwrapped from base cylinders generate involute curves. (b) Cord endpoints tied together become a belt drive which transmits a constant rotational speed ratio; knot on cord traces out involute curves associated with each of the two base cylinders

Suppose the above-mentioned cord is unwound from the upper base cylinder in Figure 2.3a and wound onto the lower base cylinder. As the cord is unwound, a point fixed on the cord (e.g., a knot) traces out the involute curve associated with the upper base cylinder, and as the cord is wound onto the lower base cylinder, the same fixed point on the cord traces out the involute curve associated with the lower base cylinder. These two involute curves are shown in Figure 2.3b. Let the two involute curves be the *working surfaces* of gear teeth mounted on their respective base cylinders, and let the cord be a belt drive. Since a belt drive transmits an exactly constant rotational speed ratio, it follows from the above geometrical argument that *a pair of meshing gears with rigid, equispaced, and perfect involute teeth would transmit an exactly constant rotational speed ratio.* Thus, the involute curve is a conjugate tooth form. It follows that all points of contact between mating involute teeth lie in the plane shown in Figure 2.3b that is tangent to the two base cylinders of the meshing gears. This plane is called the *plane of contact, plane of action, or base plane.*

In the absence of friction, the force transmitted by a pair of mating teeth is normal to the tooth surfaces. The component in the transverse plane of the force transmitted by involute teeth therefore points in the direction of – and coincides with – the straight line that is tangent to the two base cylinders in Figure 2.3b. It follows that, in the absence of friction, *the component of force in the transverse plane transmitted between meshing gears with rigid, equispaced, and perfect involute teeth always lies in the plane of contact and, therefore, is constant in both position and direction.*

It also follows from the construction of the involute curves shown in Figure 2.3a that if the distance between the shafts of a pair of meshing gears with rigid, equispaced, and perfect involute teeth is changed, the gears will continue to transmit an *exactly* constant rotational speed ratio. Thus, *errors in the distance between the shafts of meshing gears with rigid, equispaced, and perfect involute teeth do not affect their capability to transmit an exactly constant rotational speed ratio.* Such errors can be caused, for example, by elastic deformations of the shaft bearing supports and, therefore, are unavoidable. The above-described constant force transmission property of involute teeth and the insensitivity of their transmission of *uniform* rotational speeds to errors in shaft separation distance both are very desirable properties for the minimization of vibration and noise generation. In this book, it is assumed that the tooth profiles in the transverse plane of the tooth-working-surfaces are involute curves, or modifications of involute curves.

2.2 Parametric Description of Involute Helical Gear Teeth

Because the tooth-working-surfaces of a spur gear (Figure 2.1a) are parallel to the gear axis, the above-described properties of the involute curve apply directly to spur gear teeth. However, in every transverse plane, the teeth of helical gears (Figure 2.1b,c) also have involute working surfaces. This fact is most easily understood by considering the stepped gear shown in Figure 2.4. Each axial segment of the stepped gear is an involute spur gear. Now allow the axial thickness of the individual steps to shrink to infinitesimal thickness while maintaining the same axial rate of rotational advance of the steps. In the limit, the resultant gear is a helical gear (Figure 2.1b). In every transverse plane, the tooth-working-surface retains its involute form. The intersection of the involute curve with the base cylinder (Figure 2.3a) describes a perfect helix on the base cylinder. The angle between the tangent to this helix and a line of the base cylinder parallel to its axis is the *base-cylinder helix angle* ψ_b.

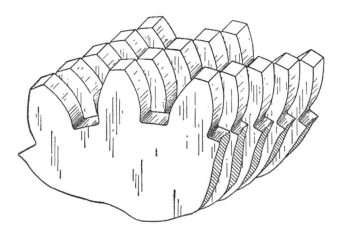

Figure 2.4 Stepped gear. As the axial thickness of the individual steps is shrunk to zero while the axial rate of rotation of advance of the steps is maintained, the resultant stepped gear becomes, in the limit, a single-helical gear (Reproduced, by permission, from Engineering Kinematics by Alvin Sloane, The Macmillan Company, 1941; republished by Dover, 1966 (Sloane, 1941))

Referring again to Figure 2.3a, in every transverse plane of either a spur gear or a helical gear, the involute working surfaces of the teeth are equally spaced around the base cylinder such that the arc-length spacing of the intersections of the involute curves with the base cylinder is Δ, the *base pitch*. If N is the number of teeth and R_b is the base-cylinder radius, then the base pitch must satisfy $N\Delta = 2\pi R_b$. Consequently, three basic parameters describe the working surfaces of an involute helical gear, the base-cylinder radius R_b, the base-helix angle ψ_b, and the base pitch Δ (or number of teeth N). For a spur gear, $\psi_b = 0$.

Referring to Figure 2.3b, the rotational speed ratio of a pair of involute gears is fixed by the ratio of the two base-cylinder radii. However, normally, gears are designed for a specific distance between the two shaft centers. Since the rotational speed ratio is fixed, once the shaft center distance is fixed the gear-meshing action is equivalent to the action of two cylinders in rolling contact, each such cylinder centered on a base-cylinder axis, with radii $R_p^{(1)}$ and $R_p^{(2)}$ which are in the exact same ratio as the base-cylinder radii, $R_b^{(1)}$ and $R_b^{(2)}$, that is,

$$\frac{R_p^{(2)}}{R_p^{(1)}} = \frac{R_b^{(2)}}{R_b^{(1)}}, \tag{2.1}$$

where each of the two superscript indices designates one of the two meshing gears. The two contacting cylinders with radii, $R_p^{(1)}$ and $R_p^{(2)}$, are called *pitch cylinders*.

The point of contact of the two pitch cylinders, which lies on the line of shaft centers, is called the *pitch point*. The plane passing through the pitch point that is perpendicular to the plane containing the two shaft centers is called the *pitch plane* (Figure 2.5). The angle ϕ between the pitch plane and the plane of contact is the *transverse pressure angle*. A change in shaft center distance will result in a change in the location of the pitch point, and a change in the pressure angle ϕ, but as explained above, the transmission of a constant rotational speed ratio is retained.

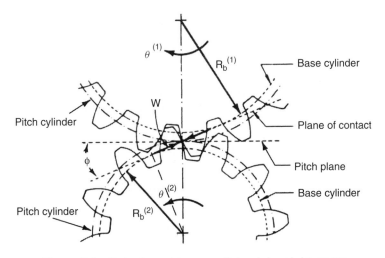

Figure 2.5   Pair of meshing parallel-axis involute gears

The arc-length spacing of the tooth-working-surfaces in the transverse plane, measured on the pitch circles, is the *circular pitch* Δ_c, which for either of the two meshing gears must satisfy

$$\Delta_c = \frac{2\pi R_p}{N}. \tag{2.2}$$

In Figure 2.5, construct a normal to the plane of contact that passes through the shaft center of the lower gear (2), shown dashed. This dashed normal intersects the plane of contact at the point of tangency between the plane of contact and the lower base cylinder. The length of the normal is the base-cylinder radius R_b, whereas the distance between the shaft center of the lower gear and the pitch point is the pitch-cylinder radius R_p. Moreover, the angle between the dashed normal and the line connecting shaft centers is ϕ, the pressure angle. Hence,

$$\cos \phi = R_b/R_p. \tag{2.3}$$

But $\Delta = 2\pi R_b/N$ and $\Delta_c = 2\pi R_p/N$; hence,

$$\frac{\Delta}{\Delta_c} = \frac{R_b}{R_p} = \cos \phi, \tag{2.4}$$

and therefore,

$$\Delta = \Delta_c \cos \phi \tag{2.5}$$

which relates the circular pitch Δ_c to the base pitch Δ and the transverse pressure angle ϕ.

The axial distance by which a helix makes exactly one full revolution around a cylinder is the lead L_h of the helix (Drago, 1988, p. 499; Colbourne, 1987, p. 318). If the base cylinder is "unwrapped" onto a plane surface, the base helix angle ψ_b, lead L_h, and base-cylinder radius R_b are related by

$$\tan \psi_b = \frac{2\pi R_b}{L_h}. \tag{2.6}$$

But the tooth locations on the pitch cylinder must possess the same lead as on the base cylinder. Let ψ denote the *pitch-cylinder helix angle*. Then,

$$\tan \psi = \frac{2\pi R_p}{L_h}. \tag{2.7}$$

Hence, from Equations (2.6) and (2.7), then Equation (2.3),

$$\tan \psi_b = \frac{R_b}{R_p} \tan \psi \tag{2.8a}$$

$$= \cos \phi \tan \psi. \tag{2.8b}$$

Remaining fundamental parameters of the meshing gear teeth in Figure 2.5 are the radii R_a of the cylinders passing through the outer ends of the teeth, which are called *addendum* cylinders and radii.

2.3 Multiple Tooth Contact of Involute Helical Gears

As described above, the arc-length spacing of the involute working surfaces in the transverse plane where they intersect the base cylinder, that is, the involute curves as in Figure 2.3a, is Δ, the base pitch. It then follows from the involute construction in Figure 2.3a,b (equally spaced knots on the string) that the spacing of the teeth in the plane of contact in the transverse plane also is the base pitch Δ, and moreover, the helix angle of the tooth contact points (Figure 2.3b) relative to a line parallel to the gear axes is the same as the helix angle ψ_b at the intersections of the involute surfaces with the base cylinders. These lines of tooth contact, and the base pitch Δ, are illustrated in the upper sketch in Figure 2.6, which is the plane of contact.

One can envision the string in Figure 2.3b to be a belt drive riding on the two base cylinders, with the equispaced lines of tooth contact drawn on this belt as illustrated in the upper portion of Figure 2.6. The solid rectangle in the upper portion of the figure represents the region of the plane of contact where the gear teeth are in actual contact. For perfect involute teeth, the locations of the sides of this rectangle are determined by where the addendum cylinders of the two gears, in the lower portion of the figure, cut the plane of contact. The length of this contact region in the transverse direction is denoted by L in the upper portion of the figure. The facewidth F, shown there, is determined by the axial facewidth of the perfect involute teeth. As the gears rotate, the fictitious belt moves through this rectangular zone of contact, and differing pairs of mating teeth come into and out of contact.

The above description, and the illustration in Figure 2.6, is an exact geometric representation of the meshing action of parallel-axis helical gears with rigid equis-paced perfect involute teeth and perfect parallel axes. It is the basis for the coordinate systems used in precision gear measurement, and for the analysis required in the remainder of this book. In the special case of spur gears, Figure 2.1a, the base helix angle ψ_b is zero, and the lines of contact shown in Figure 2.6 would be drawn parallel to the y-axis shown there, with spacing equal to the base pitch Δ.

2.4 Contact Ratios

Two geometric parameters will appear repeatedly in the following chapters. They are the transverse contact ratio, also called the profile contact ratio, and the axial contact ratio, also called the face contact ratio. The transverse contact ratio is applicable to both spur gears and helical gears. Utilizing the parameters illustrated in Figure 2.6, the transverse contact ratio Q_t is defined (Drago, 1988, p. 506; Colbourne, 1987, p. 384) by

$$Q_t \triangleq \frac{L}{\Delta} \tag{2.9}$$

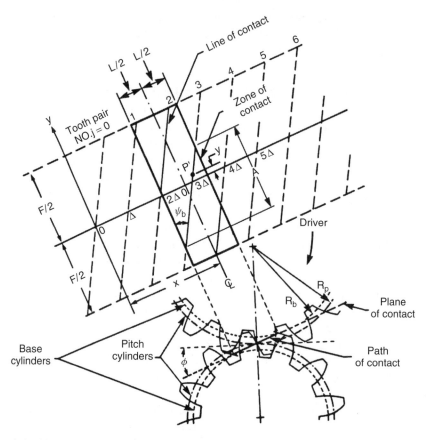

Figure 2.6 Lines of tooth-pair contact and zone of contact in plane of contact of a meshing pair of geometrically perfect involute helical gears. Generic point *P* on line of contact is denoted by *P'* because roll-distance *x* governing line-of-contact locations differs between Figures 2.6 and 3.2 (Adapted from Mark (1978))

which is the length L of the contact region in the transverse direction divided by the base pitch Δ, both "measured" in the plane of contact. Since both L and Δ are "measured" in the transverse plane, the transverse contact ratio is applicable to both spur and helical gears.

Figure 2.7a illustrates the plane of contact for a helical gear with a transverse contact ratio of exactly $Q_t = 2.0$. In every transverse section of this sketch there are exactly two teeth in contact. For a spur gear to successfully operate, the transverse contact ratio must be at least 1.0.

The axial contact ratio is the helical gear counterpart to the transverse contact ratio. It is defined as the ratio of the (axial) facewidth F divided by the axial pitch Δ_a. Both of these parameters are illustrated in Figure 2.7b, which again illustrates the lines of tooth contact in the plane of contact. The definition of axial contact ratio Q_a is (Drago,

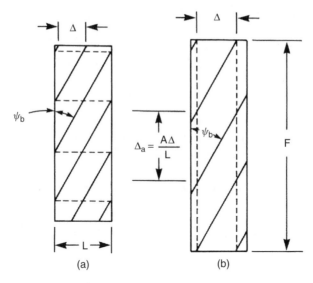

Figure 2.7 Lines of tooth-pair contact in plane of contact. (a) Illustrates transverse contact ratio $Q_t = 2.0$ and (b) illustrates axial contact ratio $Q_a = 3.0$

1988, p. 506; Colbourne, 1987, p. 385)

$$Q_a \triangleq \frac{F}{\Delta_a}. \qquad (2.10)$$

Q_t and Q_a both are dimensionless ratios.

Another form of the axial contact ratio will appear repeatedly in the derivations to follow. From Figure 2.7b one has

$$\tan \psi_b = \frac{\Delta}{\Delta_a}. \qquad (2.11)$$

From Figure 2.6, one also has

$$\tan \psi_b = \frac{L}{A}; \qquad (2.12)$$

hence, from Equations (2.11) and (2.12), there follows

$$\Delta_a = \frac{A\Delta}{L}, \qquad (2.13)$$

which is illustrated in Figure 2.7b. From Equations (2.10) and (2.13), an alternative form of the axial contact ratio is

$$Q_a = \frac{FL}{A\Delta}, \qquad (2.14)$$

where, in particular, the parameter A is illustrated in Figure 2.6. For a spur gear $A = \infty$, and therefore, $Q_a = 0$. Figure 2.7b illustrates an axial contact ratio of exactly 3.0. In every axial section, there are exactly three teeth in contact.

References

Colbourne, J.R. (1987) *The Geometry of Involute Gears*, Springer-Verlag, New York.

Drago, R.J. (1988) *Fundamentals of Gear Design*, Butterworths, Boston, MA.

Mark, W.D. (1978) Analysis of the vibratory excitation of gear systems: basic theory. *Journal of the Acoustical Society of America*, **63**, 1409–1430.

Sloane, A. (1941) *Engineering Kinematics*, Macmillan, New York. Republished by Dover, New York.

3

Mathematical Representation and Measurement of Working-Surface-Deviations

The idealized gears described in the preceding chapter would transmit exactly constant rotational speed ratios, and their teeth would be in contact everywhere on the lines of contact within the rectangular zone of contact shown in Figure 2.6, as the lines of contact pass through the zone of contact. However, the tooth-working-surfaces of real gears have manufacturing errors and intentional modifications from involute surfaces, and they and the gear bodies elastically deform under loading. The combined effect of these geometric working-surface-deviations and elastic deformations causes a fluctuating component to be superimposed on the otherwise transmission of constant rotational speed ratios. This fluctuation is the unsteady component of the transmission error, defined below.

The contribution of geometric working-surface-deviations from equispaced perfect involute surfaces is the *kinematic* (inertia and loading independent) contribution to the transmission error. Its computation requires a method to efficiently measure such working-surface-deviations and to represent from such measurements the working-surface-deviations of all N teeth on a gear from equispaced perfect involute surfaces. The required representation and measurement method is the subject of this chapter.

Performance-Based Gear Metrology: Kinematic-Transmission-Error Computation and Diagnosis,
First Edition. William D. Mark.
© 2013 John Wiley & Sons, Ltd. Published 2013 by John Wiley & Sons, Ltd.

3.1 Transmission Error of Meshing-Gear-Pairs

Independent Variable for Transmission Error

Recall that the gears in Figure 2.5, shown again in Figure 3.1, are assumed to be rotating on perfectly parallel shafts, and to have exactly equispaced rigid perfect involute working-surfaces. The rotational positions of these perfect involute gears are $\theta^{(1)}$ and $\theta^{(2)}$ as illustrated in the figure. A single variable is required to designate the rotational positions of the two gears. The most convenient such variable is the location variable x, shown in Figure 2.6, of the fictitious belt drive riding on the two base cylinders:

$$x \triangleq R_b^{(1)}\theta^{(1)} = R_b^{(2)}\theta^{(2)}, \tag{3.1}$$

where $\theta^{(1)}$ and $\theta^{(2)}$ are "measured" in radians.

Definition of Transmission Error

Suppose now that the working-surfaces of the teeth on the two gears are allowed to have geometric deviations from their equispaced perfect involute counterparts, and are allowed to elastically deform under the transmitted loading W shown in Figure 3.1, while maintaining full contact on the lines of contact shown in Figure 2.6. Further suppose that the shaft locations of the two gears remain unchanged. Let $\delta\theta^{(1)}$ and $\delta\theta^{(2)}$ denote the instantaneous rotational-position deviations of gears (1) and (2), respectively, from the positions $\theta^{(1)}$ and $\theta^{(2)}$ of their rigid perfect involute counterparts. As the gears rotate, and variable x increases, $\delta\theta^{(1)}$ and $\delta\theta^{(2)}$ will vary with x. As in Harris (1958), define the transmission error $\zeta(x)$ to be positive when it arises from *removal* of material from the equispaced perfect involute working-surfaces and from elastic deformations of those surfaces (Mark, 1982). Then

$$\zeta(x) \triangleq R_b^{(1)}\delta\theta^{(1)}(x) - R_b^{(2)}\delta\theta^{(2)}(x), \tag{3.2}$$

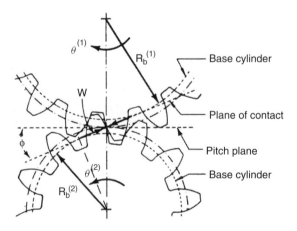

Figure 3.1 Pair of meshing parallel-axis involute gears

where the negative sign associated with the contribution from the lower gear (2) is a consequence of the sign of the positive direction of $\theta^{(2)}$. In other words, this lineal transmission error is positive when the teeth come together relative to their perfect involute counterparts.

Additive Property of Transmission-Error Contributions

The definition of transmission error given by Equation (3.2) is conceptionally equivalent to the definition provided in Gregory, Harris, and Munro (1963–1964): "The transmission error is defined, for any instantaneous angular position of one gear, as the angular displacement of the mating gear from the position it would occupy if the teeth were rigid and unmodified." However, the refinement, Equation (3.2), to this earlier definition allows the transmission error of the meshing-gear-pair to be decomposed into individual additive contributions $R_b^{(1)}\delta\theta^{(1)}$ and $R_b^{(2)}\delta\theta^{(2)}$ from each of the two meshing gears. Each of these contributions $R_b^{(1)}\delta\theta^{(1)}$ and $R_b^{(2)}\delta\theta^{(2)}$ is dependent on the deviations of the tooth-working-surfaces of that gear on all lines of contact shown in Figure 2.6 at that particular gear-pair rotational position x.

At each point on a line of contact there are two additive contributions to the transmission error from each of the mating teeth, a geometric deviation contribution and an elastic deformation contribution. The arithmetic sum of these two contributions provides the contribution of the identified point to that particular gear's transmission-error contribution. The principal subject of this book is the transmission-error contribution provided only by geometric deviations of the tooth-working-surfaces from equispaced perfect involute surfaces. The transmission-error definition provided by Equation (3.2) allows a rigorous treatment of these geometric-deviation contributions from a single gear, due to the additive form of that relationship.

At each rotational position of the gear-pair, described by the variable x, Equation (3.1), the transmission-error contribution from geometric tooth deviations is a function of the deviations on the working-surfaces along all of the lines of contact within the rectangular zone of contact illustrated in Figure 2.6. As x changes, the locations of these lines of contact on the tooth-working-surfaces also change. Two such positions of a line of contact are illustrated in Figure 3.2. As the gears

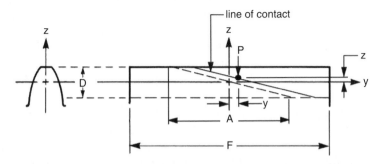

Figure 3.2 Two line of tooth-pair contact locations on working-surface of a tooth. Point P illustrates generic location on line of contact

in Figure 3.1 rotate, and variable x increases, a line of contact traverses across the full tooth-working-surface, while remaining at the same angle as illustrated in Figure 3.2. Hence, to successfully compute the transmission-error contributions from geometric deviations of the tooth-working-surfaces on a gear, a description of the geometric deviation at every point on the working-surfaces of all teeth on the gear is required. Currently available computer numerically controlled (CNC) dedicated gear-measurement machines are capable of measuring a gear in sufficient detail to provide such a description, as will be shown in the following pages. A coordinate system is required for the utilization of such measurements.

3.2 Tooth-Working-Surface Coordinate System

Figure 3.1 illustrates two teeth in contact transmitting a load W. This illustration can be regarded as a particular transverse-plane slice through the gear tooth shown in Figure 3.2. Because of the finite curvature of the two working-surfaces in contact in Figure 3.1, tooth contact takes place in each transverse plane at a single point on the working-surfaces, which would designate a single point on a line of contact in Figure 3.2. Hence, for every rotational position of the meshing gears, designated by the variable x, Equation (3.1), there exists a unique point of contact on the working-surfaces of the teeth at each *axial location y* on the line of contact in Figure 3.2.

Definition of Radial Coordinate

Figure 3.2 also illustrates a *radial coordinate z* which needs to be formally defined in relation to the contacting points on the teeth. Since all tooth contact takes place in the plane of contact shown in Figure 3.1, and the projection of points in that plane of contact onto the vertical line connecting the shaft centers is controlled by $\sin\phi$, where ϕ is the transverse pressure angle, points of contact on the tooth-working-surfaces at any axial location y can be designated by

$$z = R_b\epsilon \sin\phi + c \qquad (3.3)$$

where ϵ is involute roll angle, for example Baxter (1962, pp. 1–9), R_b is base-cylinder radius, and c is a constant, designated below. As shown in Figure 3.3a, the "roll distance" $R_b\epsilon$ uniquely designates a point on the involute curve, and therefore designates a unique point on an involute working-surface at any axial location y. The dimension of roll angle ϵ is radian measure. The projection of points on the involute curve onto the line of shaft centers, described by Equation (3.3), is readily visualized in Figure 3.3b.

The methodology described in this book computes the transmission-error contributions from geometric working-surface-deviations on *rectangular regions* on tooth-working-surfaces, as illustrated by the axial facewidth dimension F and the radial tooth-height dimension D illustrated in Figure 3.2. This rectangular "analysis region" must be chosen by the user. For mathematical reasons, it is important that the origin of the axial coordinate y be chosen at the midpoint of the axial span F

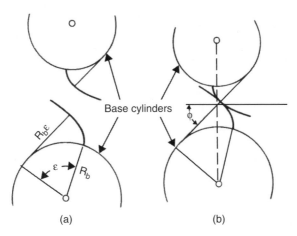

Figure 3.3   (a,b) Illustration of involute roll angle ε, roll-distance $R_b\varepsilon$, and transverse pressure angle ϕ, required for definition of radial working-surface location z given by Equations (3.3–3.6)

of this rectangular region, and the origin of the radial coordinate z be chosen at the midpoint of the radial span D.

This latter criterion sets the value of the constant c in Equation (3.3). Radial locations on tooth-working-surfaces normally are described in terms of the roll angle ϵ, Figure 3.3a. Let the roll angles at the tip and root of the radial contact region D in Figure 3.2 be denoted by ϵ_t and ϵ_r, respectively. Define

$$\epsilon_0 \triangleq \frac{\epsilon_t + \epsilon_r}{2}. \tag{3.4}$$

Then, the value of c in Equation (3.3) is

$$c = -R_b\epsilon_0 \sin \phi, \tag{3.5}$$

yielding for z in Equation (3.3),

$$z = R_b\left(\epsilon - \epsilon_0\right)\sin \phi, \tag{3.6}$$

and for the total radial span D of z,

$$D \triangleq R_b\left(\epsilon_t - \epsilon_r\right)\sin \phi. \tag{3.7}$$

There are two reasons for utilizing roll-angle ϵ to designate radial locations on a tooth: (i) equal increments in ϵ correspond to equal increments in gear rotational position, as can be seen from Figure 3.3a and (ii) gear metrology equipment normally designates radial tooth location by roll angle ϵ. *The axial coordinate y, $(-F/2) \leq y \leq (F/2)$, and the radial coordinate z, $(-D/2) \leq z \leq (D/2)$, with z defined by Equations (3.4) and (3.6), describe*

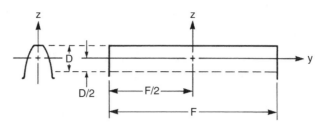

Figure 3.4 Cartesian coordinate system used to locate a generic point y, z on tooth-- working-surfaces. Origin $y=0$, $z=0$ is located at axial and radial center of chosen analysis region of axial dimension F and radial dimension D

the Cartesian coordinate system that we shall use to locate any point on the working-surface of a gear tooth. This coordinate system is illustrated in Figure 3.4. It is important to recognize that the relationship between z and roll-angle ϵ is linear, the proportionality constant being $R_b \sin \phi$, and the translation constant c described by Equations (3.3) and (3.5).

3.3 Gear-Measurement Capabilities

Present day CNC dedicated gear-measurement equipment has the capability of making precision line-scanning "lead" measurements in an axial direction, at a constant radial location z, and line-scanning "profile" measurements in a radial direction, at a constant axial location y. The locations of three such lead measurements and three such profile measurements are illustrated in Figure 3.5. The term "lead" comes from the lead of a helix, described in Section 2.2. (The term "tooth alignment" measurement is sometimes used instead of lead measurement, and the term "involute" measurement is sometimes used instead of profile measurement.) Such gear metrology equipment is designed to measure deviations of the working-surfaces of the teeth from equispaced perfect involute surfaces. It can do so with remarkable repeatability and accuracy. *If the measurement probe is set so as to measure these working-surface-deviations in a direction defined by the intersection of the plane of contact and the transverse plane (Figure 2.6), then such measurements at each working-surface location describe, exactly, the geometric deviations from the perfect equispaced involute*

Figure 3.5 Locations of three scanning lead measurements and three scanning profile measurements on a tooth-working-surface

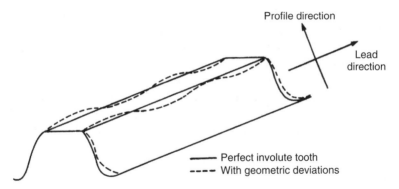

Figure 3.6 Working-surface-deviations at all locations $(-F/2) < y < (F/2)$, $(-D/2) < z$ $< (D/2)$ are required

working-surfaces described in Chapter 2, which are the deviations required for computation of the tooth-deviation contributions to the transmission error.

However, a determination of such deviations is required at *all* points y, z within the designated rectangular region shown in Figure 3.2 in order to obtain a determination of the deviations everywhere within that rectangular region, as sketched in Figure 3.6, and on all teeth. Hence, because only line-scanning measurements are available, some sort of interpolation between the line-scanning measurements is required in order to achieve a determination of the working-surface-deviations at *all* points y, z within these rectangular regions.

Clearly, closer spacing of these line-scanning measurements will enable a more accurate determination of the deviations. It is common practice to intentionally modify tooth-working-surfaces from perfect involute surfaces by providing "tip relief", sometimes "root relief", and, in the case of helical gears, some sort of "end relief", "crowning", "generated engagement relief", or "bias" modifications. Such modifications produce larger changes in the working-surfaces near the tooth tips, roots, and axial ends, with less change near the radial and axial centers of the working-surfaces. For a given measurement effort, this modification practice suggests that closer spacing of the line-scanning lead measurements near the tooth tips and roots, with sparser spacing near the radial center of a tooth, and similarly, closer spacing of the line-scanning profile measurements near the axial ends of a tooth, with sparser spacing near the axial center of a tooth, should produce a more accurate determination, by interpolation, than equally spaced line-scanning measurements. The measurement methodology, described in the following pages, utilizes such non-equally-spaced line-scanning measurements.

3.4 Common Types of Working-Surface Errors

Perfect equispaced involute teeth are manufactured relative to a base-cylinder axis. Irrespective of the type of bearings used, for example, rolling-element or journal, gears rotate relative to the centerline of a pair of bearings. Suppose the bearing axis and base-cylinder axis are perfectly parallel, but with a small eccentric offset. Suppose the axis of gear rotation during gear measurement is identical with the bearing axis. The measurement of such an otherwise perfect involute gear will exhibit a small once-per-revolution accumulated tooth spacing, that is, index error, and a small once-per-revolution linear profile error due to the effective once-per-revolution variation in the *effective* base-cylinder radius, as experienced by the measurements. Suppose now, in addition, there is a small out-of-parallel discrepancy between the bearing axis and base-cylinder axis, with the gear-measurement axis exactly coinciding with the bearing axis. In this case, there also will appear a once-per-revolution linear component in lead measurements made at any radial z location. (Such linear errors in lead measurements are almost always observed when measuring helical gears.)

In addition, various types of errors associated with specific gear cutting and finishing processes are commonly experienced (Cluff, 1992), and in particular, periodic undulation errors on the working-surfaces that are the source of so called "ghost tones" (Cluff, 1992, pp. 7, 38, 95–97; Mark, 1992a).

3.5 Mathematical Representation of Working-Surface-Deviations

For a given gear measurement effort, it was suggested above that it should be advantageous to locate line-scanning working-surface measurements more closely spaced near the tooth-tips, roots, and axial ends of the teeth, with wider spacing near the center regions of the teeth. For representation efficiency and manufacturing-error diagnostic purposes, a tooth-deviation representation method that could readily identify tooth-spacing errors and linear errors in both the axial (lead) direction and the radial (profile) direction also would be useful, as indicated above. Moreover, a working-surface representation method is required that has the capability of representing, with sufficient accuracy, *any* geometric deviations of tooth-working-surfaces from equispaced perfect involute surfaces. As described in Jackson (1941), there is a multitude of methods that can represent generic geometric deviations, but there is one method in particular that satisfies all of the above-described desirable characteristics. That method is representing tooth-working-surface-deviations by two-dimensional normalized Legendre polynomials (Mark, 1979, 1982).

Legendre Polynomials

The first few one-dimensional classical Legendre polynomials, defined over the interval $-1 \leq \xi \leq 1$, are (Jackson, 1941, pp. 46–47)

$$P_0(\xi) = 1 \qquad\qquad P_1(\xi) = \xi$$

$$P_2(\xi) = \frac{1}{2}(3\xi^2 - 1) \qquad\qquad P_3(\xi) = \frac{1}{2}(5\xi^3 - 3\xi)$$

$$P_4(\xi) = \frac{1}{8}(35\xi^4 - 30\xi^2 + 3) \qquad\qquad P_5(\xi) = \frac{1}{8}(63\xi^5 - 70\xi^3 + 15\xi). \qquad (3.8)$$

From the first two Legendre polynomials given above, all higher-order Legendre polynomials can be generated using the recurrence relation

$$nP_n(\xi) = (2n-1)\xi P_{n-1}(\xi) - (n-1)P_{n-2}(\xi), \quad n \geq 2. \tag{3.9}$$

Any well-behaved function of ξ within the interval $-1 \leq \xi \leq 1$ can be represented by a linear superposition of Legendre polynomials. (Better accuracy is obtained by using more terms.) Because $P_0(\xi) = 1$ is a constant, this Legendre term can be thought of as representing a "pure" tooth-spacing error, and because $P_1(\xi) = \xi$ is linear in ξ, it can be thought of as representing either a "pure" linear-lead error or linear-profile error, as described above, and developed in more detail below.

Each of the Legendre polynomials $P_n(\xi)$ is of degree n. They are normalized by

$$P_n(1) = 1 \quad P_n(-1) = (-1)^n \tag{3.10a,b}$$

and are orthogonal over the interval $-1 \leq \xi \leq 1$, meaning

$$\int_{-1}^{1} P_n(\xi) P_m(\xi) \, d\xi = 0, \quad n \neq m \tag{3.11}$$

With

$$\int_{-1}^{1} P_n^2(\xi) \, d\xi = \frac{2}{2n+1}. \tag{3.12}$$

Using the orthogonal property, Equation (3.11), and the normalization properties, Equations (3.10a,b), the infinite sequence of Legendre polynomials also could be generated sequentially by using the Schmidt orthogonalization process (Jackson, 1941, p. 151).

Normalized Legendre Polynomials

The above-described Legendre polynomials are functions of a single coordinate variable ξ defined within $-1 \leq \xi \leq 1$, whereas a method is required for representing tooth-working-surface-deviations over the rectangular region $(-F/2) \leq y \leq (F/2)$, $(-D/2) \leq z \leq (D/2)$ shown in Figure 3.4. Legendre polynomials meeting this requirement are $P_k(2y/F)$, $k = 0, 1, 2, \ldots$ and $P_\ell(2z/D)$, $\ell = 0, 1, 2, \ldots$ which redefine the intervals of application from $-1 \leq \xi \leq 1$ to the above-described intervals in y and z.

Also required is a normalization that will permit simple interpretations of working-surface-deviation representations and representations in the frequency domain. This normalization is accomplished with coefficients $(2k+1)^{1/2}$ and $(2\ell+1)^{1/2}$ multiplying the above-described Legendre polynomials. Therefore, define

$$\psi_{yk}(y) \triangleq (2k+1)^{1/2} P_k(2y/F), \quad (-F/2) \leq y \leq (F/2) \tag{3.13}$$

$$\psi_{z\ell}(z) \triangleq (2\ell+1)^{1/2} P_\ell(2z/D), \quad (-D/2) \leq z \leq (D/2) \tag{3.14}$$

which are the required normalized Legendre polynomials.

As shown below, the integral-square-average values of these normalized Legendre polynomials, averaged over their defined regions, are unity, and these normalized polynomials are orthogonal, as are the classical Legendre polynomials, that is, Equation (3.11). Consider using Equation (3.13),

$$\frac{1}{F} \int_{-F/2}^{F/2} \psi_{yk}(y)\, \psi_{yk'}(y) dy = \frac{1}{F}(2k+1)^{1/2}(2k'+1)^{1/2} \int_{-F/2}^{F/2} P_k(2y/F) P_{k'}(2y/F) dy. \qquad (3.15)$$

But $\xi = 2y/F$, therefore, $y = (F/2)\xi$ and $dy = (F/2)d\xi$. Hence, one has from Equation (3.15), by using Equations (3.12) and (3.11),

$$\frac{1}{F} \int_{-F/2}^{F/2} \psi_{yk}(y)\, \psi_{yk'}(y) dy = (2k+1)^{1/2}(2k'+1)^{1/2} \frac{1}{2} \int_{-1}^{1} P_k(\xi) P_{k'}(\xi) d\xi$$

$$= (2k+1)\frac{1}{2} \times \frac{2}{(2k+1)} = 1, \quad k' = k \qquad (3.16a)$$

$$= 0, \quad k' \neq k. \qquad (3.16b)$$

In exactly the same way, one has

$$\frac{1}{D} \int_{-D/2}^{D/2} \psi_{z\ell}(z)\psi_{z\ell'}(z) dz = 1, \quad \ell' = \ell \qquad (3.17a)$$

$$= 0, \quad \ell' \neq \ell. \qquad (3.17b)$$

Representation Using Two-Dimensional Normalized Legendre Polynomials

Let $\eta_{Cj}(y, z)$ denote the geometric deviation of the working-surface of generic tooth number j from an equispaced perfect involute surface described in Chapter 2. Subscript C denotes that these deviations are described as a function of the Cartesian tooth coordinates illustrated in Figure 3.4, with axial coordinate y, and the radial coordinate z defined by Equation (3.6). Deviations $\eta_{Cj}(y, z)$ are assumed measured in a direction defined by the intersection of the transverse plane and the plane of contact (Figure 3.1). We shall be concerned with the geometric deviations of all N teeth on a gear counted as $j = 0, 1, 2, \ldots, N-1$.

Any measureable tooth deviation can be represented exactly by a superposition of two-dimensional normalized Legendre polynomials of the form

$$\eta_{Cj}(y, z) = \sum_{k=0}^{\infty} \sum_{\ell=0}^{\infty} c_{j,k\ell} \psi_{yk}(y)\psi_{z\ell}(z), \qquad (3.18)$$

where the ψ functions are given by Equations (3.13) and (3.14). To obtain an expression for the expansion coefficients $c_{j,k\ell}$, we multiply both sides of Equation (3.18) by

$\psi_{yk'}(y)\psi_{z\ell'}(z)/(FD)$ and integrate over the full tooth surface, giving

$$\frac{1}{FD}\int\limits_{-D/2}^{D/2}\int\limits_{-F/2}^{F/2}\eta_{Cj}(y,z)\psi_{yk'}(y)\psi_{z\ell'}(z)dydz$$

$$=\sum_{k=0}^{\infty}\sum_{\ell=0}^{\infty}c_{j,k\ell}\frac{1}{F}\int\limits_{-F/2}^{F/2}\psi_{yk}(y)\psi_{yk'}(y)dy\frac{1}{D}\int\limits_{-D/2}^{D/2}\psi_{z\ell}(z)\psi_{z\ell'}(z)dz. \tag{3.19}$$

From the orthogonal property of the normalized Legendre polynomials Equations (3.16b) and (3.17b), it follows that the right-hand side of Equation (3.19) is zero unless both $k=k'$ and $\ell=\ell'$. When both $k=k'$ and $\ell=\ell'$, each of the two normalized integrals in the right-hand side of Equation (3.19) is unity, according to Equations (3.16a) and (3.17a); hence, when both $k=k'$ and $\ell=\ell'$, the right-hand side of Equation (3.19) is the single coefficient $c_{j,k'\ell'}$. Therefore, by removing primes from the resulting expression, one obtains a formula for the expansion coefficient $c_{j,k\ell}$, for use in Equation (3.18),

$$c_{j,k\ell}=\frac{1}{FD}\int\limits_{-D/2}^{D/2}\int\limits_{-F/2}^{F/2}\eta_{Cj}(y,z)\psi_{yk}(y)\psi_{z\ell}(z)dydz. \tag{3.20}$$

Interpretation of Expansion Coefficients

A useful interpretation of the expansion coefficients $c_{j,k\ell}$ is obtained by considering the square of the representation, Equation (3.18),

$$\eta_{Cj}^2(y,z)=\sum_{k=0}^{\infty}\sum_{\ell=0}^{\infty}c_{j,k\ell}\psi_{yk}(y)\psi_{z\ell}(z)\sum_{k'=0}^{\infty}\sum_{\ell'=0}^{\infty}c_{j,k'\ell'}\psi_{yk'}(y)\psi_{z\ell'}(z)$$

$$=\sum_{k=0}^{\infty}\sum_{k'=0}^{\infty}\sum_{\ell=0}^{\infty}\sum_{\ell'=0}^{\infty}c_{j,k\ell}c_{j,k'\ell'}\psi_{yk}(y)\psi_{yk'}(y)\psi_{z\ell}(z)\psi_{z\ell'}(z). \tag{3.21}$$

Multiplying Equation (3.21) by $1/(FD)$, integrating over the full working-surface, then using the orthogonal property, Equations (3.16b) and (3.17b), there follows

$$\frac{1}{FD}\int\limits_{-D/2}^{D/2}\int\limits_{-F/2}^{F/2}\eta_{Cj}^2(y,z)dydz$$

$$=\sum_{k=0}^{\infty}\sum_{k'=0}^{\infty}\sum_{\ell=0}^{\infty}\sum_{\ell'=0}^{\infty}c_{j,k\ell}c_{j,k'\ell'}\frac{1}{F}\int\limits_{-F/2}^{F/2}\psi_{yk}(y)\psi_{yk'}(y)dy\frac{1}{D}\int\limits_{-D/2}^{D/2}\psi_{z\ell}(z)\psi_{z\ell'}(z)dz$$

$$=\sum_{k=0}^{\infty}\sum_{\ell=0}^{\infty}c_{j,k\ell}^2\frac{1}{F}\int\limits_{-F/2}^{F/2}\psi_{yk}^2(y)dy\frac{1}{D}\int\limits_{-D/2}^{D/2}\psi_{z\ell}^2(z)dz. \tag{3.22}$$

Moreover, using the normalization property, Equations (3.16a) and (3.17a), there follows from Equation (3.22),

$$\frac{1}{FD} \int\limits_{-D/2}^{D/2} \int\limits_{-F/2}^{F/2} \eta_{Cj}^2(y,z)dydz = \sum_{k=0}^{\infty}\sum_{\ell=0}^{\infty} c_{j,k\ell}^2. \qquad (3.23)$$

These normalizations have allowed a very simple interpretation of each expansion coefficient $c_{j,k\ell}$. The left-hand side of Equation (3.23) is the mean-square value of the deviation $\eta_{Cj}(y,z)$ averaged over the working-surface area FD, Figures 3.4 and 3.6. Therefore, according to Equation (3.23), the square of the expansion coefficient $c_{j,k\ell}$ is the contribution of the expansion term $c_{j,k\ell}\, \psi_{yk}\,(y)\psi_{z\ell}(z)$ in Equation (3.18) to this mean-square deviation, as can be seen directly from Equation (3.23).

The dimension of the working-surface-deviation $\eta_{Cj}(y,z)$ is length, for example, millimeters, micrometers, or microinches, and so on. By comparing the two sides of Equation (3.23), it follows that the dimension of $c_{j,k\ell}$ is length also, in the same units as $\eta_{Cj}(y,z)$. The absolute value $\left|c_{j,k\ell}\right|$ of the expansion coefficient $c_{j,k\ell}$, Equation (3.20), is the rms contribution of the term $c_{j,k\ell}\, \psi_{yk}\,(y)\,\psi_{z\ell}(z)$ in Equation (3.18) to $\eta_{Cj}(y,z)$. This property allows one to directly numerically assess the contributions of the individual expansion terms $\psi_{yk}(y)\psi_{z\ell}(z)$ to the working-surface-deviation $\eta_{Cj}(y,z)$. Furthermore, in the next chapter, it will be shown that the Legendre polynomial normalization leading to Equation (3.23) is instrumental in enabling a simple physically-meaningful interpretation of the rotational-harmonic spectrum contributions caused by the collective deviations of the working-surfaces of all teeth on a gear from equispaced perfect involute surfaces.

Interpretation of Expansion Terms as Elementary Deviations or Errors

It is useful to regard each term $c_{j,k\ell}\, \psi_{yk}\,(y)\,\psi_{z\ell}(z)$ as an elementary deviation or error. *Each* such term is characterized by the *pair* of indices k,ℓ. Because of the orthogonal properties, Equations (3.16b) and (3.17b), no single term characterized by the index pair k,ℓ can provide a contribution to any other term characterized by another index pair, even if one of the two indices of the two pairs of indices is the same. The deviation of the working-surface of each tooth j from an equispaced perfect involute surface is the superposition of the elementary deviations described by Equation (3.18).

The first few terms in the summation, Equation (3.18), possess the simple interpretations described in Table 3.1 (Mark, 1979, 1982).

A sketch of these deviation contributions is shown in Figure 3.7. Note that our sign convention defines a positive value of tooth deviation $\eta_{Cj}(y,z)$ as "removal" of material from a perfect involute surface. Hence, the terms $k=2$, $\ell=0$ and $k=0$, $\ell=2$ showing negative values at the center of the spans therefore describe positive fullness because of the sign convention of $\eta_{Cj}(y,z)$.

Accumulated Tooth-Spacing (Index) Errors

The lowest-order term $k=0$, $\ell=0$ labeled "(accumulated) tooth-spacing deviation" in Table 3.1 requires special consideration. From Equation (3.8) we observe that

Table 3.1 Elementary working-surface-deviation classifications

$k=0, \ell=0$	(accumulated) Tooth-spacing deviation
$k=1, \ell=0$	Linear-lead deviation
$k=0, \ell=1$	Linear-profile (involute) deviation
$k=1, \ell=1$	Combined linear-lead linear-profile deviation
$k=2, \ell=0$	Lead fullness deviation
$k=0, \ell=2$	Profile (involute) fullness deviation

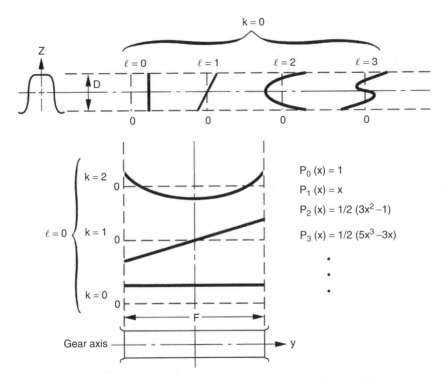

Figure 3.7 Sketch of low-order Legendre polynomials

$P_o(\xi)=1$, and therefore, from Equations (3.13) and (3.14), $\psi_{yo}(y)=1$ and $\psi_{zo}(z)=1$. Hence, from Equation (3.20) there follows, introducing a prime on the coefficient,

$$c'_{j,00} = \frac{1}{FD} \int\limits_{-D/2}^{D/2} \int\limits_{-F/2}^{F/2} \eta_{Cj}(y,z)dydz, \tag{3.24}$$

which is the average value of the deviation surface $\eta_{Cj}(y,z)$, averaged over the full tooth surface illustrated in Figure 3.4. Suppose the deviation surfaces $\eta_{Cj}(y,z), j=0, 1, 2, \ldots, N-1$ of all teeth have been measured and the coefficient $c'_{j,00}$, Equation (3.24), has been determined for every tooth. The values of $c'_{j,00}$, will be dependent on whichever tooth has been chosen as the "reference tooth". To remove the arbitrary choice of any particular tooth as the reference tooth, define $\overline{c}'_{00}$ as the average of the $c'_{j,00}$, that is,

$$\overline{c}'_{00} \triangleq \frac{1}{N} \sum_{j=0}^{N-1} c'_{j,00} \tag{3.25}$$

and

$$c_{j,00} \triangleq c'_{j,00} - \overline{c}'_{00}, \tag{3.26}$$

which removes the dependence of the term $c_{j,00}$ on whichever tooth is chosen as the reference tooth. Because the individual expansion terms $\psi_{yk}(y)_{z\ell}(z)$ are mutually orthogonal, the modification provided by Equation (3.26) has no effect on the remaining terms in Equation (3.18).

Accumulated tooth-spacing errors $c_{j,00}$ typically are an order of magnitude or more larger than other types of working-surface errors. Hence it is useful to express the mean-square deviation of a generic tooth j with its mean value $\overline{\eta}_{Cj}$ removed. Define

$$\overline{\eta}_{Cj} \triangleq \frac{1}{FD} \int_{-D/2}^{D/2} \int_{-F/2}^{F/2} \eta_{Cj}(y,z)dydz. \tag{3.27}$$

Then, from Equation (3.23),

$$\frac{1}{FD} \int_{-D/2}^{D/2} \int_{-F/2}^{F/2} \left[\eta_{Cj}(y,z) - \overline{\eta}_{Cj}\right]^2 dydz = \sum_{\substack{k=0 \\ except \\ k=0, \ell=0}}^{\infty} \sum_{\ell=0}^{\infty} c_{j,k\ell}^2 \tag{3.28}$$

where only the *single term* $k=0$, $\ell=0$ is excluded from the double summation.

Rectangular Array of Expansion Coefficients

For any particular tooth j, one can envision the expansion coefficients $c_{j,k\ell}$ being located in a rectangular array (quadrant) of points k,ℓ, $k=0, 1, 2, \ldots, \ell=0, 1, 2, \ldots$ with horizontal axis $k=0, 1, 2, \ldots$ and vertical axis $\ell=0, 1, 2, \ldots$. The sequences of coefficients on the horizontal axis, $c_{j,k0}$, for $\ell=0$, $k=0, 1, 2, \ldots$ described deviations constant in the z (radial) direction and varying in the y (axial) direction (Figure 3.4). Hence, this sequence of coefficients on the horizontal axis describes "pure" lead errors. The sequence of coefficients on the vertical axis, $c_{j,0\ell}$ for $k=0$, $\ell=0, 1, 2, \ldots$ describes deviations constant in the y (axial) direction and varying in the z (radial) direction (Figure 3.4). Hence, this sequence of coefficients on the vertical axis describes "pure" profile (involute) errors. The coefficients not on the axes, $k \neq 0$ and $\ell \neq 0$, that are interior to the k, ℓ region, describe "bias type" deviations that vary in both the axial y and radial z directions.

Since, in principle, the doubly-infinite sequence of coefficients $c_{j,k\ell}$, $k=0, 1, 2, \ldots$, $\ell=0, 1, 2, \ldots$ can be used to generate the deviation surface $\eta_{Cj}(y,z)$ exactly, as described by Equation (3.18) (Jackson, 1941, pp. 63–66; Bell, 1968, p. 57), these coefficients contain the same information as the original surface $\eta_{Cj}(y, z)$. (This concept is important because most of the analyses and calculations in this book are carried out using the expansion coefficients $c_{j, k\ell}$.)

As a practical matter, only a finite number of terms can be used in the representation of working-surface-deviations given by Equation (3.18). The representation of a generic function $f(\xi)$ by (one-dimensional) Legendre polynomials, Equations (3.8) and (3.9), satisfies the following least-squares property (Hildebrand, 1974, p. 331; Jackson, 1941, pp. 160–161): Among all polynomials $y_n(\xi)$ of degree n or less, the integrated squared error

$$\int_{-1}^{1} \left[f(\xi) - y_n(\xi) \right]^2 d\xi \tag{3.29}$$

is least when $y_n(\xi)$ is the Legendre polynomial representation of $f(\xi)$. This property is a principal reason for our use of Legendre polynomials.

Finally, we wish to emphasize one important concept. For each tooth $j, j = 0, 1, 2, \ldots,$ N -1 $c_{j, k\ell}$ uniquely characterizes a particular deviation pattern $c_{j, k\ell} \, \psi_{yk}(y)\psi_{z\ell}(z)$, as is evident from Equations (3.18) and (3.20). Hence, for each tooth j, and each k, ℓ pair of indices, the single coefficient $c_{j, k\ell}$ describes the amplitude of a unique "elementary deviation pattern" defined over the two-dimensional rectangular tooth-working-surface region $(-F/2) \leq y \leq (F/2)$ and $(-D/2) \leq z \leq (D/2)$. Therefore, apart from the amplitude $c_{j, k\ell}$ which varies with tooth number j, the *form* $\psi_{yk}(y)\psi_z\ell(z)$ of the "elementary deviation pattern" for each k, ℓ pair, is the same for every tooth $j = 0, 1, 2, \ldots,$ N -1.

3.6 Working-Surface Representation Obtained from Line-Scanning Tooth Measurements

In principle, the geometric deviation $\eta_{Cj}(y, z)$ of each tooth-working-surface $j = 0, 1, 2, \ldots,$ N -1 is characterized by the doubly-infinite sequence of expansion coefficients $c_{j, k\ell}, \ k = 0, 1, 2, \ldots, \ \ell = 0, 1, 2, \ldots$ as indicated by Equations (3.18) and (3.20). An approximation to these expansion coefficients must be obtained from currently available dedicated CNC gear-measurement machines. These machines are capable of making line-scanning profile measurements (scanning in a radial direction) and line-scanning lead measurements (scanning in an axial direction) as previously illustrated in Figure 3.5. Along each such line-scanning measurement a very accurate determination of the working-surface-deviation is obtained. However, some sort of interpolation *across* the line-scanning measurements is required to obtain an approximation to the working-surface-deviations not directly on the line-scanning measurements. (The measurements are assumed to be taken in a direction defined by the intersection of the plane of contact and transverse plane.) In principle, then, only a "dense" set of *either* profile measurements or lead measurements, not both, is required. Consequently, we shall designate as the primary set of measurements, one set of measurements, either profile or lead, possibly supplemented by the other set as secondary measurements. For spur gears, the primary set of measurements should be profile measurements (because lines of tooth contact are in an axial direction, which effectively "interpolate" or average across the profile measurements.)

For helical gears, the choice of the primary set, either profile or lead, should be made based on which set will provide the most accurate representation with the least measurement effort, that is, measurement time. Relevant factors are the type of intentional working-surface modification, the pattern of manufacturing errors on the working-surfaces, for example, the "lay" (Whitehouse, 1994, p. 78) of the surfaces, and whether the purpose of the measurements and analysis is to understand and diagnose the source of "ghost" or "phantom tones" (Cluff, 1992, pp. 7, 38, 94–97; Mark, 1992a). Criteria for determining which type of primary measurement set (profile or lead) is best, and how many of such scanning measurements are required, are discussed in the Appendices 3.A and 3.B to this chapter. If it is known that one type of measurement, either profile or lead, is more accurate than the other type, then use of the more accurate type for the primary set of measurements would almost surely be the best choice.

Choice of Lead Measurements as Primary

Suppose, based on one of the criteria of Appendices 3.A or 3.B, that lead measurements have been chosen as the primary set for representing the geometric working-surface-deviations of a helical gear. A set of four such lead measurement locations and three profile measurement locations is sketched in Figure 3.8. (In a real application, significantly more than four scanning measurements normally would be required in the primary set.)

To evaluate the expansion coefficients $c_{j,k\ell}$ of Equations (3.18) and (3.20), the double integration in Equation (3.20) is evaluated as an iterated integral, for example

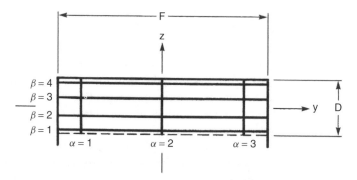

Figure 3.8 Positions of three scanning profile measurements and four scanning lead measurements located at zeros of normalized Legendre polynomials of degrees three and four, respectively

Widder (1961, pp. 185, 186). Define

$$c_{j,k.}(z) \triangleq \frac{1}{F} \int_{-F/2}^{F/2} \eta_{Cj}(y,z)\psi_{yk}(y)dy, \qquad (3.30)$$

then, from Equations (3.20) and (3.30),

$$c_{j,k\ell} = \frac{1}{D} \int_{-D/2}^{D/2} c_{j,k.}(z)\psi_{z\ell}(z)dz, \qquad (3.31)$$

which is applicable when lead measurements are chosen as the primary set. Suppose n scanning lead measurements are to be utilized in the primary set located at radial locations z_β, $\beta = 1, 2, \ldots, n$ as in Figure 3.8 where $n = 4$. For each such lead measurement, the integration in Equation (3.30) can be carried out numerically at the location z_β, that is,

$$c_{j,k.}(z_\beta) = \frac{1}{F} \int_{-F/2}^{F/2} \eta_{Cj}(y, z_\beta)\psi_{yk}(y)\, dy, \quad \beta = 1, 2, \ldots, n, \quad k = 0, 1, 2, \ldots, M. \qquad (3.32)$$

Non-smoothed lead and profile measurements should be utilized. Such digitized line-scanning measurements normally consist of several hundred measurement values distributed along the measurement. Because surface roughness is almost always present in such measurements, use of the "trapezoidal rule," for example Lanczos (1956, pp. 380–390), is most appropriate for evaluating the integrals in Equation (3.32) for each scanning lead measurement located at radial location z_β, $\beta = 1, 2, \ldots, n$. For each such scanning measurement, the integration in Equation (3.32) must be evaluated for each different normalized Legendre polynomial $\psi_{yk}(y)$, $k = 0, 1$, $2, \ldots, M$ up to some value of M limited only by the number and spacing of digitized sample points in the lead measurements.

For each value of k, the remaining integral in Equation (3.31) must be evaluated from the values of $c_{j,k.}(z_\beta)$ obtained at the n values of z_β, $\beta = 1, 2, \ldots, n$, that is,

$$c'_{j,k\ell} = \frac{1}{D} \int_{-D/2}^{D/2} c_{j,k.}(z_\beta)\psi_{z\ell}(z_\beta)dz, \qquad (3.33)$$

where a prime has been added to the coefficient $c_{j,k\ell}$ to denote that this remaining integration is only an approximation to its exact value, due to the fact that it must be approximated from only n locations of z_β within $(-D/2) \le z \le (D/2)$. In principle, these locations can be chosen freely.

If the radial locations z_β, $\beta = 1, 2, \ldots, n$ of these n lead measurements are chosen at the locations of the zeros of the normalized Legendre polynomial $\psi_{zn}(z)$ of degree

n, Equation (3.14), then Gaussian quadrature can be used to evaluate the integrals in Equation (3.33). With this specific choice of locations z_β, $\beta = 1, 2, \ldots, n$ the accuracy achieved in evaluation of the integrals, Equation (3.33), is about the same as would be achieved with twice the number of equispaced lead measurements. See for example, Lanczos (1956, pp. 396–400) and Hildebrand (1974, pp. 379–392). A particularly clear treatment of Gaussian quadrature can be found in Cheney (1982, pp. 106–111, in particular Theorem 4 on p. 110). Hence, by locating the scanning lead measurements at the locations of the zeros of a normalized Legendre polynomial $\psi_{zn}(z)$, Equation (3.14), of degree n equal to the number of scanning lead measurements to be utilized, the best a priori accuracy is achieved in evaluating the expansion coefficients $c_{j,k\ell}$ for a given measurement effort and time.

To evaluate Equation (3.33) by Gaussian quadrature, define

$$\xi = \frac{z}{D/2} = \frac{2z}{D}, \tag{3.34}$$

hence,

$$z = \frac{D}{2}\xi, \qquad dz = \frac{D}{2}d\xi. \tag{3.35a,b}$$

Then, by using Equation (3.14), Equation (3.33) becomes

$$c'_{j,k\ell} = \frac{(2\ell + 1)^{1/2}}{2} \int_{-1}^{1} c_{j,k.}(D\xi_\beta/2)P_\ell(\xi_\beta)d\xi, \quad \ell = 0, 1, 2, \ldots, n-1 \tag{3.36}$$

which allows values of $c'_{j,k\ell}$ to be evaluated only up to $\ell = n - 1$. According to Equation (8.5.4) on p. 390 of Hildebrand (1974), the Gaussian quadrature evaluation of $c'_{j,k\ell}$, Equation (3.36), is

$$c'_{j,k\ell} = \frac{(2\ell + 1)^{1/2}}{2} \sum_{\beta=1}^{n} H_\beta c_{j,k.}(z_\beta)P_\ell(2z_\beta/D)$$

$$= \frac{1}{2} \sum_{\beta=1}^{n} H_\beta c_{j,k.}(z_\beta)\psi_{z\ell}(z_\beta),$$

$$\ell = 0, 1, 2, \ldots, n-1$$

$$k = 0, 1, 2, \ldots, M. \tag{3.37}$$

From Equation (8.5.8) of Hildebrand (1974, p. 391) and Equations (8.4.18) and (8.5.3) of Hildebrand (1974, p. 390), the coefficients H_β, $\beta = 1, 2, \ldots, n$ in Equation (3.37) can be evaluated by either of the two equivalent expressions

$$H_\beta = \frac{2\left(1 - \xi_\beta^2\right)}{(n+1)^2 P_{n+1}^2(\xi_\beta)} = \frac{2}{\sum_{\ell=0}^{n-1}(2\ell + 1)P_\ell^2(\xi_\beta)}, \qquad \beta = 1, 2, \cdots, n \tag{3.38}$$

where ξ_β denotes the βth zero of the classical Legendre polynomial $P_n(\xi)$, $\beta = 1, 2, \ldots,$ n, and $P_{n+1}(\xi)$ and $P_\ell(\xi)$ are the classical Legendre polynomials of degree $n + 1$ and ℓ, respectively.

Values of the Legendre polynomials $P_n(\xi)$, $n = 0, 1, 2, \ldots,$ can be computed from Equation (3.9) and $P_0(\xi)$ and $P_1(\xi)$, Equation (3.8). Algorithms exist for computing the Legendre-polynomial zeros, for example Press *et al.* (1999, pp. 144–154). A table of Legendre-polynomial zeros and the weights H_β can be found in Stroud and Secrest (1966) for all polynomial orders from 1 to 64.

Equation (3.32) enables accurate computation of the normalized Legendre expansion coefficients $c_{j,k}$ (z_β), $k = 0, 1, 2, \ldots, M$ for each scanning lead measurement radially located at z_β (Figure 3.8). As the number n of lead measurements located at the normalized Legendre polynomial zeros z_β, $\beta = 1, 2, \ldots, n$ increases indefinitely, it is known that the errors in the Gaussian quadrature approximation, Equation (3.37), to the integration, Equation (3.31), will converge to zero for every continuous function (Cheney, 1982, p. 111). We have found that the above-described procedures work exceedingly well.

Because the accuracy achievable in transmission error computations is critically dependent on evaluation of the expansion coefficients $c_{j,k\ell}$ described by Equation (3.20), additional comment pertaining to the above-described procedure for evaluating the approximation $c'_{j,k\ell}$ to $c_{j,k\ell}$ is in order. Since each of the numerical integrations described by Equation (3.32) is an integration along a scanning lead measurement consisting of hundreds of points, these integrations can be carried out to a very high degree of accuracy – the primary accuracy limitation is that of the actual measurement data. In contrast to the integrations, Equation (3.32), the accuracy of the integrations, Equation (3.33), is limited by the number n of scanning lead measurements made.

Equation (3.30) describes the exact values of $c_{j,k}$ (z) at all values of $-(D/2) \le z \le (D/2)$ for which evaluation of Equation (3.32) will provide very accurate values of $c_{j,k}$ (z) at the n lead line-scanning locations z_β, $\beta = 1, 2, \ldots, n$. Suppose for a value of k, the exact evaluation of the function $c_{j,k}$ (z) by Equation (3.30) could be expressed by a polynomial in z of degree $2n - 1$. (This might be accomplished by an expansion in Legendre polynomials with the highest-order polynomial being of degree $2n - 1$ or less.) Then in this case, by using the Gaussian quadrature evaluation, Equation (3.37), of the coefficient $c'_{j,k\ell}$, the evaluation of $c_{j,k\ell}$ would be exact (Cheney, 1982, p. 110, Theorem 4). By requiring the n lead measurements to be made at the n zeros of the normalized Legendre polynomial $\psi_{zn}(z)$, the accuracy achieved is comparable to what would be expected by utilizing $2n$ scanning lead measurements, taken at other locations, instead of the n taken at the Legendre zeros. See, for example Lanczos (1956, pp. 396–400). Moreover, for every value of $k = 0, 1, 2, \ldots, M$ the expansion $\sum_{\ell=0}^{n-1} c'_{j,k\ell} \psi_{z\ell}(z)$ interpolates, exactly, the values $c_{j,k}$ (z_β), Equation (3.32), for all z_β, $\beta = 1, 2, \ldots, n$ (Mark, 1983).

The same considerations apply to the case where scanning profile measurements are taken as the primary set of measurements.

Choice of Profile Measurements as Primary

Consider, now, the case where profile measurements have been chosen as the primary set, based on one of the criteria described in Appendices 3.A or 3.B. The required integration procedures are exactly analogous to those described above, except that the rolls of lead measurements and profile measurements are reversed. Therefore, instead of Equation (3.30) one has for the present case of profile measurements as the primary set

$$c_{j,\cdot\ell}(y) = \frac{1}{D} \int_{D/2}^{D/2} \eta_{Cj}(y,z)\psi_{z\ell}(z)dz, \tag{3.39}$$

and instead of Equation (3.31),

$$c_{j,k\ell} = \frac{1}{F} \int_{F/2}^{F/2} c_{j,\cdot\ell}(y)\psi_{yk}(y)dy. \tag{3.40}$$

Suppose m scanning profile measurements are to be utilized. For each such profile measurement located at axial location y_α, $\alpha = 1, 2, \ldots, m$ the integration in Equation (3.39) would be carried out numerically, yielding

$$c_{j,\cdot\ell}(y_\alpha) = \frac{1}{D} \int_{D/2}^{D/2} \eta_{Cj}(y_\alpha,z)\psi_{z\ell}(z)dz, \qquad \begin{matrix} \alpha = 1,2,\cdots,m \\ \ell = 0,1,2,\cdots,N' \end{matrix} \tag{3.41}$$

for each normalized Legendre polynomial $\psi_{z\ell}$ (z), Equation (3.14), $\ell = 0, 1, 2, \ldots,$ N' where N' is limited by the number and spacing of the digitized profile sample points. The m scanning profile measurements are to be located at the m zeros of a normalized Legendre polynomial of degree m, ψ_{ym} (y), Equation (3.13). Denote these zero locations by y_α, $\alpha = 1, 2, \ldots, m$. An approximation $c'_{j,k\ell}$ to the remaining integration, Equation (3.40), for $k = 0, 1, 2, \ldots, m - 1$,

$$c'_{j,k\ell} = \frac{1}{F} \int_{-F/2}^{F/2} c_{j,\cdot\ell}(y_\alpha)\psi_{yk}(y_\alpha)dy \tag{3.42}$$

is to be evaluated by Gaussian quadrature. By a treatment that completely parallels Equations (3.34–3.38), the resultant Gaussian quadrature evaluation of Equation (3.42) is

$$c'_{j,k\ell} = \frac{1}{2} \sum_{\alpha=1}^{m} H_\alpha c_{j,\cdot\ell}(y_\alpha)\psi_{yk}(y_\alpha), \qquad \begin{matrix} k = 0,1,2,\cdots,m-1 \\ \ell = 0,1,2,\cdots,N' \end{matrix} \tag{3.43}$$

where

$$H_\alpha = \frac{2\left(1-\xi_\alpha^2\right)}{(m+1)^2 P_{m+1}^2\left(\xi_\alpha\right)} = \frac{2}{\sum_{k=0}^{m-1}(2k+1)P_k^2\left(\xi_\alpha\right)}, \qquad \alpha = 1,2,\cdots,m \tag{3.44}$$

where ξ_α denotes the αth zero of the classical Legendre polynomial $P_m(\xi)$, $\alpha = 1$, $2, \ldots, m$.

Contributions from Secondary Measurement Sets

If lead measurements are chosen as the primary set, then there would be no way to capture short wavelength behavior and surface roughness in the radial direction unless some profile measurements also are made. In such situations, a set of m (fewer) profile measurements also would be taken as the secondary set, located at the zeros of a normalized Legendre polynomial $\psi_{ym}(y)$. Conversely, if profile measurements are chosen as the primary set, then to capture short wavelength behavior and surface roughness in the axial direction, a set of n (fewer) lead measurements also would be taken as the secondary set, located at the zeros of a normalized Legendre polynomial $\psi_{zn}(z)$. If lead measurements constitute the secondary set of measurements, then Equations (3.32) and (3.37) yield the expansion coefficients $c'_{j,k\ell}$ for that set. If profile measurements constitute the secondary set, then Equations (3.41) and (3.43) yield the expansion coefficients $c'_{j,k\ell}$ for that set. How to handle the resultant redundant sets of coefficients is addressed below.

In principle, the expansion coefficients $c'_{j,k\ell}$, of the two-dimensional expansion, Equation (3.18) of the working-surface-deviation $\eta_{Cj}(y, z)$ of tooth j, will fill an entire quadrant $k = 0, 1, 2, \ldots; \ell = 0, 1, 2, \ldots$ at the integer locations k, ℓ. The Gaussian quadrature approximations fill only a region of this k, ℓ quadrant as illustrated in Figure 3.9. Figure 3.9a illustrates the $c'_{j,k\ell}$ locations obtained when lead measurements are the primary measurement set, and Figure 3.9b illustrates the $c'_{j,k\ell}$ locations when profile measurements are the primary set.

In each of Figures 3.9a,b, the region near the origin $k = 0$, $\ell = 0$ has (redundant) contributions from both the primary and secondary sets of measurements. In principal, the values of these redundant coefficients from each of the two measurement sets should be the same, but of course, they will differ somewhat. If there is prior knowledge pertaining to tooth-working-surface imperfections, such as the "lay" of the surface (Whitehouse, 1994, p. 78), this knowledge might suggest use of expansion coefficients $c'_{j,k\ell}$ obtained from one measurement set or the other in the region where coefficients are available from both sets of measurements. If no such prior knowledge is available, use of the diagonals shown in Figure 3.10 to designate which coefficients to use is the best a priori choice. Coefficient locations falling on the diagonal should be chosen from the primary measurement set (which has the larger number of line-scanning measurements).

For each measured tooth j, Equation (3.18) can be used to regenerate the working-surface-deviation $\eta_{Cj}(y, z)$ by using the available expansion coefficients $c'_{j,k\ell}$ from both measurement sets, as described above and illustrated, for example, in Figure 3.10. Examples of these regenerated surfaces are shown in the following pages. Consequently, these expansion coefficients contain all available information pertaining to the geometric deviation of the working-surface of each tooth j from a perfect involute surface. The various computational algorithms developed in the following chapters use these expansion coefficients to characterize the tooth-working-surface-deviations.

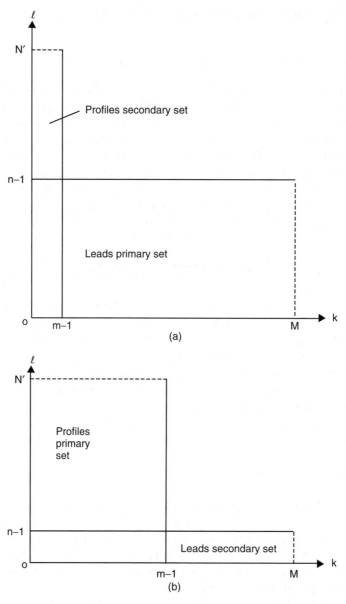

Figure 3.9 Regions in k, ℓ plane occupied by profile and lead measurement expansion coefficients $c'_{j,k\ell}$. (a) lead measurements are primary set, (b) profile measurements are primary set

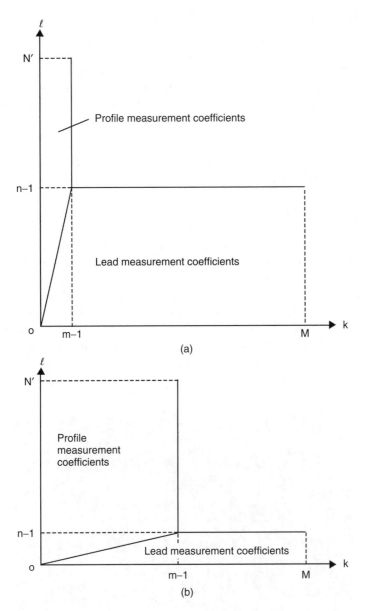

Figure 3.10 Regions in k,ℓ plane where coefficients $c'_{j,k\ell}$ from profile measurements and lead measurements are suggested to be utilized

Mark (1983) describes a rather complicated procedure titled "Reconstruction of Deviations of a Tooth Surface in Two Dimensions". Because of the two-dimensional orthogonal property of the expansion functions $\psi_{yk}(y)\psi_{z\ell}(z)$, Equations (3.13) and (3.14), the complicated two-dimensional reconstruction procedure described there is entirely unnecessary, and offers no advantage over the above-described method of representing tooth deviations from line-scanning measurements of tooth-working-surfaces.

3.7 Example of Working-Surface Generations Obtained from Line-Scanning Measurements

To illustrate use of the above-described method for measurement of gear-tooth deviations, computation of the normalized Legendre-polynomial expansion coefficients $c'_{j,k\ell}$, and regeneration of tooth-working-surface-deviations, the (smaller) pinion of the helical-gear-pair illustrated in Figure 3.11 was measured in detail. This pinion has $N = 38$ teeth. Involute roll-angle values ε_r and ε_t near the tooth-root and tip locations, respectively, were identified that determined the radial region of interest of the tooth deviations. From these roll-angle values, transverse pressure angle ϕ, and base-circle radius R_b, the Cartesian radial coordinate z was determined by Equations (3.4) and (3.6), and the total radial span of interest D was determined by Equation (3.7). Using these quantities, the locations of 15 lead-measurement roll-angle and z coordinate values were computed from the locations of the zeros

Figure 3.11 The working-surface-deviations of all $N = 38$ teeth on the smaller helical pinion were measured using 15 line-scanning lead measurements and 7 line-scanning profile measurements located at normalized Legendre-polynomial zeros (From Mark and Reagor (2001), Reproduced by permission of the American Gear Manufacturers Association)

of a normalized Legendre polynomial of degree 15. These 15 line-scanning lead measurements were the primary set of measurements. The axial range F of interest was chosen, and this value of F allowed the locations of the seven line-scanning profile measurements to be determined at the zeros of a normalized Legendre polynomial of degree 7. These seven profile measurements were the secondary set of measurements. (The 15 Legendre-zero locations in the radial z direction and the seven Legendre-zero locations in the axial y direction were scaled to the radial span of interest D and the axial span of interest F, respectively.) All $N = 38$ teeth were measured as described above.

Individual Tooth Deviations

From the above-described lead and profile measurements, the normalized Legendre polynomial expansion coefficients were computed as described in the preceding section of this chapter. This computation was carried out for each of the $N = 38$ teeth on the pinion. For two consecutive teeth on the pinion, labeled tooth $j = 0$ and $j = 1$, the working-surface-deviations were regenerated from the expansion coefficients $c'_{j,k\ell}$ by using Equation (3.18). These regenerated working-surface-deviations are displayed in Figures 3.12 and 3.13.

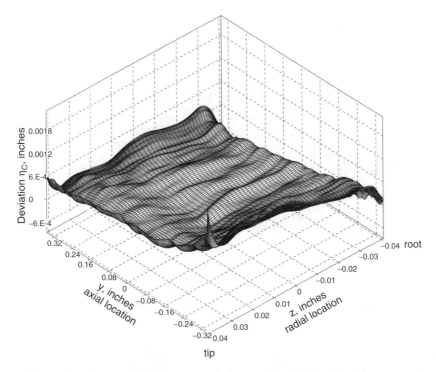

Figure 3.12   Working-surface-deviations of tooth $j = 0$ of 38 tooth pinion re-generated by Equation (3.18) from Legendre expansion coefficients $c'_{0,k\ell}$ obtained from line-scanning measurements made on tooth $j = 0$. All dimensions in inches (From Mark and Reagor (2001), Reproduced by permission of the American Gear Manufacturers Association)

Each of the two working-surfaces displays significant structure, which differs between the two surfaces. These differences are real, and are attributable to manufacturing deviations that differ from one tooth to another. (These differences will be diagnosed and explained in a later chapter.) This pinion was hobbed, but not "finished" after hobbing.

The visible edge on the left side of the surfaces, "rolled-up," is the tip of the teeth and the right side of the surfaces, "rolled down," is the root of the teeth. (Recall that our sign convention is: positive deviation is removal of material from perfect involute surfaces.) The Legendre-zero sampling of the line-scanning lead and profile measurements, and the Gaussian-quadrature integrations utilized in computation of the expansion coefficients $c'_{j,k\ell}$, has enabled regeneration of the working-surface-deviations by Equation (3.18) with the remarkable detail illustrated in Figures 3.12 and 3.13. In particular, the close spacing of the scanning lead measurements taken at the Legendre polynomial zeros near the ends of their intervals of definition has allowed the accurate representation of the tooth surfaces shown in Figures 3.12 and 3.13 near the tooth tips and tooth roots.

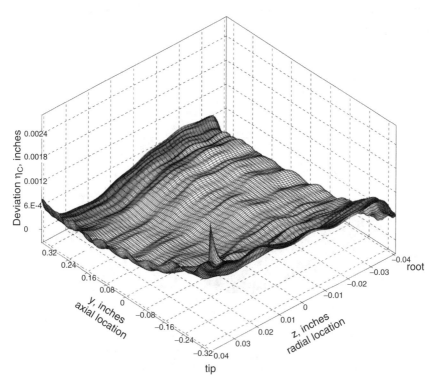

Figure 3.13 Working-surface-deviations of tooth $j = 1$ of 38 tooth pinion re-generated by Equation (3.18) from Legendre expansion coefficients $c'_{1,k\ell}$ obtained from line-scanning measurements made on tooth $j = 1$. All dimensions in inches (From Mark and Reagor (2001), Reproduced by permission of the American Gear Manufacturers Association)

Average Deviation Surfaces

The working-surfaces of power transmission gears normally are modified from the involute tooth form to compensate for elastic deformations and tooth-spacing errors. In virtually all applications, the modifications are intended to be the same for every tooth on a gear under consideration. It therefore is of considerable interest to the design engineer to determine how close the actual measured tooth modifications are to those intended by the designer.

Such modifications of the tooth-working-surfaces are intended to reduce or minimize the *tooth-meshing-harmonics* of the vibration excitation and noise caused by meshing-gear-pairs. It will be shown in the next chapter that it is the deviation surface formed by the average of all teeth $j = 0,1, \ldots, N-1$ on a gear that, together with tooth/gearbody elastic deformations, causes the tooth-meshing-harmonic contributions to the transmission error. Therefore, the quantity to be compared with the desired working-surface modifications is the average deviation surface,

$$\bar{\eta}_C(y,z) \triangleq \frac{1}{N} \sum_{j=0}^{N-1} \eta_{Cj}(y,z). \tag{3.45}$$

Inserting Equation (3.18) into Equation (3.45), interchanging the order of summation, and defining

$$\bar{c}'_{k\ell} \triangleq \frac{1}{N} \sum_{j=0}^{N-1} c'_{j,k\ell}, \qquad \begin{aligned} k &= 0,1,2,\cdots \\ \ell &= 0,1,2,\cdots \end{aligned} \tag{3.46}$$

there follows,

$$\bar{\eta}_C(y,z) = \sum_{k=0}^{\infty} \sum_{\ell=0}^{\infty} \bar{c}'_{k\ell} \psi_{yk}(y) \psi_{z\ell}(z), \tag{3.47}$$

where the double summation in Equation (3.47) includes all of the approximate expansion terms $k = 0, 1, 2, \ldots; \ell = 0, 1, 2, \ldots$ as illustrated in Figure 3.10. Equation (3.47) describes generation of the average deviation surface $\bar{\eta}_C(y,z)$ by using, for each k, ℓ pair, the average over all N teeth of the approximate expansion coefficient $c'_{j,k\ell}$ obtained by Equation (3.46). This procedure is a very computationally efficient method for obtaining the average working-surface-deviation.

The approximate expansion coefficients $c'_{j,k\ell}$ were generated from the line-scanning measurements of all 38 teeth on the pinion shown in Figure 3.11. For each k, ℓ pair, these 38 expansion coefficients were averaged as shown in Equation (3.46). The resultant average working-surface-deviation then was regenerated as shown by Equation (3.47). This average working-surface-deviation is displayed in Figure 3.14. Figure 3.14 shows that most of the tooth-to-tooth variations in manufacturing deviations, illustrated in Figures 3.12 and 3.13, are not found in the average deviation surface displayed in Figure 3.14. It is such an average deviation surface that a designer would want to compare with the working-surface modifications he or she had prescribed.

In the next chapter, it is shown that the tooth-to-tooth variations in working-surface-deviations, as illustrated in Figures 3.12 and 3.13, are the primary cause of the rotational-harmonic contributions to gear kinematic transmission errors.

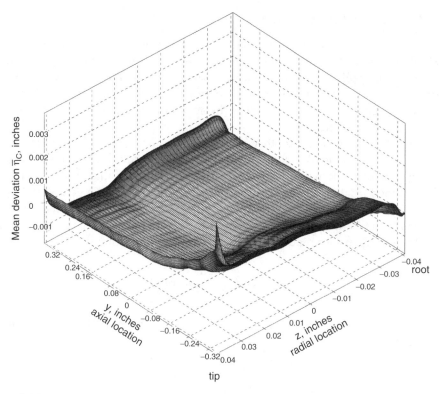

Figure 3.14 Average deviation surface of all $N = 38$ working-surfaces of 38 tooth pinion computed by Equation (3.47) from the average values, Equation (3.46), of the Legendre expansion coefficients. All dimensions in inches (From Mark and Reagor (2001), Reproduced by permission of the American Gear Manufacturers Association)

Appendix 3.A. Method for Estimating Required Number of Primary Line-Scanning Measurements Based on Surface-Roughness Criteria

The measurement criterion described below is useful for obtaining an accurate representation of working-surface-deviations with amplitudes larger than the rms surface roughness. Use of this rms criterion should enable accurate representation of form deviations, such as intentional working-surface modifications, and other relatively long-wavelength deviations with amplitudes larger than the rms surface roughness.

Yet, some generating-type manufacturing operations are known to produce working-surface "undulation errors" (Cluff, 1992, pp. 7, 38, 94–97; Mark, 1992a) with amplitudes smaller than the rms surface roughness that can cause unacceptable "ghost-tone" vibration excitations and noise, even for automotive gears. A working-surface measurement method is described in Appendix 3.B that will provide an accurate representation of such undulation errors producing "ghost-tone" rotational harmonics of rotational-harmonic numbers no larger than a prescribed harmonic number n used to determine the required number of primary line-scanning measurements.

RMS Criterion Method

1. Identify a typical gear tooth and choose the rectangular analysis region on the tooth-working-surfaces to be measured. The radial range of this analysis region is designated by the roll-angle values ε_t and ε_r of Equations (3.4–3.7). The axial range of this region is designated by F.

2. Begin determination if lead or profile measurements are to be chosen as the primary measurement set. To make this determination, ideally measure three scanning lead measurements radially located at the zeros z_β, $\beta = 1, 2, 3$ of the normalized Legendre polynomial $\psi_{z3}(z)$ of degree 3, Equation (3.14), and ideally measure three scanning profile measurements axially located at the zeros y_α, $\alpha = 1, 2, 3$ of the normalized Legendre polynomial $\psi_{y3}(y)$ of degree 3, Equation (3.13). The locations of six such scanning measurements are illustrated in Figure 3.5. For each of the three scanning lead measurements located at z_β, $\beta = 1, 2, 3$ compute and store the entire available set of expansion coefficients $c_{j,k.}(z_\beta)$, Equation (3.32), by numerical integration, for $k = 0, 1, 2, \ldots, M$ up to a value of M limited only by the number of samples and sampling interval utilized in obtaining the measurements $\eta_{Cj}(y, z_\beta)$, $\beta = 1, 2, 3$. Carry out the comparable computation of the expansion coefficients $c_{j,\cdot\ell}(y_\alpha)$, Equation (3.41), $\alpha = 1, 2, 3$ by numerical integration of Equation (3.41) for $\ell = 1, 2, \ldots, N'$ up to a value of N' limited only by the sampling of the three profile measurements $\eta_{Cj}(y_\alpha, z)$, $\alpha = 1, 2, 3$. For $M \to \infty$, the expansion coefficients $c_{j,k.}(z_\beta)$ obtained from the lead measurements satisfy the generalized Parseval relation (Jackson, 1941, p. 216) applied to Equation (3.32),

$$
\frac{1}{F} \int_{-F/2}^{F/2} \eta_{Cj}^2(y, z_\beta) dy = \sum_{k=0}^{\infty} c_{j,k.}^2(z_\beta) , \qquad \beta = 1, 2, 3 \tag{3.A.1}
$$

and for $N' \to \infty$, the expansion coefficients $c_{j,\cdot\ell}(y_\alpha)$ obtained from the profile measurements satisfy the comparable relation applied to Equation (3.41),

$$
\frac{1}{D} \int_{-D/2}^{D/2} \eta_{Cj}^2(y_\alpha, z) dz = \sum_{\ell=0}^{\infty} c_{j,\cdot\ell}^2(y_\alpha) , \qquad \alpha = 1, 2, 3. \tag{3.A.2}
$$

Equations (3 A.1) and (3.A.2) are the one-dimensional counterparts of the two-dimensional relation, Equation (3.23).

3. For $\beta = 1, 2, 3$ numerically compute and store the left side of Equation (3.A.1) for the three lead measurements, and for $\alpha = 1, 2, 3$ numerically compute and store the left side of Equation (3.A.2) for the three profile measurements. Then compute

and plot as a function of K for the lead measurements

$$\left[\frac{1}{F} \int_{-F/2}^{F/2} \eta^2_{Cj}(y, z_\beta)dy - \sum_{k=0}^{K} c^2_{j,k} (z_\beta) \right]^{1/2}, \quad \beta = 1, 2, 3 \qquad (3.A.3)$$

and compute and plot as a function of L' for the profile measurements

$$\left[\frac{1}{D} \int_{-D/2}^{D/2} \eta^2_{Cj}(y_\alpha, z)dz - \sum_{\ell=0}^{L'} c^2_{j,\ell} (y_\alpha) \right]^{1/2}, \quad \alpha = 1, 2, 3. \qquad (3.A.4)$$

According to Equations (3.A.1) and (3.A.2) the above two quantities, Equations (3.A.3) and (3.A.4), will become vanishing small as $K \to \infty$ and $L' \to \infty$, respectively. For finite values of K and L', Equations (3.A.3) and (3.A.4) represent the rms "errors" in truncating the normalized Legendre expansions using terms only up to $k = K$ and $\ell = L'$, respectively. Each of the quantities of Equations (3.A.3) and (3.A.4) will decay very rapidly for small K and L', but for larger K and L' will decay much more slowly.

In forming the expansion coefficients $c'_{j,k\ell}$, Equations (3.36) and (3.37), one is effectively interpolating *across* the scanning lead measurements. In forming the coefficients $c'_{j,k\ell}$ by Equation (3.43), one is effectively interpolating *across* the scanning profile measurements. In such effective interpolations we cannot realistically hope to capture surface-roughness deviations, but we can aim to obtain accurate representations of form and, hopefully, waviness deviations (Whitehouse, 1994, pp. 102–114), and especially, intentional working-surface modifications. Therefore, for each value of $\beta = 1, 2, 3$ determine the smallest value of K for which the expression (3.A.3) is approximately equal to the rms surface roughness. Record the largest of these three smallest values of K, say K_m. Do the same for the three smallest values of L' from the expression (3.A.4) and record the largest of these three values of L', say L'_m. If the largest of the three K values K_m is smaller than the largest of the three L' values L'_m, then to minimize measurement time, use line-scanning *profile* measurements as the primary measurement set. Otherwise, use line-scanning lead measurements as the primary measurement set.

If scanning profile measurements are chosen as the primary measurement set, choose the number m of profile measurements to be equal to (or larger than) $K_m + 1$, that is, $m \geq K_m + 1$, where these scanning profile measurements are to be located at the zeros of a normalized Legendre polynomial $\psi_{ym}(y)$ of degree $m \geq K_m + 1$. Similarly, if scanning lead measurements are chosen as the primary measurements set, choose the number n of lead measurements to be equal to (or larger than) $L'_m + 1$, that is, $n \geq L'_m + 1$, to be located at the zeros of a normalized Legendre polynomial $\psi_{zn}(z)$ of degree $n \geq L'_m + 1$.

4. According to Lanczos (1956, pp. 371, 372), the resultant expansions using these effective interpolations across the line-scanning measurements will result in

accuracies comparable to those formed by *integrations* to obtain the expansion coefficients, which were used in forming the expansion coefficients in Equations (3.A.3) and (3.A.4). To fully understand this argument, it is necessary to recognize that, except for a reordering of operations, which will result in an *improvement* in accuracy, computation of the approximate expansion coefficients, $c'_{j,k\ell}$ by the Gaussian quadrature formulas, Equations (3.37) and (3.43) is *exactly* equivalent to interpolation between the line-scanning measurements, as is proved in Mark (1983, Equation (20)) and mentioned in Hildebrand (1974, p. 467).

5. In order to capture surface roughness characteristics in the direction across a primary set of line-scanning measurements, a secondary set of fewer line-scanning measurements is to be utilized.

6. Because the rms amplitudes of undulation errors, that are the source of ghost or phantom tones, often are smaller than rms surface roughness, when the purpose of the measurements and analysis is to understand and diagnose ghost tones, another requirement for choosing the number of primary line-scanning measurements is more appropriate, as is described next.

Appendix 3.B. Method for Estimating Required Number of Primary Line-Scanning Measurements for Case of Known Ghost-Tone Rotational-Harmonic Number

Ordinarily, the simultaneous multiple tooth contact of tooth-meshing action causes an averaging action of tooth-working-surface-deviations in such a manner that the transmission-error contribution from such deviations is much smaller than the amplitudes of the working-surface-deviations. However, some generating-type manufacturing operations cause undulation errors that experience negligible averaging-action attenuation from this multiple tooth contact (Cluff, 1992, pp. 7, 38, 94–97). The resultant transmission-error amplitude is essentially the same as the undulation-error amplitude on the tooth-working-surfaces. Such undulation errors with amplitudes smaller than the rms surface roughness on tooth-working-surfaces are known to be sources of unacceptable vibrations and noise – even for automotive gears. When the rotational-harmonic number of such unattenuated "ghost-tone" harmonics is known, it can be used to determine the required number of primary line-scanning measurements, as described below.

A reasonably full analysis of the transmission-error contributions resulting in such ghost tones is provided in Mark (1992a), and summarized in Section 6.6 of this book. The undulation-error cause of such ghost tones are sinusoidal-type errors on the tooth-working-surfaces. As shown by Equations (91a–c) of Mark (1992a), for such errors to experience no attenuation from the averaging action of simultaneous multiple tooth contact, two conditions must be satisfied. The direction of the lines of constant phase of these sinusoidal-type errors must coincide with the direction of lines of contact on tooth-working-surfaces, Equation (91b) of Mark (1992a), and the rotational-harmonic number n of the resultant ghost tones must satisfy Equation (95)

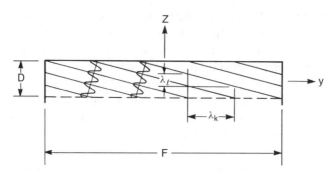

Figure 3.B.1   ``Pure'' undulation error on tooth-working-surface illustrating axial-wavelength λ_k and radial-wavelength λ_ℓ

of Mark (1992a), that is,

$$n\lambda_k = N\Delta_a,$$ (3.B.1a)

where λ_k is the wavelength of the sinusoidal undulation error "measured" in an axial direction (Figure 3.B.1), Δ_a is the axial pitch (Figure 2.7), and N is the number of teeth. But from Equation (2.11) and then Equation (2.6) there follows $\Delta_a = \Delta L_h/(2\pi R_b)$, where L_h is the lead of the helix "measured" on the base cylinder. Hence, from Equation (3.B.1a), there follows

$$n\lambda_k = \frac{N\Delta}{2\pi R_b}L_h = L_h.$$ (3.B.1b)

Therefore, the requirement of Equations (3.B.1a,b) is that there be an exact integer number n of axial wavelengths λ_k in one rotation of the gear.

From Equations (2.14) and (2.13), one has

$$\frac{F}{Q_a} = \frac{A\Delta}{L} = \Delta_a$$ (3.B.2)

which, when combined with Equation (3.B.1a) yields

$$\frac{F}{\lambda_k} = \frac{n}{N}\,Q_a,$$ (3.B.3)

where F/λ_k is the number of axial cycles of the sinusoidal undulation error within the axial facewidth F of the contact region, n is the resultant "ghost tone" rotational-harmonic number, N is the number of gear teeth, and Q_a is axial contact ratio of the contact region. Equation (3.B.3) is satisfied by a sinusoidal undulation error that has experienced no attenuation from gear meshing action, resulting in a ghost tone of rotational-harmonic number n.

A relation analogous to Equation (3.B.3) exists for the radial component of a sinusoidal undulation error experiencing no attenuation. From Equations 90, 91a, 51, and 48 of Mark (1992a), there follows

$$\frac{n}{N} = \rho_{k\ell}q_k = \frac{D}{A}\frac{\lambda_k}{\lambda_\ell}\frac{\Delta_a}{\lambda_k} = \frac{D}{A}\frac{\Delta_a}{\lambda_\ell},$$ (3.B.4)

where D is radial depth of the tooth contact region, Equation (3.7), A as shown in Figure 2.6, Δ_a is axial pitch, Figure 2.7, and λ_ℓ is radial wavelength, Figure 3.B.1. Utilizing Equations (2.13) and (2.9), there follows from Equation (3.B.4),

$$\frac{n}{N} = \frac{D}{A}\frac{A}{L}\frac{\Delta}{\lambda_\ell} = \frac{D}{Q_t \lambda_\ell}, \tag{3.B.5}$$

and therefore,

$$\frac{D}{\lambda_\ell} = \frac{n}{N}Q_t, \tag{3.B.6}$$

which is the radial analog of the axial relation, Equation (3.B.3). D/λ_ℓ is the number of radial cycles of the undulation error within the radial depth D of the tooth contact region, and Q_t is the transverse contact ratio of that contact region. Sinusoidal undulation errors that experience no attenuation from tooth-meshing action satisfy both of the Equations (3.B.3) and (3.B.6).

Because each class of line-scanning measurements, profile or lead, is continuous in one direction, for undulation errors unattenuated by multiple tooth contact, either of Equation (3.B.3) or (3.B.6) may be utilized in choosing the primary set of measurements. Normally, this choice would be made to minimize the number of line-scanning measurements required to accurately characterize the undulation error, which would be determined by the smaller of Equation (3.B.3) or (3.B.6), where n is the rotational-harmonic number of the ghost tone. If $Q_t > Q_a$ there are more radial cycles (D/λ_ℓ); hence, it is desirable to utilize line-scanning profile measurements in this case, and therefore Equation (3.B.3) would be utilized to determine the required number of profile measurements (not fewer than one) to accurately capture the axial fluctuation of the undulation. Conversely, if $Q_a > Q_t$, there are more axial cycles (F/λ_k); hence, it is desirable to utilize line-scanning lead measurements in this case, and therefore Equation (3.B.6) would be utilized to determine the required number of lead measurements to accurately capture the radial fluctuation of the undulation. In either of the above-two cases, the number of measurements, determined as described above, would be the primary set of measurements. The required interpolation to take place is then *across* the above-described primary set of line-scanning measurements. Since the undulation error is a sinusoid, the remaining problem is to determine the minimum number of line-scanning measurements located at the zeros of a normalized Legendre polynomial that, when interpolated across these measurements, will accurately represent a sinusoid of arbitrary phase with F/λ_k or D/λ_ℓ cycles.

Minimum Number of Line-Scanning Measurements Required for Accurate Representation of Undulation Errors

Since the above-described line-scanning measurements are made at the locations of the zeros of normalized Legendre polynomials, the problem is to determine the number of Legendre zeros required to accurately interpolate (i.e., represent) a sinusoid of arbitrary phase with a specified number of cycles (F/λ_k or D/λ_l) within the representation interval. The interpolation is carried out using the normalized Legendre polynomials of Equation (3.13) or (3.14), as described in Mark (1983, Equations (8)–(21)). Because Gaussian quadrature and interpolation yield identical results (Mark, 1983), the method is essentially that leading to Equations (3.37) and (3.43).

An extensive numerical study was carried out to determine the above interpolation requirements. Sinusoids with m cycles for every integer m between $m=1$ and $m=26$ were represented by interpolation as described above, each with three different phases. The goal was to determine the minimum number of Legendre-zero interpolation points required to accurately interpolate the sinusoids with no visible difference between the sinusoid being represented and the result obtained by interpolation. The required number of Legendre zeros n' as a function of the number of full cycles m of the sinusoid being interpolated was found to be capable of representation by the simple formula

$$n' = m\pi + k \tag{3.B.7}$$

where k is a function of m, and the computed value of n' is to be rounded off to the closest integer. The values of k as a function of m are given by Table 3.B.1. To determine k, the number of cycles computed by either of Equation (3.B.3) or (3.B.6) should be rounded *up* to the nearest integer m for use in the table.

There is a simple explanation of our use of π in the formula (3.B.7). Within the interval $-1 < x < 1$, $\sin(m\pi x + \varphi)$ has m full cycles and hence $2m$ zeros. Therefore, the distance in x between zeros is $1/m$. Within this same interval $-1 < x < 1$, the zeros of a Legendre polynomial are most densely spaced near the endpoints -1 and 1, and least densely spaced at the center $x=0$. As $n' \to \infty$, it can be shown from Szego (1939, p. 311) that at the center $x=0$ of the interval $-1 < x < 1$, the distance between Legendre polynomial zeros is π/n' for a Legendre polynomial of degree n'. For successful interpolation of a sinusoid, this Legendre zero spacing cannot be larger than the spacing between the sinusoid zeros of $1/m$. Equating these two spacings gives $(\pi/n') = (1/m)$ or $n' = m\pi$. One cannot expect a Legendre polynomial of order lower than $n' = m\pi$ to accurately interpolate $\sin(m\pi x + \varphi)$. The additional constant k in Equation (3.B.7) is required to account for the general phase of a sinusoid and the fact that the above argument pertaining to Legendre zeros is valid for asymptotically large values of n'. (A slightly improved asymptotic formula for Legendre zero spacing at $x=0$ is $(2\pi)/(2n'+1)$.)

Figure 3.B.2 illustrates the representation of a sine wave with $m=8$ full cycles by Legendre-polynomial interpolation of samples of the sine wave taken at the locations of the zeros of a Legendre polynomial of degree $n'=32$. The number of $n'=32$ sample points was computed by Equation (3.B.7) with a value of $k=7$ obtained from Table 3.B.1, yielding $n' = 8\pi + 7 = 32.13$, which rounded off to the nearest integer, gives $n' = 32$. The upper portion of Figure 3.B.2 is the Legendre polynomial representation of the sine wave generated using the 32 sample values shown in the lower portion of the figure. The (remarkable) Legendre polynomial representation of the upper curve, obtained by interpolation of the sample values, is a virtually exact representation of the sine wave shown in the lower curve.

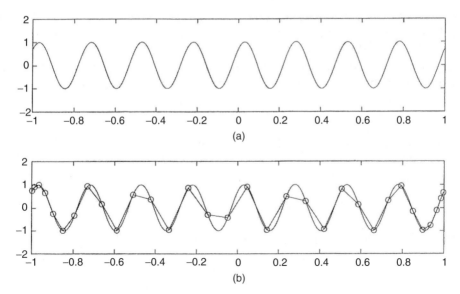

Figure 3.B.2 Upper curve (a) is Legendre polynomial reconstruction of the sinusoid, where the Legendre polynomial expansion coefficients were evaluated by Gaussian quadrature using only the 32 sample values of the sinusoid from the lower curve (b). Lower curve (b) shows 32 discrete samples of a sinusoid with eight full cycles. Abscissa sample locations are the zero locations of a Legendre polynomial of degree 32. (From Mark and Reagor (2001), Reproduced by permission of the American Gear Manufacturers Association)

As pointed out in Hildebrand (1974, p. 467) and shown explicitly in Mark (1983, Equation (20)), reconstruction of the sine wave (upper figure) from the encircled sample points (lower figure) shown in Figure 3.B.2 can be regarded either as an interpolation between the sample points, or as a reconstruction of the sine wave utilizing Gaussian quadrature. However, when viewing the locations of the encircled sample points, one normally would conclude that there exists no smooth interpolation between these sample points (in the conventional interpretation of interpolation). Hence, the Gaussian quadrature reconstruction of this remarkable result would seem to be the preferred interpretation.

In contrast to $n' = \pi m$, the asymptotic formula for Nyquist sampling is $n' = 2m$. Therefore, interpolation of sinusoids using Legendre polynomials, with samples taken at Legendre zeros, is about 50% less *asymptotically* efficient than interpolation of sinusoids using Nyquist sampling. However, the strongest deviations of gear tooth-working-surfaces from perfect involute surfaces generally occur near the radial and axial ends of the working-surfaces. As can be seen from the lower part of Figure 3.B.2, it is in these endpoint regions where the Legendre polynomial zeros (sample points) are most densely spaced, thereby providing more accurate interpolations in these regions where the strongest deviations from involute surfaces are found. Furthermore, the lowest-order term in a two-dimensional Legendre-polynomial representation is a constant, which is a tooth-spacing error, and the next-order terms are linear terms, which very efficiently represent deviations of tooth-working-surfaces caused by the frequently encountered skewness between gear cutting/finishing base-cylinder axes and gear-measurement or operating axes of rotation. These properties of Legendre polynomials enable very efficient representation of the above-described error types.

Example Calculation

Consider a helical gear with the following characteristics. The rectangular analysis region of assumed contact has been chosen. The axial Q_a and transverse Q_t contact ratios of this contact region are

$$Q_a = 1.1645 \qquad\qquad Q_t = 1.6094. \qquad (3.B.8)$$

The number of teeth on the gear is

$$N = 59. \qquad (3.B.9)$$

A "ghost" tone has been observed at rotational-harmonic number

$$n = 144. \qquad (3.B.10)$$

We wish to compute the number of profile or lead measurements, made at the locations of Legendre polynomial zeros, in order to fully represent by the above-described interpolation the undulation-error sinusoid responsible for the ghost tone at rotational-harmonic number $n = 144$.

Consider first computation of the required number of profile measurements. From Equation (3.B.3), the number m of axial cycles of the undulation error is

$$m = \frac{n}{N} Q_a = \frac{144}{59} \times 1.1645 = 2.84 \ axial\ cycles. \qquad (3.B.11)$$

If this number is rounded to 3, we observe from Table 3.B.1 that $k = 5$. Therefore, utilizing Equation (3.B.7), the number n' of profile measurements required to accurately represent this undulation harmonic, by interpolation across these profile measurements, is

$$n' = 2.84\pi + 5 = 13.9. \qquad (3.B.12)$$

Therefore, 14 profile measurements located at Legendre zeros will accurately interpolate this undulation sinusoid.

Table 3.B.1 Values of k for use in Equation (3.B.7)

m	1	2	3	4	5	6	7	8	9	10 to infinity
k	3	4	5	5	5	6	6	7	7	8

Now consider computation of the required number of lead measurements. From Equation (3.B.6), the number m of radial cycles of the undulation error is

$$m = \frac{n}{N}Q_t = \frac{144}{59} \times 1.6094 = 3.93 \; radial \; cycles. \qquad (3.B.13)$$

If this number is rounded to 4, we observe from Table 3.B.1 that $k = 5$ in this case also. Therefore, utilizing Equation (3.B.7), the number n' of lead measurements required to accurately represent this undulation harmonic, by interpolation across these lead measurements, is

$$n' = 3.93\pi + 5 = 17.3. \qquad (3.B.14)$$

Therefore, 17 lead measurements located at Legendre zeros will accurately interpolate this undulation sinusoid.

Because it normally is desirable to minimize the number of measurements, in this example one would choose the 14 profile measurements as the primary set of measurements. A significantly smaller number of lead measurements should suffice as the secondary set of measurements.

Because of the symmetry of Equations (3.B.3) and (3.B.6), only a single calculation is required to determine the number of primary measurements for a helical gear, by utilizing the smaller of Q_a or Q_t. For a spur gear, $Q_a = 0$, and for a constant amplitude undulation error, only a single profile measurement would be required, in principle, to capture the undulation error, in the case of a spur gear.

References

Baxter, M.L. (1962) Basic theory of gear-tooth action and generation, in *Gear Handbook*, Chapter 1 (ed. D.W. Dudley), 1st edn, McGraw-Hill, New York, pp. 1–1, 1–21.

Bell, W.W. (1968) *Special Functions for Scientists and Engineers*, D. Van Nostrand, London. Republished by Dover, Mineola, NY.

Cheney, E.W. (1982) *Introduction to Approximation Theory*, 2nd edn, Chelsea Publishing Company, New York.

Cluff, B.W. (ed.) (1992) *Gear Process Dynamics*, 7th edn, American Pfauter Limited Partnership, Loves Park, IL.

Gregory, R.W., Harris, S.L., and Munro, R.G. (1963–1964) Dynamic behavior of spur gears. *Proceedings of the Institution of Mechnical Engineers*, **178**, 207–218.

Harris, S.L. (1958) Dynamic loads on the teeth of spur gears. *Proceedings of the Institution of Mechnical Engineers*, **172**, 87–100.

Hildebrand, F.B. (1974) *Introduction to Numerical Analysis*, 2nd edn, McGraw-Hill, New York. Republished by Dover, New York.

Jackson, D. (1941) *Fourier Series and Orthogonal Polynomials*, The Mathematical Association of America, Buffalo, New York. Republished by Dover, Mineola, NY.

Lanczos, C. (1956) *Applied Analysis*, Prentice-Hall, Englewood Cliffs, NJ. Republished by Dover, New York.

Mark, W.D. (1979) Analysis of the vibratory excitation of gear systems. II: tooth error repre-sentations, approximations, and application. *Journal of the Acoustical Society of America*, **66**, 1758–1787.

Mark, W.D. (1982) Gear noise excitation, in *Engine Noise: Excitation, Vibration, and Radiation* (eds R. Hickling and M.M. Kamal), Plenum Press, New York, pp. 55–93.

Mark, W.D. (1983) Analytical reconstruction of the running surfaces of gear teeth using stan-dard profile and lead measurements. *ASME Journal of Mechanisms, Transmissions, Automation in Design*, **105**, 725–735.

Mark, W.D. (1992a) Contributions to the vibratory excitation of gear systems from periodic undulations on tooth running surfaces. *Journal of the Acoustical Society of America*, **91**, 166–186.

Mark, W.D. and Reagor, C.P. (2001) *Performance-Based Gear-Error Inspection, Specification, and Manufacturing-Source Diagnostics*, AGMA Technical Paper 01FTM6, American Gear Manufacturing Association, Alexandria, Virginia.

Press, W.H., Teukolsky, S.A., Vetterling, W.T., and Flannery, B.P. (1999) *Numerical Recipes in Fortran 77: The Art of Scientific Computing*, 2nd edn, Vol. **1** of Fortran Numerical Recipes, Cambridge University Press, Cambridge.

Stroud, A.H. and Secrest, D. (1966) *Gaussian Quadrature Formulas*, Prentice-Hall, Englewood Cliffs, NJ.

Szego, G. (1939) *Orthogonal Polynomials*, American Mathematical Society, Providence, RI.

Whitehouse, D.J. (1994) *Handbook of Surface Metrology*, Institute of Physics Publishing, Bristol and Philadelphia.

Widder, D.V. (1961) *Advanced Calculus*, 2nd edn, Prentice-Hall, Englewood Cliffs, NJ. Repub-lished by Dover, Mineola, NY.

4

Rotational-Harmonic Analysis of Working-Surface Deviations

Because frequency values of vibration/noise sources remain unchanged by transmission through linear time-invariant structural paths, vibration/noise analyses and measurements often are carried out in the frequency domain. Dominant frequencies arising from meshing gears are low-order rotational-harmonic frequencies, tooth-meshing harmonic frequencies, so-called sideband frequencies, and ghost-tone frequencies. In this chapter, it is shown how the rotational-harmonic frequency content of tooth-working-surface geometric deviations from equispaced perfect involute surfaces can be computed from detailed working-surface measurements before the attenuation arising from gear-meshing action takes place. It also is shown how the working-surface manufacturing deviations responsible for generation of any particular harmonic can be computed, which is a powerful manufacturing-source diagnostic tool.

4.1 Periodic Sequence of Working-Surface Deviations at a Generic Tooth Location

Consider a generic point P located on the line of contact shown in Figure 3.2 at Cartesian coordinates y, z locating that point on the tooth-working-surface. Because that point is on the line of contact, it also is in the plane of contact as shown in Figure 2.6. As the gears rotate and the variable x, Equation (3.1), advances a distance exactly equal to the base pitch Δ, the tooth adjacent to this earlier tooth occupies exactly the same location in the zone of contact in Figure 2.6 as the earlier tooth had occupied, and the point P occupies exactly the same location in the y, z tooth coordinates shown in Figure 3.2 on the new tooth as on the earlier tooth. Hence, each

Performance-Based Gear Metrology: Kinematic-Transmission-Error Computation and Diagnosis,
First Edition. William D. Mark.
© 2013 John Wiley & Sons, Ltd. Published 2013 by John Wiley & Sons, Ltd.

such generic location y, z on successive teeth is contacted by mating gear teeth at intervals in x exactly equal to the base pitch Δ (assuming perfect involute geometry). Thus, the geometric deviations on successive teeth at *each* such y, z location constitute a *discrete* sequence of values located at intervals in x separated by the base pitch Δ. This sequence of values is periodic with period in x of $N\Delta$, the base-circle circumference, where N is the number of teeth.

4.2 Heuristic Derivation of Rotational-Harmonic Contributions

Because the above-described sequence of values is periodic in x with period $N\Delta$, we normally would expect to be able to represent it by using Fourier series, for example, Gaskill (1978, p. 108) and Lanczos (1956, p. 254). Consider a function $f(x)$ periodic in x with period $N\Delta$, the base-circle circumference. Its complex Fourier series coefficients $\hat{f}(n)$ can be expressed as

$$\hat{f}(n) = \frac{1}{N\Delta} \int_0^{N\Delta} f(x)\exp\left[-i2\pi nx/(N\Delta)\right] dx, \qquad n = 0, \pm 1, \pm 2, \cdots \qquad (4.1)$$

from which $f(x)$ can be exactly reconstructed by

$$f(x) = \sum_{n=-\infty}^{\infty} \hat{f}(n)\exp\left[i2\pi nx/(N\Delta)\right] \qquad (4.2)$$

which is periodic in x with period $N\Delta$. Equations (4.1) and (4.2) are valid for periodic functions $f(x)$ defined for all values of x, but as described above, the concern here is a periodic function defined only at equispaced discrete intervals in x of $\delta x = \Delta$, the base pitch. This observation suggests that we replace the integration in Equation (4.1) by a summation, and the differential dx by $\delta x = \Delta$. Letting j denote tooth number, we have for these discrete values of x, $x=j\Delta$, yielding the discrete counterpart to Equation (4.1) as

$$\hat{f}(n) = \frac{1}{N\Delta} \sum_{j=0}^{N-1} f(j\Delta)\exp\left[-i2\pi nj\Delta/(N\Delta)\right] \Delta$$

$$= \frac{1}{N} \sum_{j=0}^{N-1} f(j\Delta)\exp\left(-i2\pi nj/N\right), \qquad n = 0, \pm 1, \pm 2, \cdots \qquad (4.3)$$

where the upper limit becomes $N-1$ in order to avoid counting the same tooth $j=0$ twice. The function $f(x)$ is evaluated in Equation (4.3) only at the discrete equispaced locations $x=j\Delta, j=0, 1, 2, \ldots, N-1$ which are the locations on the working surfaces of the N teeth, as described above. A careful formal derivation of Equation (4.3) is provided in Appendix 4.A.

Although Equation (4.3) was heuristically arrived at, it is, in fact, the discrete counterpart to the complex Fourier series, that is, the discrete finite Fourier transform (Cooley *et al.*, 1969, 1972). Its inverse transform (Cooley *et al.*, 1969, 1972) is

$$f(j\Delta) = \sum_{n=0}^{N-1} \hat{f}(n)\exp(i2\pi nj/N), \qquad j = 0, 1, \cdots, N-1 \tag{4.4}$$

where we again emphasize that $f(j\Delta)$ is evaluated only at the discrete values of $x = j\Delta$, $j = 0, 1, \ldots, N-1$. Just as Fourier series deals with periodic functions defined on a continuous variable x, and the Fourier integral transform deals with functions defined over $-\infty < x < \infty$, the discrete finite Fourier transform and its inverse, Equations (4.3) and (4.4), are the counterparts to the complex Fourier series, Equations (4.1) and (4.2), but defined for periodic functions designated only at equispaced discrete intervals in x. Because gears are circular with equispaced teeth, the discrete finite Fourier transform is an essential *exact* mathematical tool for the analysis dealt with herein. Its properties are fully derived (probably for the first time) in Cooley *et al.* (1969, 1972), which is an excellent reference.

4.3 Rotational-Harmonic Contributions from Working-Surface Deviations

As pointed out at the beginning of this chapter, our concern is deviations of the working surfaces of all N teeth at a generic point P on the working surfaces, as illustrated in Figures 3.2 and 2.6. The location of this generic point is described by the Cartesian coordinate y, z. Hence, the tooth deviations $\eta_{Cj}(y, z)$ at generic working-surface location y, z at the discrete (tooth) values $j = 0, 1, 2, \ldots, N-1$ constitute the sequence of discrete values at locations $x = j\Delta$ whose transform, Equation (4.3), is now formed,

$$\hat{\eta}_C(n; y, z) \triangleq \frac{1}{N} \sum_{j=0}^{N-1} \eta_{Cj}(y, z)\exp(-i2\pi nj/N), \qquad n = 0, \pm 1, \pm 2, \cdots. \tag{4.5}$$

Because the tooth deviations $\eta_{Cj}(y, z)$ differ at different working-surface locations y, z (Figure 3.2), the transform values $\hat{\eta}_C(n; y, z)$ differ for the different locations y, z.

Euler's formula (Hildebrand, 1976, p. 543) enables us to interpret the periodic character of the exponential function in Equation (4.5),

$$\exp(i\theta) = \cos\theta + i\sin\theta; \tag{4.6}$$

hence, using the fact that the cosine function is an even function of its argument and the sine function is an odd function of its argument, the exponential function in Equation (4.5) can be expressed as

$$\exp(-i2\pi nj/N) = \cos(2\pi nj/N) - i\sin(2\pi nj/N). \tag{4.7}$$

Since $sin\ (0) = 0$ and $cos\ (0) = 1$, the term $n = 0$ in Equation (4.5) is the average value of $\eta_{Cj}(y, z)$ averaged over all N teeth,

$$\hat{\eta}_C(0; y, z) = \frac{1}{N} \sum_{j=0}^{N-1} \eta_{Cj}(y, z). \tag{4.8}$$

The first periodic term in Equation (4.5) is $n = \pm 1$. For $n = \pm 1$, both the cosine and sine terms move through 2π radians as tooth number j progresses from 0 to N, which is one revolution around the gear. Hence, the terms $n = \pm 1$ are the fundamental *rotational*-harmonic terms. These terms represent a single sinusoidal cycle around the gear. The terms $n = \pm 2$ represent two sinusoidal cycles in one revolution around the gear, and so on. Hence, the various terms n of $\hat{\eta}_C(n; y, z)$, Equation (4.5), constitute a decomposition of the tooth deviations of all N teeth at location y, z into sinusoidal terms. The low-order terms $n = 0, \pm 1, \pm 2, \ldots$ of the discrete finite Fourier transform (DFT), Equations (4.3) and (4.5), behave in a very similar manner to the comparable terms of the Fourier series coefficients, Equation (4.1).

Contributions from Mean-Working-Surface Deviations

However, due to the discrete uniform spacing Δ of the gear teeth, the higher-order terms behave very differently from the low-order terms. Because the tooth spacing is Δ and the gear has N teeth, the particular rotational harmonic $\pm n$ that has a period equal to the tooth-spacing period Δ is $n = \pm N$, as one can observe from the exponential term in Equation (4.5) evaluated at rotational-harmonic number $n = N$,

$$\begin{aligned} \exp(-i2\pi Nj/N) &= \exp(-i2\pi j) \\ &= \cos(2\pi j) - i \sin(2\pi j) \\ &= 1 \end{aligned} \tag{4.9}$$

which has a period of 2π radians as tooth number j incrementally advances to $j = 1, 2, 3, \ldots$. Moreover, it follows directly from Equation (4.9) that for any integer $j = 1, 2, \ldots$ one has for harmonic number $n = N$, $exp\ (-i2\pi Nj/N) = 1$, and therefore, from Equation (4.5) and then Equation (4.8),

$$\hat{\eta}_C(N; y, z) = \frac{1}{N} \sum_{j=0}^{N-1} \eta_{Cj}(y, z) = \hat{\eta}_C(0; y, z). \tag{4.10}$$

The term $n = N$ of the DFT, Equation (4.5), which is the *tooth-meshing fundamental harmonic*, is, simply, the average deviation surface, averaged over all N teeth on the gear, as shown by Equation (4.10). This average deviation surface is a function of the Cartesian y, z coordinates (Figure 3.2).

Let $p = 0, \pm 1, \pm 2, \ldots$ denote *tooth-meshing* harmonic number, where $p = \pm 1$ is the above-described tooth-meshing *fundamental* harmonic. For any integer value p, tooth-meshing harmonic $p = 1, 2, 3, \ldots$ is rotational harmonic $n = pN$. For tooth-meshing harmonic p, the exponential term in the DFT, Equation (4.5) is

$$\exp(-i2\pi pNj/N) = \exp(-i2\pi pj)$$
$$= \cos(2\pi pj) - i\sin(2\pi pj) = 1 \tag{4.11}$$

again, since p and j are integers. Therefore, for all tooth-meshing harmonic numbers $n = \pm pN$, $p = 0, 1, 2, \ldots$ there follows from the DFT, Equation (4.5),

$$\hat{\eta}_C\left(pN; y, z\right) = \frac{1}{N}\sum_{j=0}^{N-1}\eta_{Cj}(y, z), \quad p = 0, \pm 1, \pm 2, \cdots \tag{4.12}$$

which is the average working-surface, a function of y and z, averaged over all teeth on the gear. Therefore, the characteristic of the working-surface geometric deviations from equispaced perfect involute surfaces (not including the attenuating effects of gear-meshing action) that determines their contributions to the tooth-meshing harmonics of the transmission error is the average deviation surface described by Equation (4.12). Ordinarily, intentional modifications of these surfaces from the involute are a principal component of these average deviation surfaces.

Contributions from Tooth-to-Tooth Variations in Working Surfaces

Let us now consider the remaining rotational-harmonic contributions $n \neq pN$ of the DFT, Equation (4.5). Consider any rotational harmonic $n' = n + pN$, $p = 0, \pm 1, \pm 2, \ldots$. For this harmonic, the exponential in Equation (4.5) is

$$\exp[-i2\pi(n + pN)j/N] = \exp(-i2\pi nj/N)\exp(-i2\pi pj)$$
$$= \exp(-i2\pi nj/N) \tag{4.13}$$

according to Equation (4.11). But n appears in the right-hand side of Equation (4.5) only in the exponential term. *Therefore, it follows from* Equation (4.13) *that the DFT, Equation (4.5), is periodic in rotational harmonic n with period in n equal to $n = N$, the tooth-meshing harmonic spacing.*

Using the DFT notation of Equation (4.5), the inverse DFT, Equation (4.4), is

$$\eta_{Cj}\left(y, z\right) = \sum_{n=0}^{N-1}\hat{\eta}_C(n; y, z)\exp(i2\pi nj/N), \quad j = 0, 1, \cdots, N - 1 \tag{4.14}$$

which is periodic in j with period N, the number of teeth, as the exponential indicates. Let us now decompose the tooth-working-surface deviations into their mean or average component

$$\bar{\eta}_C\left(y, z\right) \triangleq \frac{1}{N}\sum_{j=0}^{N-1}\eta_{Cj}\left(y, z\right), \tag{4.15}$$

and the difference of the individual working-surface deviations from the mean deviation surface,

$$\epsilon_{Cj}(y,z) \triangleq \eta_{Cj}(y,z) - \bar{\eta}_C(y,z), \qquad j = 0,1,2,\cdots,N-1. \tag{4.16}$$

Because it normally is the goal to manufacture gears with *identical* equispaced working surfaces, the contribution $\epsilon_{Cj}(y,z)$ normally is wholly composed of manufacturing errors (and surface roughness). (There also can be a manufacturing-error contribution to the mean-working-surface deviation $\bar{\eta}_C(y,z)$.)

Incorporating Equation (4.16) into Equation (4.14), there follows

$$\bar{\eta}_C(y,z) + \epsilon_{Cj}(y,z) = \sum_{n=0}^{N-1} \hat{\eta}_C(n;y,z)\exp(i2\pi nj/N)$$

$$= \hat{\eta}_C(0;y,z) + \sum_{n=1}^{N-1} \hat{\eta}_C(n;y,z)\exp(i2\pi nj/N) \tag{4.17}$$

because exp $(0) = 1$. But from Equations (4.15) and (4.10) we have

$$\bar{\eta}_C(y,z) = \hat{\eta}_C(0;y,z), \tag{4.18}$$

and therefore from Equations (4.17) and (4.18),

$$\epsilon_{Cj}(y,z) = \sum_{n=1}^{N-1} \hat{\eta}_C(n;y,z)\exp(i2\pi nj/N). \tag{4.19}$$

Hence, it follows from Equations (4.19), (4.12), and (4.15) and the above-described periodic behavior of the DFT with period N in n, that all rotational-harmonic contributions $\hat{\eta}_C(n;y,z)$ of the DFT, Equation (4.5), except those at $n = pN, p = 0,\pm1,\pm2,\ldots$ are caused by manufacturing-error differences $\epsilon_{Cj}(y,z)$ of the tooth-working-surfaces from the mean-working-surface deviation $\bar{\eta}_C(y,z)$. These rotational-harmonic amplitudes do not include the attenuating effects resulting from the gear-meshing action, which will be treated in the following chapter.

Introduction of Normalized-Legendre-Polynomial Expansions of Working-Surface Deviations

In the previous chapter, a method was described for representing the tooth-working-surface geometric deviations $\eta_{Cj}(y,z)$ by a superposition of two-dimensional normalized Legendre polynomials, Equation (3.18), and it was shown there how the expansion coefficients $c_{j,k\ell}$ of the individual two-dimensional Legendre-polynomial terms in Equation (3.18) could be evaluated from line-scanning lead measurements by Equation (3.37) and from line-scanning profile measurements by Equation (3.43), with the aid of Figures 3.9 and 3.10. The expansion coefficients $c_{j,k\ell}$, $k = 0,1,2,\ldots$;

$\ell = 0, 1, 2, \ldots$ contain all information required for a complete description of the working-surface deviations of each individual tooth j.

By incorporating Equation (3.18) into Equation (4.5), we obtain

$$\hat{\eta}_C(n; y, z) = \frac{1}{N} \sum_{j=0}^{N-1} \left[\sum_{k=0}^{\infty} \sum_{\ell=0}^{\infty} c_{j,k\ell} \psi_{yk}(y) \psi_{z\ell}(z) \right] \exp(-i2\pi nj/N)$$

$$= \sum_{k=0}^{\infty} \sum_{\ell=0}^{\infty} \left[\frac{1}{N} \sum_{j=0}^{N-1} c_{j,k\ell} \exp(-i2\pi nj/N) \right] \psi_{yk}(y) \psi_{z\ell}(z). \tag{4.20}$$

Define for $k = 0, 1, 2, \ldots; \ell = 0, 1, 2, \ldots$ the DFT of $c_{j,k\ell}$,

$$B_{k\ell}(n) \triangleq \frac{1}{N} \sum_{j=0}^{N-1} c_{j,k\ell} \exp(-i2\pi nj/N) \tag{4.21a}$$

$$= B_{k\ell}(n + pN), \qquad p = 0, \pm 1, \pm 2, \cdots \tag{4.21b}$$

according to Equation (4.13). Then, from Equations (4.20) and (4.21a,b) there follows

$$\hat{\eta}_C(n; y, z) = \sum_{k=0}^{\infty} \sum_{\ell=0}^{\infty} B_{k\ell}(n) \psi_{yk}(y) \psi_{z\ell}(z), \qquad n = 0, \pm 1, \pm 2, \cdots \tag{4.22a}$$

$$= \hat{\eta}_C(n + pN; y, z), \qquad p = 0, \pm 1, \pm 2, \cdots. \tag{4.22b}$$

For each Legendre term $k\ell$, the quantity $B_{k\ell}(n)$, Equation (4.21), is the DFT of the normalized Legendre expansion coefficient $c_{j,k\ell}$, $j = 0, 1, 2, \ldots, N-1$. $B_{k\ell}(n)$ is periodic in rotational-harmonic n with period N, the tooth-meshing harmonic spacing. According to Equation (4.22b), for each location y, z on the tooth-working-surfaces, the DFT, Equation (4.5), of $\eta_{Cj}(y, z)$ also is periodic in rotational-harmonic number n with the same tooth-meshing harmonic number spacing N. This very important property will be explained in later sections of the book.

4.4 Rotational-Harmonic Spectrum of Mean-Square Working-Surface Deviations

Divide both sides of Equation (3.23) by the number of teeth N, and sum the results over all N teeth,

$$\frac{1}{NFD} \sum_{j=0}^{N-1} \int_{-D/2}^{D/2} \int_{-F/2}^{F/2} \eta_{Cj}^2(y, z) \, dy \, dz = \frac{1}{N} \sum_{j=0}^{N-1} \sum_{k=0}^{\infty} \sum_{\ell=0}^{\infty} c_{j,k\ell}^2$$

$$= \sum_{k=0}^{\infty} \sum_{\ell=0}^{\infty} \frac{1}{N} \sum_{j=0}^{N-1} c_{j,k\ell}^2. \tag{4.23}$$

Parseval's theorem for the DFT, Equation (4.21a), is (Cooley, Lewis, and Welch, 1969, 1972, Equation (17))

$$\frac{1}{N}\sum_{j=0}^{N-1}c_{j,k\ell}^2 = \sum_{n=0}^{N-1}\left|B_{k\ell}(n)\right|^2. \tag{4.24}$$

However, from the periodic property of $B_{k\ell}(n)$, Equation (4.21b), there follows

$$\sum_{n=0}^{N-1}\left|B_{k\ell}(n)\right|^2 = \sum_{n=n'}^{n'+N-1}\left|B_{k\ell}(n)\right|^2. \tag{4.25}$$

Combining Equations (4.24) and (4.25) with Equation (4.23), we have (Mark, 1982; Mark and Reagor, 2001)

$$\frac{1}{NFD}\sum_{j=0}^{N-1}\int_{-D/2}^{D/2}\int_{-F/2}^{F/2}n_{Cj}^2(y,z)dydz = \sum_{n=n'}^{n'+N-1}\sum_{k=0}^{\infty}\sum_{\ell=0}^{\infty}\left|B_{k\ell}(n)\right|^2, \quad n'=0,\pm1,\pm2,\cdots. \tag{4.26}$$

Since FD is the analysis area of the tooth-working-surfaces (Figure 3.2), and N is the number of teeth, the left-hand side of Equation (4.26) is the mean-square deviation of the working surfaces, averaged over all N teeth on the gear. It then follows that

$$\sum_{k=0}^{\infty}\sum_{\ell=0}^{\infty}\left|B_{k\ell}(n)\right|^2, \quad n' \le n \le (n'+N-1) \tag{4.27}$$

is the rotational-harmonic decomposition of the mean-square tooth deviations, averaged over the working surfaces of all N teeth on the gear under consideration. This rotational-harmonic spectrum is periodic in n with period N equal to the number of teeth N, which also is the spacing in n of the tooth-meshing harmonics $n = pN$, $p = 0, \pm1, \pm2, \ldots$.

Because the expansion coefficients $c_{j,k\ell}$ of Equation (3.20) are real, it follows directly from Equation (4.21a) that

$$B_{k\ell}(-n) = B_{k\ell}^*(n), \tag{4.28}$$

where the asterisk denotes the complex conjugate, and therefore,

$$\left|B_{k\ell}(-n)\right|^2 = \left|B_{k\ell}(n)\right|^2 \tag{4.29}$$

for every integer n and all $k\ell$ pairs. Hence, the rotational-harmonic spectra (Equation (4.27)) are "two-sided" spectra, which are even functions of rotational-harmonic number n. For example, for $N = $ odd, choose for the lower limit of the sum in Equation (4.26), $n' = -(N-1)/2$, which yields the symmetric upper limit, $n'+N-1 = (N-1)/2$. For $N = $ even with $n' = -(N/2)$, the upper limit becomes $n'+N-1 = (N/2)-1$. Because it is customary to display such two-sided even spectra only for positive harmonic numbers $n = 1, 2, 3, \ldots$, we *define* the "one-sided" mean-square rotational-harmonic spectrum of the tooth-working-surface deviations as

$$G_\eta(n) \triangleq \begin{cases} \sum\limits_{k=0}^{\infty} \sum\limits_{\ell=0}^{\infty} |B_{k\ell}(0)|^2, & n = 0 \\ 2 \sum\limits_{k=0}^{\infty} \sum\limits_{\ell=0}^{\infty} |B_{k\ell}(n)|^2, & n = 1, 2, 3, \cdots. \end{cases} \qquad (4.30\text{a,b})$$

From the periodic property of $B_{k\ell}(n)$, Equation (4.21b), except for $n = 0$, the spectrum $G_\eta(n)$ possesses the comparable "one-sided" periodic property

$$G_\eta(n) = G_\eta(n + pN), \qquad p = 0, 1, 2, \cdots; \qquad n = 1, 2, 3, \cdots. \qquad (4.31)$$

It follows directly from the left-hand side of Equation (4.26) that the dimension of $|B_{k\ell}(n)|^2$ is the same as the dimension of $\eta_{Cj}^2(y, z)$, which is length-squared. $G_\eta(n)$, Equation (4.30), has the same dimension. Because it is easier to assess quantities of dimension "length" rather than "length-squared," the computed mean-square tooth-deviation spectra will be displayed as the square-root of Equation (4.30), that is, $[G_\eta(n)]^{1/2}$.

As defined by Equation (4.21a), $B_{k\ell}(n)$ is complex. Formulas for the real quantities $|B_{k\ell}(n)|^2$ and $G_\eta(n)$, involving only real numbers, are derived in Appendix 4.B.

Finally, it should be pointed out that the *relative* simplicity of the mean-square spectrum results of Equations (4.26) and (4.30) is a consequence of Equation (3.23) which, in turn, is a consequence of the normalization factors $(2k+1)^{1/2}$ and $(2\ell+1)^{1/2}$ utilized in the definitions (3.13) and (3.14) of the normalized Legendre polynomials, and of their orthogonal property.

Contributions from Mean and Tooth-to-Tooth Variations of Tooth-Working-Surface Deviations

It follows directly from Equations (4.12), (4.22a), and (4.26) that the contributions to the tooth-meshing harmonics $n = pN, p = 1, 2, 3, \ldots$ of the spectrum, Equation (4.30b),

$$G_\eta(pN) = 2 \sum_{k=0}^{\infty} \sum_{\ell=0}^{\infty} |B_{k\ell}(pN)|^2, \qquad p = 1, 2, 3, \cdots \qquad (4.32)$$

arise from the mean-tooth-surface deviation $\bar{\eta}_C(y, z)$, Equation (4.15). Moreover, it follows directly from Equations (4.16), (4.19), and (4.22a) that the contributions to the non-tooth-meshing rotational harmonics $n \neq pN$ of the spectrum, Equation (4.30b), arise from the geometric tooth-to-tooth variations $\epsilon_{Cj}(y, z)$, Equation (4.16), of the working surfaces from the mean-working-surface deviation $\bar{\eta}_C(y, z)$; that is, from manufacturing errors and surface roughness. Hence, the rms value of these tooth-working-surface variations $\epsilon_{Cj}(y, z)$ from the mean working-surface, evaluated over the analysis area FD of all N working surfaces, can be evaluated from Equations (4.26) and (4.30b) as

$$\text{rms variation of } \epsilon_{Cj}(y,z) = \left\{ \sum_{n=1}^{N-1} \sum_{k=0}^{\infty} \sum_{\ell=0}^{\infty} |B_{k\ell}(n)|^2 \right\}^{1/2} \tag{4.33a}$$

$$= \frac{1}{\sqrt{2}} \left[\sum_{n=1}^{N-1} G_\eta(n) \right]^{1/2} . \tag{4.33b}$$

Similarly, it follows from Equations (4.18), (4.22), (4.26), and (4.30b) that the rms value of the deviation of the mean working-surface $\bar{\eta}_C(y,z)$, Equation (4.15), from a perfect involute surface, evaluated over the analysis area FD, is

$$\text{rms deviation of } \bar{\eta}_C(y,z) = \left\{ \sum_{k=0}^{\infty} \sum_{\ell=0}^{\infty} |B_{k\ell}(pN)|^2 \right\}^{1/2} , \quad p = 0, \pm 1, \pm 2, \cdots \tag{4.33c}$$

$$= \frac{1}{\sqrt{2}} \left[G_\eta(pN) \right]^{1/2} \quad , \quad p = 1, 2, \cdots \tag{4.33d}$$

Alternative proofs of Equations (4.33a) and (4.33c) are provided in Appendix 4.C.

Hence, the mean deviation $\bar{\eta}_C(y,z)$, Equation (4.15), of the working surfaces provides contributions only to the tooth-meshing harmonics $n = pN$, $p = 0, \pm 1, \pm 2, \ldots$, and the tooth-to-tooth variations, $\epsilon_{Cj}(y,z)$, $j = 0, 1, \ldots, N-1$, Equation (4.16), of the working surfaces provide contributions only to the non-tooth-meshing rotational harmonics n, $n \neq pN$.

Comment on "Aliasing"

When utilizing equispaced sampling of *continuous* signals, the DFT, Equation (4.3), is commonly used to generate the frequency spectrum of the signal (by implementing the fast Fourier transform algorithm). In such applications of the DFT, actual spectrum contributions of the continuous signal from frequencies beyond the Nyquist frequency $n = N/2$ appear in the DFT spectrum as contributions to frequencies $n < N/2$. This behavior is referred to as "aliasing," for example, Hamming (1973, p. 14). (To avoid such aliasing, continuous signals must be low-pass filtered before sampling in order to eliminate all spectrum contributions beyond the Nyquist frequency $n = N/2$.)

In contrast to this discrete sampling of continuous signals, in the present application of the DFT, Equation (4.3), the "signal", that is, the gear teeth, are actually discrete, and the DFT, Equation (4.3), is the exact mathematical tool required to carry out the spectrum analysis of geometric working-surface deviations, and to predict their transmission error contributions. (This fact first became apparent to this writer in the derivation leading to Equation (135) of Mark (1978)). Although the teeth are discrete, the deviations of the working surfaces are continuous functions of the tooth coordinates y, z, which is accurately represented by the two-dimensional expansion in normalized Legendre polynomials, Equation (3.18), and is incorporated in the above derivation of the spectrum, that is, in Equation (4.20). The importance and utility of the spectrum, Equation (4.30), will become more apparent in the treatment of the next chapter, where it will be shown that all rotational-harmonic frequencies of this spectrum can be present in transmission-error spectra.

Example Rotational-Harmonic Spectrum of Mean-Square Working-Surface Deviations

Equation (4.30) provides the mean-square rotational-harmonic spectrum in terms of the DFT, $B_{k\ell}(n)$, and Equation (4.21a) provides $B_{k\ell}(n)$ in terms of the normalized-Legendre-polynomial expansion coefficients $c_{j,k\ell}$ of the individual teeth $j = 0, 1, \ldots,$ $N - 1$ described by Equation (3.20). These coefficients are evaluated from line-scanning tooth measurements by Equations (3.37) and (3.43), with the aid of Figures 3.9 and 3.10.

All of the teeth on the smaller helical gear shown in Figure 3.11 were measured as described in Section 3.7 of the previous chapter, yielding the expansion coefficients $c'_{j,k\ell}$ of Equations (3.37) and (3.43). From these expansion coefficients the rms rotational-harmonic spectrum $[G(n)]^{1/2}$ was computed as explained above. It is displayed in Figure 4.1 using a logarithmic amplitude scale. The measured gear has $N = 38$ teeth, and we observe that the spectrum in Figure 4.1 is perfectly periodic in rotational-harmonic number n with period N, the number of teeth, as described by Equation (4.31). The tooth-meshing harmonic contributions at $n = pN$, $p = 1, 2, 3, \ldots$ are strong, as one might expect. Additional strong harmonics are observed at rotational-harmonic numbers $n = 1, 18, 19, 20,$ and 37, which continue to repeat at higher harmonic numbers $n > N$ due to the spectrum periodic property.

The spectrum shown in Figure 4.1 is symmetric about the midpoint harmonic location $n = N/2$, and the corresponding higher-harmonic locations $n = pN + N/2$, $p = 1, 2, 3, \ldots$, Equation (4.31). *It is a general property of all such spectra.*

The transmission-error contributions of the strong rotational-harmonic amplitudes located at harmonic numbers $n = 1, 18, 19, 20, 37,$ and 38 in Figure 4.1 will be discussed in the next chapter. The geometric deviations of the tooth-working-surfaces causing such rotational harmonics are discussed next.

4.5 Tooth-Working-Surface Deviations Causing Specific Rotational-Harmonic Contributions

It is shown below that once any harmonic of the rotational-harmonic spectrum $G_\eta(n)$ is identified to be of interest, such as those at $n = 19$ or 20 in Figure 4.1, the unique geometric manufacturing-error pattern on the tooth-working-surfaces that is the cause of the identified rotational harmonic can be computed. This capability is a powerful tool for diagnosing the manufacturing source of the identified rotational harmonic, since it should generally be possible to relate the kinematic properties of manufacturing operations to any such solved-for manufacturing-error pattern on the working surfaces. (However, gear-meshing action generally will attenuate some

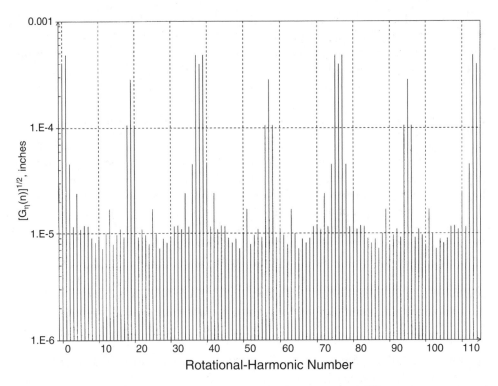

Figure 4.1 Rotational harmonic rms spectrum $(G_\eta(n))^{1/2}$, square-root of Equation (4.30b), of working-surface deviations of 38 tooth pinion in Figure 3.11. Logarithemic amplitude scale in inches (From Mark and Reagor (2001), Reproduced by permission of the American Gear Manufacturers Association)

rotational-harmonic amplitudes much more than others; hence, those rotational-harmonic amplitudes contributing strongly to the transmission error can be identified only after this attenuating action has been computed, as described in the next chapter, or unless they are identified experimentally, as in vibration or noise measurements.)

Derivation of Working-Surface-Deviation Rotational-Harmonic Contributions

The inverse DFT, Equation (4.4), applied to the DFT, $B_{k\ell}(n)$, Equation (4.21a), is

$$c_{j,k\ell} = \sum_{n=0}^{N-1} B_{k\ell}(n)\exp(i2\pi nj/N), \quad j = 0, 1, \cdots, N-1 \qquad (4.34)$$

where the $c_{j,k\ell}$ are the (real) normalized Legendre expansion coefficients, Equation (3.20). Using Euler's formula, Equation (4.6), one has

$$\exp(\pm i2\pi nj/N) = \cos(2\pi nj/N) \pm i\sin(2\pi nj/N) \qquad (4.35)$$

since the cosine function is an even function of its argument and the sine function is an odd function. If we define the even and odd contributions to $B_{k\ell}(n)$, Equation (4.21a), as

$$B_{k\ell}(n)_e \triangleq \frac{1}{N} \sum_{j=0}^{N-1} c_{j,k\ell} \cos(2\pi nj/N) \qquad (4.36)$$

and

$$B_{k\ell}(n)_o \triangleq \frac{1}{N} \sum_{j=0}^{N-1} c_{j,k\ell} \sin(2\pi nj/N), \qquad (4.37)$$

both of which are real, since $c_{j,k\ell}$ is real, there then follows from Equation (4.21a) and Equations (4.35–4.37),

$$B_{k\ell}(n) = \frac{1}{N} \sum_{j=0}^{N-1} c_{j,k\ell} \left[\cos(2\pi nj/N) - i\sin(2\pi nj/N)\right]$$

$$= B_{k\ell}(n)_e - i\, B_{k\ell}(n)_o. \qquad (4.38)$$

Hence, using Equations (4.35) and (4.38), the summand in Equation (4.34) is

$$B_{k\ell}(n)\exp(i2\pi nj/N) = \left[B_{k\ell}(n)_e - iB_{k\ell}(n)_o\right] \left[\cos(2\pi nj/N) + i\sin(2\pi nj/N)\right]$$

$$= B_{k\ell}(n)_e \cos(2\pi nj/N) + B_{k\ell}(n)_o \sin(2\pi nj/N)$$

$$+i\left[B_{k\ell}(n)_e \sin(2\pi nj/N) - B_{k\ell}(n)_o \cos(2\pi nj/N)\right]. \qquad (4.39)$$

But the normalized Legendre expansion coefficient $c_{j,k\ell}$, Equation (3.20), is real. Therefore, the summation over n of the imaginary component of Equation (4.39) in Equation (4.34) must vanish identically, leaving

$$c_{j,k\ell} = \sum_{n=0}^{N-1} \left[B_{k\ell}(n)_e \cos(2\pi nj/N) + B_{k\ell}(n)_o \sin(2\pi nj/N)\right], \qquad (4.40)$$

which involves only real quantities. However, $B_{k\ell}(n)$ in Equation (4.34) is periodic in n with period N, as shown by Equation (4.21b), as is $exp\,(i2\pi nj/N)$, as in Equation (4.13). Hence, the summations in Equations (4.34) and (4.40) can be replaced by $\sum_{n=n'}^{n'+N-1}$. Substituting Equation (4.40) into Equation (3.18), introducing these more general summation limits, and interchanging the order of summations, there follows

$$\eta_{Cj}(y,z) = \sum_{n=n'}^{n'+N-1} \left\{\sum_{k=0}^{\infty}\sum_{\ell=0}^{\infty} \left[B_{k\ell}(n)_e \cos(2\pi nj/N) + B_{k\ell}(n)_o \sin(2\pi nj/N)\right] \psi_{yk}(y)\psi_{z\ell}(z)\right\},$$

$$j = 0,1,2,\cdots, N-1; \quad n' = 0, \pm 1, \pm 2, \cdots \qquad (4.41)$$

Equation (4.41) is an expression for the working-surface deviation of tooth j, expressed as a superposition of the rotational-harmonic contributions of the working-surface deviations. The quantity within the square brackets in Equations (4.41) and (4.40) is an even function of n, a two-sided spectrum, as before. Therefore, except for the tooth-meshing harmonic contributions, $n = pN$, $p = 0, 1, 2, \ldots$ the "one-sided" contribution, $n > 0$, of each rotational harmonic $n \neq pN$ to the working-surface deviation of a generic tooth j at Cartesian coordinate location y, z is, for example, see the comments following Equation (4.29),

$$\eta_{Cj}(y, z; n) = 2 \sum_{k=0}^{\infty} \sum_{\ell=0}^{\infty} \left[B_{k\ell}(n)_e \cos(2\pi nj/N) + B_{k\ell}(n)_o \sin(2\pi nj/N) \right] \psi_{yk}(y)\psi_{z\ell}(z),$$

$$j = 0, 1, 2, \cdots, N-1; \quad n = 1, 2, \cdots; n \neq pN. \tag{4.42}$$

For the tooth-meshing harmonics, $n = pN, p = 0, 1, 2, \ldots$ we have for the trigonometric terms in Equation (4.41),

$$\cos(2\pi pNj/N) = \cos(2\pi pj) = 1 \tag{4.43}$$

and

$$\sin(2\pi pNj/N) = \sin(2\pi pj) = 0. \tag{4.44}$$

In this case, there is only one term $n = pN$ within the summation over n in Equation (4.41). Hence, using Equations (4.43) and (4.44), the contribution of a generic tooth-meshing harmonic $n = pN$, $p = 1, 2, \ldots$ to the working-surface deviation of the tooth j at Cartesian coordinate location y, z is

$$\eta_{Cj}(y, z; pN) = \sum_{k=0}^{\infty} \sum_{\ell=0}^{\infty} B_{k\ell}(pN)_e \, \psi_{yk}(y)\psi_{z\ell}(z), \quad p = 0, 1, 2, \cdots \tag{4.45}$$

But from Equations (4.36) and (4.43), and following that, Equation (3.18), there follows from Equation (4.45)

$$\eta_{Cj}(y, z; pN) = \sum_{k=0}^{\infty} \sum_{\ell=0}^{\infty} \frac{1}{N} \sum_{j=0}^{N-1} c_{j,k\ell} \, \psi_{yk}(y)\psi_{z\ell}(z) \tag{4.46a}$$

$$= \frac{1}{N} \sum_{j=0}^{N-1} \left[\sum_{k=0}^{\infty} \sum_{\ell=0}^{\infty} c_{j,k\ell} \, \psi_{yk}(y)\psi_{z\ell}(z) \right] \tag{4.46b}$$

$$= \frac{1}{N} \sum_{j=0}^{N-1} \eta_{Cj}(y, z) \tag{4.46c}$$

which is the average working-surface deviation, averaged over all N teeth, again confirming that this average working-surface deviation is the geometric working-surface-deviation contribution arising from the tooth-meshing harmonics, $p = 1, 2, 3, \ldots$

Equation (4.42) is an expression for the contribution to the working-surface deviation $\eta_{Cj}(y,z)$ of tooth j that arises from rotational-harmonic number n of the working-surface deviations, $n \neq pN$. Equations (4.45) and (4.46) are equivalent expressions for the working-surface deviation contribution that arises from the tooth-meshing harmonics $n = pN$, $p = 1, 2, 3, \ldots$ of the working-surface deviations.

Comment: Clearly, the manufacturing-error pattern on the tooth-working-surfaces is the *physical cause* of the resultant rotational-harmonic spectrum. But, in order to diagnose the manufacturing source of any particular rotational harmonic n, the inverse DFTs of Equations (4.4) and (4.34) have been used, leading us to describe working-surface deviations as arising from rotational-harmonic contributions, which is a reversal of physical cause and effect. There is a unique correspondence between the spectra described by Equations (4.42), (4.45), and (4.46) and the working-surface deviations on all of the N teeth.

4.6 Discussion of Working-Surface-Deviation Rotational-Harmonic Contributions

By comparing Equation (4.42) with Equation (3.18), we observe that the term $2[\ldots]$ of the summation in Equation (4.42) is the expansion coefficient in normalized Legendre polynomials of the contribution of rotational harmonic n to the deviation surface $\eta_{Cj}(y,z)$ of tooth j. This expansion coefficient $2[\ldots]$ is generally different from every tooth $j = 0, 1, \ldots, N-1$ and is a function of rotational-harmonic number n.

Now consider the contributions $\eta_{Cj}(y, z; n)$, Equation (4.42), to the working-surface deviation $\eta_{Cj}(y, z)$ of tooth j that arise from rotational harmonics $n' = pN \pm n$. From Equation (4.36) we observe that the first term within the square bracket in Equation (4.42) involves the product of two terms $\cos(2\pi nj/N)$, and from Equation (4.37), the second term within the same square bracket involves the product of two terms $\sin(2\pi nj/N)$. Using elementary trigonometric identities, one has

$$\cos\left[2\pi(pN \pm n)j/N\right] = \cos(2\pi nj/N), \quad p = 0, 1, 2, \cdots \qquad (4.47)$$

and

$$\sin\left[2\pi(pN \pm n)j/N\right] = \pm\sin(2\pi nj/N), \quad p = 0, 1, 2, \cdots \qquad (4.48)$$

Therefore, it follows from Equations (4.42), (4.47), and (4.48) that

$$\eta_{Cj}(y, z; pN \pm n) = \eta_{Cj}(y, z; n), \quad p = 0, 1, 2, \cdots \quad n = 1, 2, \cdots \qquad (4.49)$$

that is, the contribution to the working-surface deviations $\eta_{Cj}(y,z)$ of tooth j that arises from rotational harmonic $pN \pm n$ is identical to that arising from rotational harmonic n of the working-surface deviations.

The result of Equation (4.49) is significant for the following reason. Gear manufacturing errors such as accumulated tooth-spacing errors, involute slope errors, helix-angle errors, and so on often exhibit low-order rotational-harmonic contributions $n = 1, 2, \ldots$ in their rotational-harmonic decompositions, Equation (4.5). Such errors are a cause of the low-order rotational-harmonic contributions to transmission-error, vibration, and noise spectra. However, when such low-order rotational-harmonic contributions are observed, so-called sideband rotational-harmonic contributions around the tooth-meshing harmonic locations $n = pN$ also are frequently observed at harmonic locations $pN \pm 1$, $pN \pm 2$, and so on. It will be shown in the following chapters that these so-called sideband contributions are a consequence of the periodic character, Equation (4.49), of the contributions to the working-surface deviations $\eta_{Cj}(y, z)$ arising from the low-order rotational harmonics n. This same periodic behavior is an essential factor in the generation of "ghost tones". Additional explanation of this behavior will be provided in Chapters 5–7.

Examples of Working-Surface Deviations Causing Specific Rotational-Harmonic Contributions

Upon earlier examination of Figure 4.1, we observed strong error-spectrum rotational-harmonic amplitudes at harmonic numbers $n = 1, 18, 19, 20$, and 37, in addition to the tooth-meshing fundamental $p = 1$ at rotational-harmonic number $n = N = 38$, which is the number of teeth on the smaller gear of Figure 3.11. However, taking $n = 1$ in Equation (4.49) with $N = 38$, the number of teeth, we observe for $p = 1$,

$$\eta_{Cj}(y, z; 39) = \eta_{Cj}(y, z; 37) = \eta_{Cj}(y, z; 1). \tag{4.50}$$

Hence, the same collective error pattern on the gear teeth $j = 0, 1, 2, \ldots, 37$ causes rotational-harmonic numbers $n = 1, 37$, and 39. Taking $n = 18$ in Equation (4.49) with $N = 38$, we observe for $p = 1$,

$$\eta_{Cj}(y, z; 20) = \eta_{Cj}(y, z; 18), \tag{4.51}$$

that is, the same collective error pattern causes rotational-harmonic numbers $n = 18$ and 20. Therefore, non-duplicating selective error patterns are shown below arising from rotational-harmonic numbers $n = 1, 19$, and 20.

The contribution from rotational-harmonic number $n = 38$, which is the tooth-meshing fundamental, was already displayed in Figure 3.14. As indicated by Equation (4.46), this average working-surface deviation is the geometric-deviation contribution to each of the tooth-meshing harmonics $n = pN$, $p = 1, 2, 3, \ldots$.

The contributions for the remaining rotational harmonics $n = 1, 19$, and 20 were computed using Equation (4.42). The same set of working-surface measurements of the smaller gear in Figure 3.11 was used in these computations as was used in the generation of Figures 3.12–3.14 described in Section 3.7.

Equation (4.42) is an expansion in the two-dimensional normalized Legendre polynomials of Equations (3.13) and (3.14) over the rectangular working-surface region $-(F/2) \le y \le (F/2), (-D/2) \le z \le (D/2)$ of the chosen tooth j, thereby providing a representation over this region of the contribution to the working-surface deviation of tooth j provided by rotational harmonic n of the working-surface deviations. Evaluation of Equation (4.42) requires evaluation of $B_{k\ell}(n)_e$ and $B_{k\ell}(n)_o$, which are defined by Equations (4.36) and (4.37), respectively. These functions of n require, for each $k\ell$ pair, the Legendre expansion coefficient $c_{j,k\ell}$, Equation (3.20), for every tooth $j = 0, 1, \ldots, N-1$, which are evaluated from the line-scanning lead measurements by Equation (3.37) and the line-scanning profile measurements by Equation (3.43).

Figure 4.2 displays the computation by Equation (4.42) of the working-surface deviations provided by rotational harmonic $n = 1$ on four approximately equally

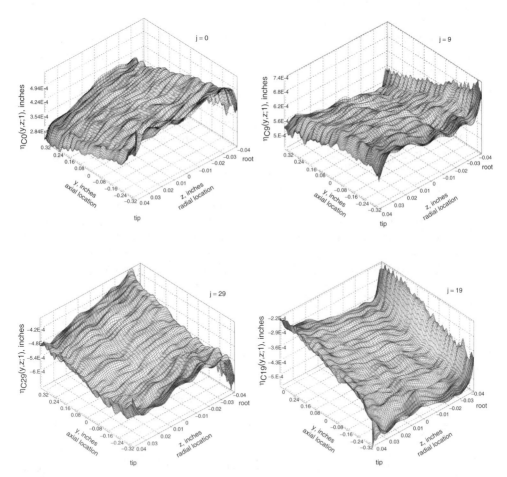

Figure 4.2 Computation by Equation (4.42) of the contribution from rotational harmonic $n = 1$ to the working-surface deviations of four approximately equally-spaced teeth, $j = 0, 9, 19$, and 29 of the 38 tooth pinion in Figure 3.11. All dimensions in inches (From Mark and Reagor (2001), Reproduced by permission of the American Gear Manufacturers Association)

spaced teeth, $j = 0, 9, 19$, and 29 of the $N = 38$ tooth gear. Because rotational harmonic $n = 1$ describes a single sinusoidal cycle around the gear of the working-surface deviations, only the deviations on four approximately equally-spaced teeth were chosen to be displayed. The horizontal axes to the left and right describe the axial y and radial z working-surface locations, respectively, as defined in Section 3.2. The vertical axis describes the deviation from a perfect involute surface of this rotational-harmonic contribution $n = 1$. This deviation direction is described in Section 3.5 with the aid of Figure 3.1. The total span of each vertical axis in Figure 4.2 is determined by the range of the computed deviations. Hence, its origin is not displayed. These coordinate axes are the same as those used in Figures 3.12–3.14.

Except for the behavior in the tooth-root region to the right, and to a lesser extent the tip region to the left, the behavior of the four displays in Figure 4.2 approximates a wobbling plane surface, with a one-cycle wobble. This behavior can be explained by a lack of parallel between the tooth generation (base cylinder) axis and the axis of rotation of the gear measurements. The total vertical span of the deviations ranges from about 270 μin. (6.9 μm) for tooth number $n = 29$ to about 350 μin. (8.9 μm) for tooth numbers $j = 0$ and 19. The plane-surface contributions to the four surfaces shown in Figure 4.2 are provided by the normalized Legendre polynomial terms $k = 1, \ell = 0$ and $k = 0, \ell = 1$ of Equations (3.18) and (4.42).

The dominant component of this $n = 1$ once-per-revolution contribution is the vertical displacement (not graphically displayed) of the mean value of each of the four surfaces shown in Figure 4.2. For tooth numbers $j = 0, 9, 19$, and 29, this mean vertical displacement is about 385, 590, −395, and −510 μin., respectively. This component is the $k = 0, \ell = 0$ (accumulated tooth-spacing error) two-dimensional normalized Legendre term of Equations (3.18) and (4.42). This accumulated tooth-spacing error component arises from an eccentric displacement between the tooth generation (base-cylinder) axis and the axis of rotation of the gear measurements. The rms contribution of rotational harmonic $n = 1$ to the working-surface deviations can be seen from Figure 4.1 to be about 480 μin. (12 μm).

Figure 4.3 displays the computation by Equation (4.42) of the working-surface deviations provided by rotational harmonic $n = 19$. Figure 4.3a shows the deviations of the even-number teeth, $j = 0, 2, 4, \ldots$, and Figure 4.3b shows those of the odd-numbered teeth, $j = 1, 3, 5, \ldots$. Consider the two trigonometric terms in Equation (4.42) for $n = 19$ and $N = 38$. For the sine term we have

$$\sin(2\pi 19j/38) = \sin(\pi j) = 0 \tag{4.52}$$

and for the cosine term

$$\cos(2\pi 19j/38) = \cos(\pi j) = (-1)^j. \tag{4.53}$$

Therefore, for $n = 19$ and $N = 38$, Equation (4.42) reduces to

$$\eta_{Cj}(y, z; 19) = (-1)^j 2 \sum_{k=0}^{\infty} \sum_{\ell=0}^{\infty} B_{k\ell}(19)_e \, \psi_{yk}(y) \, \psi_{z\ell}(z), \quad j = 0, 1, 2, \cdots, 37 \tag{4.54}$$

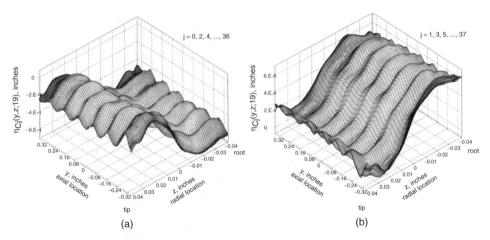

Figure 4.3   Computation by Equation (4.42) of the contribution from rotational harmonic $n = 19$ to the working-surface deviations of teeth $j = 0, 2, 4, \ldots, 36$ and teeth $j = 1, 3, 5, \ldots 37$ of the 38 tooth pinion in Figure 3.11. All dimensions in inches (From Mark and Reagor (2001), Reproduced by permission of the American Gear Manufacturers Association)

which shows that the contribution of rotational harmonic $n = 19$ to every second tooth must be identical, and moreover, the contributions to adjacent teeth are the same except for a reversal in arithmetic sign. This behavior is clearly displayed in Figure 4.3. According to Figure 4.1, the rms contribution of rotational harmonic $n = 19$ to the working-surface deviations is about 290 μin. (7.4 μm).

As can be seen from Figure 4.4, the errors shown in Figure 4.3 are hobbing errors. The gear under investigation was cut by hobbing, and not finished after hobbing.

Figure 4.5 displays the computation by Equation (4.42) of the working-surface deviations provided by rotational harmonic $n = 20$ on two adjacent teeth, $j = 0$ and $j = 1$. The errors shown there are approximately sinusoidal, with a period of $(N/n) = (38/20) = 1.9$ tooth spacings. Therefore, the error on every adjacent tooth is approximately that on the previous tooth, but with opposite sign, and slightly retarded phase. From Figure 4.1, the rms contribution of rotational harmonic $n = 20$ to the working-surface deviations is a shade over 100 μin. (2.5 μm).

The manufacturing errors shown in Figure 4.5 are classical "undulation errors" (Cluff, 1992, pp. 7, 38, 91–97; Merritt, 1971, p. 82). Using the rotational-harmonic number ($n = 20$ in this case) it almost always is possible to identify a kinematic element in the manufacturing operation that is the cause of such errors.

The rms manufacturing-error contribution of rotational harmonic $n = 19$ is about $(290/100) \approx 3$ times as large as that from rotational harmonic $n = 20$. However, in the next chapter, it will be shown that the rms contribution to the transmission error of rotational harmonic $n = 19$ is only about $1/4$ of that provided by rotational harmonic $n = 20$. This comparison is a dramatic example of the dominant importance of the manufacturing-error *pattern* on the tooth-working-surfaces in transmission-error generation, vibration excitation, and noise.

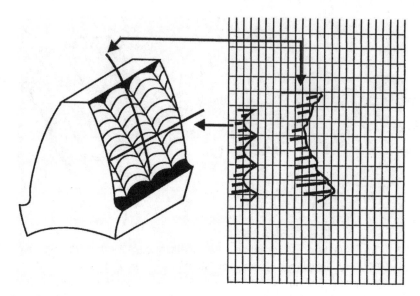

Figure 4.4 Helical gear tooth topography as "seen" by one lead and one involute trace by an inspection machine (From Cluff (1992, p. 250), Reproduced by permission of American Pfauter Limited Partnership)

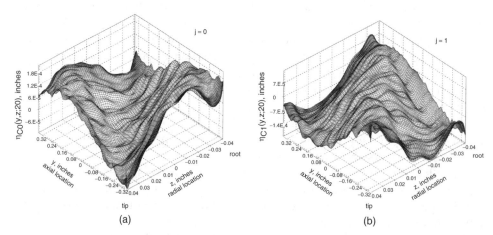

Figure 4.5 Computation by Equation (4.42) of the contribution from rotational harmonic $n = 20$ to the working-surface deviations of adjacent teeth $j = 0$ and $j = 1$ of the 38 tooth pinion in Figure 3.11. All dimensions in inches (From Mark and Reagor (2001), Reproduced by permission of the American Gear Manufacturers Association)

Appendix 4.A. Formal Derivation of Equation (4.3)

Let $f_j(x), j = 0, 1, 2, \ldots, N - 1$ denote the sequence of tooth working-surface deviations at a generic working-surface location y, z on tooth numbers $j = 0, 1, 2, \ldots, N - 1$ as described in the main text. The individual working surfaces are separated from each other by an interval $\delta x = \Delta$. Consider the Fourier series coefficients $\hat{f}(n)$, Equation (4.1), of the discrete sequence of normalized working-surface deviations $f_j(x)\Delta$, $j = 0, 1, 2, \ldots, N-1$. Because Fourier series is normally defined for periodic functions of a continuous variable, utilizing it to represent the above-described discrete periodic sequence requires use of the Dirac delta function (Lanczos, 1956, p. 222; Gaskill, 1978, p. 56), defined as

$$\delta(x) = \begin{cases} \infty, & x = 0 \\ 0, & x \neq 0 \end{cases} \tag{4.A.1a}$$

together with

$$\int_{-\infty}^{\infty} \delta(x)dx = \int_{-\epsilon/2}^{\epsilon/2} \delta(x)dx = 1 \tag{4.A.1b}$$

for any small positive ϵ. Consequently, for any continuous function $f(x)$, Equations (4.A.1a,b) imply that

$$\int_{-\infty}^{\infty} f(x)\delta(x - \xi)dx = f(\xi) \int_{-\infty}^{\infty} \delta(x - \xi)dx \tag{4.A.2a}$$

$$= f(\xi) \tag{4.A.2b}$$

because $\delta(x)$ is infinitely narrow in x, that is, Equation (4.A.1a), and has unit area, Equation (4.A.1b).

Consider a single revolution of duration $N\Delta$ in x of the discrete sequence of working-surface deviations $f_j(x)$, $j = 0, 1, 2, \ldots, N-1$. Using the above-described Dirac delta function, this sequence, normalized by the constant Δ, can be represented as

$$f(x) = \sum_{j=0}^{N-1} f_j(x)\Delta\delta(x - j\Delta) \tag{4.A.3}$$

which is zero everywhere except at the values $x = 0, \Delta, 2\Delta, \ldots, (N-1)\Delta$ according to Equation (4.A.1a). Substituting Equation (4.A.3) into the expression (4.1) for the Fourier series coefficients, we obtain

$$\hat{f}(n) = \frac{1}{N\Delta} \int_{0}^{N\Delta} \sum_{j=0}^{N-1} f_j(x)\Delta\delta(x - j\Delta)\exp\left[-i2\pi nx/(N\Delta)\right]dx \tag{4.A.4a}$$

$$= \frac{1}{N} \sum_{j=0}^{N-1} \int_{0-\epsilon}^{N\Delta-\epsilon} f_j(x)\delta(x - j\Delta)\exp\left[-i2\pi nx/(N\Delta)\right]dx \tag{4.A.4b}$$

$$= \frac{1}{N} \sum_{j=0}^{N-1} f_j(j\Delta)\exp\left[-i2\pi nj\Delta/(N\Delta)\right] \tag{4.A.4c}$$

$$= \frac{1}{N} \sum_{j=0}^{N-1} f_j(j\Delta)\exp\left(-i2\pi nj/N\right), \quad n = 0 \pm 1, \pm 2, \cdots \tag{4.A.4d}$$

where the infinitesimal value ϵ was introduced in Equation (4.A.4b) to accommodate the tooth deviation for $j=0$, and the property Equation (4.A.2b) was used in going from Equation (4.A.4b) to Equation (4.A.4c). Equation (4.A.4d) is the discrete Fourier transform, Equation (4.3).

Appendix 4.B. Formulas for $|B_{k\ell}(n)|^2$ and $G_\eta(n)$ Involving Only Real Quantities

From Equation (4.21a), we have for the DFT, $B_{k\ell}(n)$,

$$B_{k\ell}(n) = \frac{1}{N} \sum_{j=0}^{N-1} c_{j,k\ell} \exp\left(-i2\pi nj/N\right)$$

$$= \frac{1}{N} \sum_{j=0}^{N-1} c_{j,k\ell} \left[\cos\left(2\pi nj/N\right) - i\sin(2\pi nj/N)\right]$$

$$= \frac{1}{N} \sum_{j=0}^{N-1} c_{j,k\ell} \cos\left(2\pi nj/N\right) - i\frac{1}{N} \sum_{j=0}^{N-1} c_{j,k\ell}\sin\left(2\pi nj/N\right)$$

$$= RB_{k\ell}(n) - iIB_{k\ell}(n), \tag{4.B.1}$$

where R and I designate real and imaginary contributions. But

$$|B_{k\ell}(n)|^2 = B_{k\ell}(n)B_{k\ell}^*(n) \tag{4.B.2}$$

where the asterisk denotes the complex conjugate. Therefore, from Equations (4.B.2) and (4.B.1),

$$|B_{k\ell}(n)|^2 = [RB_{k\ell}(n)]^2 + [IB_{k\ell}(n)]^2$$

$$= \left[\frac{1}{N} \sum_{j=0}^{N-1} c_{j,k\ell} \cos\left(2\pi nj/N\right)\right]^2 + \left[\frac{1}{N} \sum_{j=0}^{N-1} c_{j,k\ell}\sin(2\pi nj/N)\right]^2 \tag{4.B.3}$$

which involves only real quantities. The coefficients $c_{j,k\ell}$ are the working-surface deviation expansion coefficients defined by Equation (3.20) and evaluated from the line-scanning lead and profile measurements by Equations (3.37) and (3.43) with the aid of Figures 3.9 and 3.10. The one-sided rotational-harmonic spectrum of the mean-square working-surface deviations $G_n(n)$, Equation (4.30), is obtained by summing the quantities of Equation (4.B.3) using the available expansion coefficients $c_{j,k\ell}$ as indicated in Figure 3.10.

Appendix 4.C. Alternative Proofs of Equations (4.33a) and (4.33c)

Define

$$\bar{c}_{k\ell} \triangleq \frac{1}{N} \sum_{j=0}^{N-1} c_{j,k\ell} \tag{4.C.1a}$$

$$= B_{k\ell}(0) \tag{4.C.1b}$$

by Equation (4.21a), since $exp(0) = 1$. Then, from Equations (4.15), (3.18), and (4.C.1a),

$$\bar{\eta}_{Cj}(y,z) = \sum_{k=0}^{\infty} \sum_{\ell=0}^{\infty} \bar{c}_{k\ell} \psi_{yk}(y) \psi_{z\ell}(z). \tag{4.C.2}$$

From Equations (4.16), (3.18), and (4.C.2),

$$\epsilon_{Cj}(y,z) = \sum_{k=0}^{\infty} \sum_{\ell=0}^{\infty} \left(c_{j,k\ell} - \bar{c}_{k\ell} \right) \psi_{yk}(y) \psi_{z\ell}(z). \tag{4.C.3}$$

Applying Equation (3.23) to $\epsilon_{Cj}^2(y,z)$ yields

$$\frac{1}{FD} \int_{-D/2}^{D/2} \int_{-F/2}^{F/2} \epsilon_{Cj}^2(y,z)\, dy dz = \sum_{k=0}^{\infty} \sum_{\ell=0}^{\infty} \left(c_{j,k\ell} - \bar{c}_{k\ell} \right)^2. \tag{4.C.4}$$

Form the average of Equation (4.C.4) in the same manner as in Equation (4.23),

$$\frac{1}{NFD} \sum_{j=0}^{N-1} \int_{-D/2}^{D/2} \int_{-F/2}^{F/2} \epsilon_{Cj}^2(y,z)\, dy dz = \sum_{k=0}^{\infty} \sum_{\ell=0}^{\infty} \frac{1}{N} \sum_{j=0}^{N-1} \left(c_{j,k\ell} - \bar{c}_{k\ell} \right)^2. \tag{4.C.5}$$

But

$$\frac{1}{N} \sum_{j=0}^{N-1} \left(c_{j,k\ell} - \bar{c}_{k\ell} \right)^2 = \frac{1}{N} \sum_{j=0}^{N-1} \left(c_{j,k\ell}^2 - 2\bar{c}_{k\ell} c_{j,k\ell} + \bar{c}_{k\ell}^2 \right)$$

$$= \left(\frac{1}{N} \sum_{j=0}^{N-1} c_{j,k\ell}^2 \right) - \bar{c}_{k\ell}^2. \tag{4.C.6}$$

Then, from Equations (4.C.6), (4.24), and (4.C.1b),

$$\frac{1}{N} \sum_{j=0}^{N-1} \left(c_{j,k\ell} - \bar{c}_{k\ell} \right)^2 = \sum_{n=1}^{N-1} \left| B_{k\ell}(n) \right|^2, \tag{4.C.7}$$

and from Equations (4.C.5) and (4.C.7),

$$\frac{1}{NFD} \sum_{j=0}^{N-1} \int_{-D/2}^{D/2} \int_{-F/2}^{F/2} \epsilon_{Cj}^2(y,z)\, dy dz = \sum_{n=1}^{N-1} \sum_{k=0}^{\infty} \sum_{\ell=0}^{\infty} \left| B_{k\ell}(n) \right|^2. \tag{4.C.8}$$

The square-root of the left-hand side of Equation (4.C.8) is the rms variation of $\epsilon_{Cj}(y,z)$, averaged over the analysis region FD and over all N teeth. Equation (4.33a) is the square root of Equation (4.C.8).

From Equations (4.18) and (4.22a), and then Equations (4.C.1a,b), we have

$$\bar{\eta}_C(y,z) = \sum_{k=0}^{\infty} \sum_{\ell=0}^{\infty} B_{k\ell}(0)\, \psi_{yk}(y)\, \psi_{z\ell}(z) \tag{4.C.9a}$$

$$= \sum_{k=0}^{\infty} \sum_{\ell=0}^{\infty} \bar{c}_{k\ell} \psi_{yk}(y)\, \psi_{z\ell}(z). \tag{4.C.9b}$$

Applying Equation (3.23) to Equation (4.C.9b), and then applying Equation (4.21b) for $n = 0$ to Equation (4.C.9a), there follows directly

$$\frac{1}{FD} \int_{-D/2}^{D/2} \int_{-F/2}^{F/2} \left[\bar{\eta}_C(y,z) \right]^2 dy dz = \sum_{k=0}^{\infty} \sum_{\ell=0}^{\infty} \left(\bar{c}_{k\ell} \right)^2 \tag{4.C.10a}$$

$$= \sum_{k=0}^{\infty} \sum_{\ell=0}^{\infty} \left| B_{k\ell}(pN) \right|^2, \quad p = 0, \pm 1, \pm 2, \cdots. \tag{4.C.10b}$$

Equation (4.33c) is the square root of Equation (4.C.10b).

References

Cluff, B.W. (ed.) (1992) *Gear Process Dynamics*, 7th edn, American Pfauter Limited Partnership, Loves Park, IL.

Cooley, J.W., Lewis, P.A.W., and Welch, P.D. (1969) The finite fourier transform. *IEEE Trans. Audio Electroaustic.*, **AU-17**, 77–85. Reprinted in Rabiner, L.R. and Rader, C.M. (eds) (1972) *Digital Signal Processing*, IEEE Press, New York, pp. 251–259.

Gaskill, J.D. (1978) *Linear Systems, Fourier Transforms, and Optics*, John Wiley & Sons, Inc., New York.

Hamming, R.W. (1973) *Numerical Methods for Scientists and Engineers*, 2nd edn, McGraw-Hill, New York. Republished by Dover, Mineola, NY.

Hildebrand, F.B. (1976) *Advanced Calculus for Applications*, 2nd edn, Prentice-Hall, Englewood Cliffs, NJ.

Lanczos, C. (1956) *Applied Analysis*, Prentice-Hall, Englewood Cliffs, NJ. Republished by Dover, New York.

Mark, W.D. (1978) Analysis of the vibratory excitation of gear systems: basic theory. *Journal of the Acoustical Society of America*, **63**, 1409–1430.

Mark, W.D. (1982) Gear noise excitation, in *Engine Noise: Excitation, Vibration, and Radiation* (eds R. Hickling and M.M. Kamal), Plenum Press, New York, pp. 55–93.

Mark, W.D. and Reagor, C.P. (2001) *Performance-Based Gear-Error Inspection, Specification, and Manufacturing-Source Diagnostics*, AGMA Technical Paper 01FTM6, American Gear Manufacturing Association, Alexandria, Virginia.

Merritt, H.E. (1971) *Gear Engineering*, John Wiley & Sons, Inc., New York.

5

Transmission-Error Spectrum from Working-Surface-Deviations

To compute the transmission error contribution $R_b^{(\cdot)} \delta\theta^{(\cdot)}(x)$ from a single gear, $(\cdot) = (1)$ or (2), defined in Equation (3.2), it is necessary to take into account the attenuating effects on the tooth deviations of the subject gear $(\cdot)$ that arise from the meshing action with a mating gear. In this chapter, it is shown that this attenuating action is a linear operation consisting of averaging the tooth deviations of the subject gear $(\cdot)$ over all lines of tooth contact located at each instantaneous value of roll distance x, Equation (3.1). Such linear operations are amenable to a "transfer function" approach in the frequency (rotational harmonic) domain. The resultant transfer functions $\hat{\phi}_{k\ell}(n/N)$ that characterize this averaging action are functions of the ratio n/N of rotational harmonic n divided by the number of teeth N; these transfer functions differ for each different Legendre term $k\ell$ of the tooth-deviation representation, Equation (3.18). In addition to their dependence on n/N and $k\ell$, the mesh transfer functions $\hat{\phi}_{k\ell}(n/N)$ are dependent only on the transverse Q_t and axial Q_a contact ratios of the rectangular working-surface analysis region. This formulation provides complete separation of the representation of geometric working-surface-deviations from the representation of attenuating effects arising from meshing action with a mating gear. The action of these attenuating effects on the dominant rotational harmonics of the spectrum as shown in Figure 4.1 is presented. The formulation allows computation of the complex Fourier series coefficients of the geometric-deviation transmission-error contributions of the subject gear $(\cdot)$, thereby permitting computation of these transmission-error contributions in the "time" (roll-distance) domain x, by using Fourier series.

Performance-Based Gear Metrology: Kinematic-Transmission-Error Computation and Diagnosis,
First Edition. William D. Mark.
© 2013 John Wiley & Sons, Ltd. Published 2013 by John Wiley & Sons, Ltd.

5.1 Transmission-Error Contributions from Working-Surface-Deviations

Transmission Error of Gear-Pair

A careful derivation of the transmission error $\zeta(x)$ arising from a meshing generic pair of helical gears is provided in Chapter 7. The transmission error is a function of the "roll-distance" variable x defined by Equation (3.1). There are three additive contributions to $\zeta(x)$, an elastic-deformation contribution $\zeta_W(x)$ which is loading dependent, and contributions $\zeta^{(1)}(x)$ and $\zeta^{(2)}(x)$ arising from geometric deviations of the tooth-working-surfaces of meshing gears (1) and (2) from equispaced perfect involute surfaces. The elastic deformation contribution $\zeta_W(x)$ includes the additive deformation contributions from each of the two meshing gears. Apart from changes in the zones of tooth contact on the working-surfaces, the geometric-deviation contributions $\zeta^{(1)}(x)$ and $\zeta^{(2)}(x)$ are only very weakly dependent on the loading carried by the gears. The total transmission error of a meshing-gear-pair is the superposition of these contributions:

$$\zeta(x) = \zeta_W(x) + \zeta^{(1)}(x) + \zeta^{(2)}(x). \tag{5.1}$$

Geometric Deviation Contribution from Single Gear

The subject of this book is the transmission-error geometric-deviation contribution $\zeta^{(\cdot)}(x)$ arising from a single generic gear of a meshing pair, $(\cdot) = (1)\ or\ (2)$, which can be expressed as

$$\zeta^{(\cdot)}(x) = \frac{1}{K_M(x)} \sum_j \tilde{\eta}_{Kj}(x) \tag{5.2}$$

where $K_M(x)$ is the total mesh stiffness and $\sum_j \tilde{\eta}_{Kj}(x)$ is the summation of the local-stiffness-weighted geometric working-surface-deviations, summed over the lines of contact of all teeth in contact of the gear under consideration, as defined below. Both of the above-described quantities, $K_M(x)$ and $\sum_j \tilde{\eta}_{Kj}(x)$ are functions of the roll-distance variable x, Equation (3.1), which describes the instantaneous rotational position of the gears.

The total mesh stiffness $K_M(x)$ is the summation over all tooth pairs j in contact of the stiffness $\tilde{K}_{Tj}(x)$ of the individual tooth pairs,

$$\tilde{K}_{Tj}(x) \triangleq \int_{y_{Aj}(x)}^{y_{Bj}(x)} K_{Tj}(x,y)d\ell, \tag{5.3}$$

where $K_{Tj}(x,y)$ is the tooth-pair stiffness per unit length of line of contact and $d\ell$ is the differential length of line of contact, $d\ell = sec\ \psi_b dy$ (Figure 2.6). Hence, the total mesh stiffness is

$$K_M(x) \triangleq \sum_j \tilde{K}_{Tj}(x).$$ (5.4)

Figure 2.6 illustrates such lines of contact within the rectangular zone of contact, where the location of a generic point P on each line of contact is a function of axial location y for each value of gear-pair rotational position x (Figure 3.2). The limits of integration $y_{Aj}(x)$ and $y_{Bj}(x)$ in Equation (5.3) designate the endpoints of the individual lines of contact, which are functions of roll distance x. The tilde (squiggle) symbol $\tilde{\ }$ located above η in Equation (5.2) and above K in Equations (5.3) and (5.4), and elsewhere, designates the integral over the line of contact of the quantity it sits above. Its similarity to the integral sign motivated its use.

The local stiffness-weighted geometric working-surface-deviations in Equation (5.2) are defined by

$$\tilde{\eta}_{Kj}(x) \triangleq \int_{y_{Aj}(x)}^{y_{Bj}(x)} K_{Tj}(x,y)\, \eta_j(x,y)\, d\ell$$ (5.5)

where $K_{Tj}(x,y)$, the integration limits, and $d\ell$ are defined as above, and $\eta_j(x,y)$ is the working-surface-deviation from an equispaced perfect involute surface of tooth j of the gear $(\cdot)$ of interest on the line of contact (Figures 2.6 and 3.2) at the particular location determined by the coordinate location x, y. At each coordinate location x, y, $\eta_j(x,y)$ is "measured" in a direction determined by the intersection of the plane of contact and transverse plane (plane of paper in lower portion of Figure 2.6). Apart from the assumption of a rectangular zone of contact in Figure 2.6, which is equivalent to the rectangular contract region on tooth-working-surfaces (Figure 3.2), the above formulation of Equations (5.1–5.5) is essentially "exact," as one can verify from the derivation carried out in Chapter 7.

Key Simplifying Assumption

However, by making the additional assumption that the local tooth-pair stiffness per unit length of line of contact is a constant independent of local line of contact location, that is,

$$K_{Tj}(x,y) = K, \text{a constant,}$$ (5.6)

it is symbolically shown below that the geometric-deviation contribution $\zeta^{(\cdot)}(x)$ to the transmission error becomes independent of tooth-pair stiffness and, therefore, of mesh loading. Consider the total mesh stiffness $K_M(x)$, Equation (5.4), and denote by $\ell_j(x)$ the length of contact line of tooth pair j within the rectangular zone of contact in Figure 2.6. Then, from Equations (5.3) and (5.6) there follows symbolically,

$$\tilde{K}_{Tj}(x) = K \int_{y_{Aj}(x)}^{y_{Bj}(x)} d\ell = K\ell_j(x)$$ (5.7)

where

$$\ell_j(x) \triangleq \int_{y_{Aj}(x)}^{y_{Bj}(x)} d\ell \tag{5.8}$$

is the length of contact line of tooth-pair j, and from Equations (5.4) and (5.7),

$$K_M(x) = K \sum_j \ell_j(x) = K\ell(x) \tag{5.9}$$

where

$$\ell(x) \triangleq \sum_j \ell_j(x) \tag{5.10}$$

is the total length of all lines of contact within the zone of contact (Figure 2.6) at roll distance value x. Moreover, from Equations (5.5) and (5.6),

$$\tilde{\eta}_{Kj}(x) = K \int_{y_{Aj}(x)}^{y_{Bj}(x)} \eta_j(x,y)d\ell = K\tilde{\eta}_j(x) \tag{5.11}$$

where

$$\tilde{\eta}_j(x) \triangleq \int_{y_{Aj}(x)}^{y_{Bj}(x)} \eta_j(x,y)d\ell. \tag{5.12}$$

Inserting Equations (5.9) and (5.11) into Equation (5.2), there follows for the case where the local tooth-pair stiffness per unit length of line of contact is a constant, Equation (5.6),

$$\zeta^{(\cdot)}(x) = \frac{1}{K\ell(x)} K \sum_j \tilde{\eta}_j(x) = \frac{1}{\ell(x)} \sum_j \tilde{\eta}_j(x) \tag{5.13}$$

where $\ell(x)$ and $\tilde{\eta}_j(x)$ are defined by Equations (5.10) and (5.12), respectively. This result is independent of local tooth-pair stiffness K, Equation (5.6).

Interpretation of Geometric-Deviation Transmission-Error Contribution

The interpretation of Equation (5.13) is very simple: *The tooth-working-surface geometric-deviation contribution to the instantaneous value of the transmission error at gear-pair rotational position x, from either gear (1) or (2) of a meshing pair, is the integrated value of the working-surface-deviations, integrated over all lines of tooth contact, divided by the total length of these lines of contact. This ratio is, simply, the average deviation of the working-surfaces at "instant" x, averaged over the total length of the lines of tooth contact.*

This average deviation is a generally fluctuating function of the instantaneous rotational position x of the gear-pair. This interpretation is valid for any shape of contact zone. However, it is an approximation because of the assumption, Equation (5.6), that the local tooth-pair stiffness per unit length of line of contact is a constant.

The enormous value of utilizing the above constant stiffness approximation, Equation (5.6), is that it permits evaluation of the contributions to the transmission error from the geometric working-surface-deviations of all teeth on a gear by utilizing only measurements made on the teeth of the single gear of interest. These transmission-error contributions include contributions to all rotational harmonics caused by the working-surface-deviations, including working-surface-deviation contributions to the tooth-meshing harmonics. In the following pages, it is shown how the transmission-error contribution $\zeta^{(\cdot)}(x)$, Equation (5.13), can be computed, in both frequency domain and time domain, from working-surface measurements made on a gear of interest (as described in Chapter 3). Because this transmission-error contribution is independent of mesh forces and inertia properties, it is the *kinematic contribution* of gear $(\cdot)$ to the transmission error (Merritt, 1971, p. 84).

5.2 Fourier-Series Representation of Transmission-Error Contributions from Working-Surface-Deviations

Because a gear is circular with base-circle circumference $N\Delta$, the transmission-error contribution $\zeta^{(\cdot)}(x)$ from working-surface-deviations is periodic with fundamental period $N\Delta$ in the "roll-distance" variable x, Equation (3.1). Denote the complex Fourier series coefficients of $\zeta^{(\cdot)}(x)$ by $\alpha_n^{(\cdot)}$,

$$\alpha_n^{(\cdot)} = \frac{1}{N\Delta} \int_{-N\Delta/2}^{N\Delta/2} \zeta^{(\cdot)}(x) \exp\left[-i2\pi nx/(N\Delta)\right] dx, \quad n = 0, \pm 1, \pm 2, \cdots \tag{5.14}$$

from which $\zeta^{(\cdot)}(x)$ can be generated by

$$\zeta^{(\cdot)}(x) = \sum_{n=-\infty}^{\infty} \alpha_n^{(\cdot)} \exp\left[i2\pi nx/(N\Delta)\right]. \tag{5.15}$$

Hence, if the full set of expansion coefficients $\alpha_n^{(\cdot)}$, $n = 0, \pm 1, \pm 2, \cdots$ are determined, the transmission-error contribution $\zeta^{(\cdot)}(x)$ in the "time" domain can be generated from these coefficients by using Equation (5.15).

Because the derivation of these Fourier-series expansion coefficients of the working-surface-deviation contributions to the transmission error is less straightforward than the material covered up to this juncture, the derivation of these expansion coefficients has been delegated to Chapter 7 in order not to interrupt the main flow of the work. The general form of that derivation will allow for "exact" evaluation of the Fourier series coefficients of $\zeta^{(\cdot)}(x)$, Equation (5.2), including the local stiffness-weighting effects described in Section 5.1. A plausibility argument explaining the resulting form of these Fourier-series expansion coefficients is given below.

Linear-System Interpretation of Gear-Meshing Action

We can regard the tooth-working-surface-deviations of all teeth $j = 0, 1, 2, \ldots, N-1$ on a gear as the "input" to a (somewhat unusual) linear system described by Equations (5.2–5.5). The system input contribution from each tooth j is the working-surface-deviations $\eta_{Cj}(y, z)$ described as a function of the Cartesian coordinates y, z over $(-F/2) \leq y \leq (F/2)$, $(-D/2) \leq z \leq (D/2)$, as defined in Section 3.2 and illustrated by Figure 3.2. A simple coordinate transformation relates these Cartesian coordinate locations to the coordinates x, y utilized in Equations (5.2–5.5). The stiffness-weighted integration of the tooth deviations along the lines of contact (Figure 3.2) described by Equation (5.5), followed by the summation over all teeth j in contact and division by the total mesh stiffness $K_M(x)$, Equation (5.2), describes the linear "system" operations on the working-surface-deviations.

Because at least one tooth pair is always in contact (Figure 2.6) at every value of roll-distance variable x, the transmission error contribution $\zeta^{(\cdot)}(x)$ generally is a continuous function of x, and because $\zeta^{(\cdot)}(x)$ is periodic in x with period $N\Delta$, the base-circle circumference, it is described in the frequency domain by Fourier series (Equations (5.14) and (5.15)). Nevertheless, because the teeth are discrete and spaced in x at intervals of Δ, the base pitch, it was illustrated by Equations (4.1) and (4.3) that the discrete counterpart to forming the Fourier series coefficients is the discrete finite Fourier transform (DFT) (Equation (4.3)). When the working-surface-deviation of each tooth j was represented by the superposition of normalized two-dimensional Legendre polynomials (Equation (3.18)), it was shown that the DFT of the working-surface-deviations $\eta_{Cj}(y, z)$ (Equation (4.5)), became the result (Equation (4.20)), and then Equation (4.22), where the strength $c_{j,k\ell}$ of each Legendre term $k\ell$ as a function of tooth number j was transformed to rotational harmonic number strength $B_{k\ell}(n)$ by Equation (4.21a). The resultant representation of the DFT, $\hat{\eta}_C(n; y, z)$ given by Equation (4.22a), corresponds to the DFT of the summation over j in Equation (5.2). That leaves the smoothing action provided by the integration in Equation (5.5) and the division by the total mesh stiffness $K_M(x)$ in Equation (5.2). These two operations are characterized in the frequency domain by "mesh transfer functions," which necessarily differ for each different Legendre term $\psi_{yk}(y) \, \psi_{z\ell}(z)$ in Equation (4.22a). These "mesh transfer functions" are designated by the symbol $\hat{\phi}_{k\ell}(n/N)$. In addition to their dependence on Legendre term $k\ell$, they are functions of the ratio of rotational harmonic number n divided by the number of teeth N. Hence, the final form of the complex Fourier series coefficients (Equation (5.14)), of the tooth-working-surface-deviation contributions of a generic helical gear $(\cdot)$ to the transmission error is

$$\alpha_n^{(\cdot)} = \sum_{k=0}^{\infty} \sum_{\ell=0}^{\infty} B_{k\ell}(n) \hat{\phi}_{k\ell}(n/N), \quad n = 0, \pm 1, \pm 2, \cdots \tag{5.16}$$

which the reader might compare with Equation (4.22a).

For each Legendre $k\ell$ pair, the *form* $\psi_{yk}(y)\psi_{z\ell}(z)$ of the deviation contribution in Equation (4.22a) is the same for every tooth $j = 0, 1, \ldots, N-1$ on a gear, since its amplitude $c_{j,k\ell}$ is contained in $B_{k\ell}(n)$ (Equation (4.21a)). Consequently, $\hat{\phi}_{k\ell}(n/N)$ characterizes, for each $k\ell$ pair, the attenuation of $\psi_{yk}(y)\psi_{z\ell}(z)$ that arises from the computed meshing action with a mating gear. After the Fourier series coefficients, Equation (5.16), have been computed, the transmission-error contributions $\zeta^{(\cdot)}(x)$ from the tooth-working-surface-deviations can be computed as a function of roll distance x, Equation (3.1), by Equation (5.15).

The above explanation was carried out in an attempt to explain the functional form of Equation (5.16). Its rigorous derivation is carried out in Chapter 7. It was first derived in Mark (1978, Equations (134), (135)) where the necessity of using a representation such as the expansion in two-dimensional Legendre polynomials was shown (Mark, 1978, Equation (119)).

The result, Equation (5.16), is particularly useful when the assumption, Equation (5.6), of constant local tooth-pair stiffness per unit length of line of contact is made. In such cases, $\zeta^{(\cdot)}(x)$ is independent of local tooth-pair stiffness and total mesh stiffness, as shown by Equation (5.13). When this assumption is made, each "mesh transfer function" $\hat{\phi}_{k\ell}(n/N)$, for each $k\ell$ pair, is dependent only on the ratio n/N, and the transverse and axial contact ratios Q_t and Q_a, Equations (2.9) and (2.10), respectively, based on the rectangular analysis region utilized in the tooth-working-surface measurements. *Hence, apart from harmonic number dependence n/N, $\hat{\phi}_{k\ell}(n/N)$ is a function only of the well-recognized gear design parameters Q_t and Q_a, and is independent of working-surface-deviations. On the other hand, the function $B_{k\ell}(n)$, defined by Equation (4.21a), is determined only from the Legendre expansion coefficients $c_{j,k\ell}$ of the deviations of the individual working-surfaces $\eta_{Cj}(y,z)$, as determined by Equation (3.20). These expansion coefficients completely characterize the tooth-working-surface-deviations.* Therefore, the first term $B_{k\ell}(n)$ in Equation (5.16) characterizes the working-surface-deviations of all teeth on the gear $(\cdot)$, and the second term $\hat{\phi}_{k\ell}(n/N)$ characterizes the attenuation of these working-surface-deviations provided by the averaging action of the integrations of the working-surface-deviations along the lines of contact (Figure 3.2) and the averaging action of the multiple teeth in contact (Figure 2.6). *In other words, the second term $\hat{\phi}_{k\ell}(n/N)$ in Equation (5.16) characterizes the computed attenuating effects on the tooth-working-surface-deviations of the gear under consideration that would arise from the meshing action with a mating gear, if the gear under consideration were meshing with a mating gear. Therefore, a more descriptive term for $\hat{\phi}_{k\ell}(n/N)$ is "mesh-attenuation function".* Approximate and exact formulas and algorithms for computing the mesh-attenuation functions $\hat{\phi}_{k\ell}(n/N)$ are derived and summarized in Chapters 7 and 8, respectively.

5.3 Rotational-Harmonic Spectrum of Mean-Square Mesh-Attenuated Working-Surface-Deviations

The mesh-attenuated working-surface-deviations are the transmission error contributions from those deviations. Parseval's formula (Korn and Korn, 1961, Sec. 4.11-4) for Fourier series, Equations (5.14) and (5.15), is

$$\frac{1}{N\Delta} \int_{-N\Delta/2}^{N\Delta/2} \left[\zeta^{(\cdot)}(x)\right]^2 dx = \sum_{n=-\infty}^{\infty} \left|\alpha_n^{(\cdot)}\right|^2 \qquad (5.17a)$$

$$= \left|\alpha_o^{(\cdot)}\right|^2 + 2\sum_{n=1}^{\infty} \left|\alpha_n^{(\cdot)}\right|^2, \qquad (5.17b)$$

since $\zeta^{(\cdot)}(x)$ in Equation (5.14) is real. The left-hand side of Equation (5.17) is the mean-square transmission-error contribution, averaged over the base-circle circumference $N\Delta$, and the right-hand side of Equation (5.17) describes the decomposition of this mean-square transmission-error contribution into its squared rotational-harmonic amplitudes. Therefore, define from Equation (5.17b) the "one-sided" mean-square rotational-harmonic spectrum of the mesh-attenuated tooth-working-surface-deviations as

$$G_\zeta(n) \triangleq \begin{cases} \left|\alpha_o^{(\cdot)}\right|^2, & n = 0 \\ 2\left|\alpha_n^{(\cdot)}\right|^2, & n = 1, 2, 3, \cdots \end{cases} \qquad (5.18a)$$

$$= \begin{cases} \left|\sum_{k=0}^{\infty}\sum_{\ell=0}^{\infty} B_{k\ell}(0)\hat{\phi}_{k\ell}(0)\right|^2, & n = 0 \\ 2\left|\sum_{k=0}^{\infty}\sum_{\ell=0}^{\infty} B_{k\ell}(n)\hat{\phi}_{k\ell}(n/N)\right|^2, & n = 1, 2, 3, \cdots \end{cases} \qquad (5.18b)$$

which follows from Equation (5.16). Equation (5.18b) provides an expression for the "one-sided" mean-square rotational-harmonic spectrum of the tooth-working-surface-deviation contribution of the subject gear $(\cdot)$ to the transmission error. It is a function of the DFT, Equation (4.21), of the normalized Legendre expansion coefficients and of the "mesh-attenuation functions" $\hat{\phi}_{k\ell}(n/N)$.

Justification of the Term "Mesh-Attenuation Function"

The mean-square value of the tooth-working-surface-deviations was given by the left-hand side of Equation (4.26), and its (periodic) rotational-harmonic spectrum was given by Equation (4.30). The mean-square value of the transmission error contributions from these working-surface-deviations is given by the left-hand side of Equation (5.17), and its rotational harmonic spectrum is given by Equation (5.18b). Note the difference in locations of absolute value indicators between Equations (4.30) and (5.18b).

Applying the Cauchy-Schwartz inequality (Korn and Korn, 1961, Sec.1.3-3) to the right-hand side of Equation (5.18a), there follows directly

$$\left|\sum_{k=0}^{\infty}\sum_{\ell=0}^{\infty} B_{k\ell}(n)\hat{\phi}_{k\ell}(n/N)\right|^2 \le \left[\sum_{k=0}^{\infty}\sum_{\ell=0}^{\infty} \left|B_{k\ell}(n)\right|^2\right]\left[\sum_{k=0}^{\infty}\sum_{\ell=0}^{\infty} \left|\hat{\phi}_{k\ell}(n/N)\right|^2\right] \qquad (5.19)$$

which relates the right-hand side of Equation (5.18b) to that of Equation (4.30). The equality in Equation (5.19) can hold for any rotational harmonic n only if the ratio $B_{k\ell}(n)/\hat{\phi}_{k\ell}(n/N)$ is a constant (Beckenbach and Bellman, 1965, pp. 2, 3) independent of k and ℓ, in which case, the spectrum $B_{k\ell}(n)$ and mesh-attenuation function $\hat{\phi}_{k\ell}(n/N)$ are perfectly correlated. But for any rotational harmonic n where the equality in Equation (5.19) holds, $\hat{\phi}_{k\ell}(n/N)$ cannot legitimately be regarded as a mesh-*attenuation* function unless

$$\sum_{k=0}^{\infty}\sum_{\ell=0}^{\infty}\left|\hat{\phi}_{k\ell}(n/N)\right|^2 \leq 1, \qquad (5.20)$$

which can be seen by comparing Equation (5.19) with Equations (4.30) and (5.18b). By Equation (6.50), it is shown in closed form, for the case where the total mesh stiffness $K_M(x)$ and local tooth-pair stiffness $K_{Tj}(x,y)$ both are assumed to be constant, that the linear operations consisting of the integration over the lines of contact in Equation (5.5) and the summation over all lines of contact in Equation (5.2), both of which are averaging (attenuating) operations yield, exactly

$$\sum_{k=0}^{\infty}\sum_{\ell=0}^{\infty}\left|\hat{\phi}_{k\ell}(n/N)\right|^2 = 1, \qquad (5.21)$$

which when inserted into Equation (5.19) gives

$$\left|\sum_{k=0}^{\infty}\sum_{\ell=0}^{\infty}B_{k\ell}(n)\hat{\phi}_{k\ell}(n/N)\right|^2 \leq \sum_{k=0}^{\infty}\sum_{\ell=0}^{\infty}\left|B_{k\ell}(n)\right|^2, \qquad (5.22)$$

thereby showing that the mesh-attenuated mean-square spectrum, Equation (5.18), cannot possess contributions at any given rotational harmonic n larger than the contribution of that harmonic to the non-mesh-attenuated spectrum (Equation (4.30)).

It follows directly from the left-hand side of Equation (5.17) that the dimension of the right-hand side, and of $G_\zeta(n)$, is the same as the dimension of $[\zeta^{(\cdot)}(x)]^2$, which is length-squared. Because it is easier to assess quantities of dimension "length" rather than "length-squared," the mean-square transmission-error spectra will be displayed as the square-root of Equation (5.18), that is, $[G_\zeta(n)]^{1/2}$.

5.4 Example of Rotational-Harmonic Spectrum of Mean-Square Mesh-Attenuated Working-Surface-Deviations

Equation (5.18b) for $G_\zeta(n)$ is the mesh-attenuated mean-square rotational harmonic spectrum counterpart to Equation (4.30) for $G_\eta(n)$, which describes the mean-square rotational harmonic spectrum of the tooth-working-surface-deviations (without mesh attenuation). The square root $[G_\zeta(n)]^{1/2}$ of the mesh-attenuated counterpart to its non-mesh-attenuated counterpart $[G_\eta(n)]^{1/2}$ of Figure 4.1 is shown in Figure 5.1b, also using a logarithmic amplitude scale. It is instructive to understand the highly selective attenuation of the dominant rotational harmonic amplitudes in Figure 4.1 (also displayed in Figure 5.1a) that are computed and displayed in Figure 5.1b.

Rotational Harmonic n = 1 Tooth-Spacing-Error Contribution

Consider, first, rotational harmonic $n = 1$. The amplitudes of $n = 1$ in Figures 5.1a and 5.1b appear to be virtually identical, indicating negligible computed attenuation of the working-surface-deviation contributions causing this lowest-order rotational harmonic. As explained in Section 4.5 and shown in Figure 4.2, the dominant working-surface-deviation contribution to $n = 1$ is a sinusoidal one-cycle accumulated tooth-spacing error caused by an eccentricity between the locations of the axis of rotation in gear measurement and the base-cylinder axis used in tooth generation. Because the variation of this one-cycle sinusoidal error is negligible over the roll-distance span of $(Q_t + Q_a)\Delta$ where any single tooth would be in contact with a mating-gear-tooth, the computed attenuation of this one-cycle accumulated tooth-spacing error is negligible; hence, the amplitudes of $n = 1$ in Figures 5.1a and 5.1b are virtually identical, indicating negligible computed attenuation.

The most rational definition of an (accumulated) tooth-spacing error is the average deviation of the working-surface of a tooth, averaged over the rectangular region $(-F/2) \le y \le (F/2), (-D/2) \le z \le (D/2)$, where this average deviation is the measured distance in roll-distance x from the location that working-surface would occupy if the teeth were exactly equally spaced (see Equations (3.24–3.26)). This accumulated tooth-spacing error is the term $k = 0, \ell = 0$ in the Legendre expansion, Equation (3.18), as can be seen from Equations (3.8), (3.13), (3.14), and (3.20), where $c_{j,00}$ describes this accumulated tooth-spacing error for tooth j, before the adjustment described by Equations (3.24–3.26) is made. As will be shown in Chapter 6, the mesh-attenuation function for tooth-spacing errors satisfies $\hat{\phi}_{00}(n/N) \approx 1$ for small n/N, so for this accumulated tooth-spacing harmonic contribution $n = 1$, Equations (4.30) and (5.18b) predict that no attenuation of the accumulated tooth-spacing harmonic $n = 1$ would be caused by meshing with a mating gear.

Rotational Harmonic n = 1 Linear-Lead and Linear-Profile-Error Contributions

Figure 4.2 also exhibits a once-per-revolution wobble of an approximately plane surface, as described in Section 4.5. The Legendre terms $k = 0, \ell = 1$ and $k = 1, \ell = 0$ describe this plane-surface wobble. Their respective mesh-attenuation functions $\hat{\phi}_{01}(n/N)$ and $\hat{\phi}_{10}(n/N)$ strongly attenuate this $n = 1$ once-per-revolution contribution, as illustrated in Chapter 6. However, their contribution to rotational-harmonic amplitude $n = 1$ in Figure 5.1a is very small in comparison to that caused by the $k = 0, \ell = 0$ accumulated tooth-spacing contribution.

Rotational Harmonics n = 2 and n = 4 Tooth-Spacing-Error Contributions

Comparison of the amplitudes of rotational harmonics $n = 2$ and 4 in Figures 5.1a and 5.1b show only modest attenuation of these two additional low-order harmonics. Because all Legendre error types except accumulated tooth-spacing errors $k = 0, \ell = 0$ are strongly attenuated in this very-low harmonic number region, these

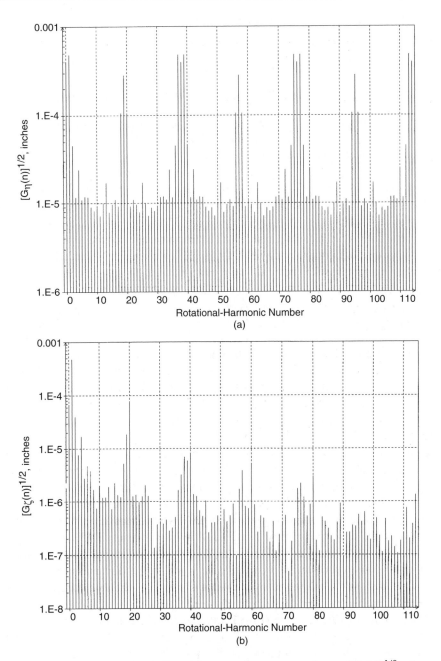

Figure 5.1 (a) Repeat of rotational harmonic rms spectrum $(G_\eta(n))^{1/2}$, shown in Figure 4.1, of non-mesh-attenuated working-surface-deviations. (b) Rotational harmonic rms spectrum $(G_\zeta(n))^{1/2}$, square-root of Equation (5.18b), of mesh-attenuated working-surface-deviations. Both spectra computed from measurements of 38 tooth pinion in Figure 3.11. Logarithmic amplitude scales in inches (From Mark and Reagor (2001), Reproduced by permission of the American Gear Manufacturers Association)

two additional harmonics $n=2$ and 4 in Figure 5.1 would indicate significant two-cycle and four-cycle rotational harmonic contributions of accumulated tooth-spacing errors.

Rotational Harmonic n = 19 Hobbing-Error Contributions

Rotational harmonic $n=19$ is the next strong rotational harmonic shown in Figure 5.1b. The manufacturing errors causing $n=19$ on even-numbered and odd-numbered teeth were shown in Figure 4.3, and are attributable to hobbing errors, as shown in Figure 4.4. Careful inspection of the rms amplitude of rotational harmonic $n=19$ in Figure 5.1b shows an rms amplitude of about 19 µin. (0.48 µm). Because the rms amplitude of this same harmonic $n=19$ of the non-attenuated working-surface-deviations displayed in Figure 5.1a was shown to be about 290 µin. (7.4 µm), the computed effect on rotational harmonic $n=19$ of the attenuation caused by gear-meshing action is about $(290/19) \approx 15$, or about a 20 log $15 \approx 24$ decibel reduction. This factor of 15 reduction in rms amplitudes is a substantial reduction.

Rotational Harmonic n = 20 Undulation/Ghost-Tone-Error Contributions

The next strong rotational harmonic in Figure 5.1b is $n=20$. Careful inspection of Figure 5.1b shows that the rms amplitude of rotational harmonic $n=20$ is about 80 µin. (2 µm). The rms amplitude of this same $n=20$ harmonic shown in Figure 5.1a was seen to be just slightly over 100 µin. (2.54 µm). Hence, this $n=20$ rotational harmonic was computed to be attenuated by a factor of only about $(100/80)=1.25$, or about 20 log $1.25 \approx 2$ db, virtually no attenuation.

A gear manufacturing error of $2\frac{1}{2}$ µm is not generally considered to be exceptionally large, but a transmission-error harmonic with rms amplitude of 2 µm is a significant vibration and noise producer. The manufacturing-error pattern on the two teeth shown in Figure 4.5, which is the cause of rotational harmonic $n=20$, explains why this particular manufacturing-error pattern experiences almost no attenuation.

In the italicized description in Section 5.1 following Equation (5.13), it was explained that the transmission-error contribution arising from working-surface-deviations, at any instantaneous roll-distance location x, is "the average deviation of the working-surfaces at "instant" x, averaged over the total length of the lines of tooth contact". For there to be negligible attenuation then, as in the case of rotational harmonic $n=20$, the manufacturing-error contribution *to rotational harmonic* $n=20$ must be essentially the same at all points on all lines of tooth contact at every roll distance location x, that is, at every rotational position of the gear. From Figure 4.5 we can see how the manufacturing-error contributions shown there on two adjacent teeth can satisfy this condition. The error on each tooth has the general appearance of slightly over one period of a sine wave. That error is almost constant along diagonal lines, for example, the trough in the left tooth $j=0$ in Figure 4.5. The direction of this trough is the direction of the lines of contract on the working-surface of the teeth, as illustrated by Figure 3.2. Hence, since the error is essentially the same at all points in the direction of this line of contact, negligible averaging of the errors along these lines of contact will take place.

Because this particular rotational harmonic is $n = 20$, there are 20 cycles of this error contribution around the full circumference of the gear. The gear has 38 teeth. Therefore, in passing from one tooth to an adjacent tooth, a pure sinusoidal error will exhibit a phase shift of $(20/38) = 0.526 \ldots$ cycles, that is, slightly over $180°$. This is exactly the behavior displayed in Figure 4.5. Hence, the manufacturing error displayed in Figure 4.5 approximates a pure sinusoidal error pattern with an integral number of 20 cycles around the circumference of the gear, always constant in the direction of the lines of tooth contact. These are the conditions required for negligible attenuation arising from the meshing action with a matting gear (Mark, 1992a, Equations (91b), (91c), and (95)), as described in Section 6.6.

The strong amplitude of rotational harmonic $n = 20$ in Figure 5.1b is commonly referred to as a "ghost tone," and the error pattern shown on the two teeth in Figure 4.5 is typical of ghost-tone error patterns. Such "undulation errors" are caused by generating type manufacturing operations (Drago, 1988; Dudley, 1984).

Even though rotational harmonic numbers $n = 19$ and 20 in Figures 5.1a and 5.1b are adjacent harmonics, the working-surface manufacturing errors causing these two harmonics, shown in Figures 4.3 and 4.5 respectively, have completely different characteristics, and the attenuation factors of 15 for $n = 19$ versus 1.25 for $n = 20$ are dramatically different. In particular, in contrast to the manufacturing errors in Figure 4.5 causing $n = 20$, those shown in Figure 4.3 causing $n = 19$ are strongly attenuated by the computed meshing action with a mating gear, which arises from the averaging effect of the errors along lines of tooth contact, as described by Equations (5.12) and (5.13) and the accompanying italicized discussion.

Simple rms working-surface-error metrics are inadequate to characterize vibration/noise generating properties. Because the square-root of the sum of the squares of the individual rotational harmonic line amplitudes shown in Figure 5.1a describes the rms geometric deviations of the tooth-working-surfaces, we can regard the individual rotational-harmonic line amplitudes shown there as representative of the nonattenuated manufacturing errors on the tooth-working-surfaces. The above-described contrasting attenuation factors of 15 versus 1.25 of rotational harmonics $n = 19$ and 20 illustrates the fact that *simple* rms working-surface error descriptors do not directly relate to the vibration/noise generating properties of manufacturing errors, because as shown above, these vibration/noise generating properties are very strongly dependent on the detailed manufacturing error *patterns* on the working-surfaces of all of the teeth. The nonattenuated rms errors of $n = 19$ were seen to be $(290/100) \approx 3$ times as large as the nonattenuated errors of $n = 20$, yet the attenuated errors of $n = 20$ were seen to be $(80/19) \approx 4$ times as large as the attenuated errors of $n = 19$.

Rotational Harmonic $n = 38$ Tooth-Meshing Fundamental Contributions

Because the subject pinion shown in Figure 3.11 has 38 teeth, rotational harmonic $n = 38$ of the mesh-attenuated spectrum of Figure 5.1b is the manufacturing-deviation contribution to the *tooth-meshing fundamental* harmonic. The deviation surface causing this harmonic is the average deviation surface described by Equations (3.45) and (3.47), and displayed in Figure 3.14. In an operating gear-pair transmitting significant loading, tooth elastic deformations would be superimposed on the average deviation surface of the teeth, causing a significantly stronger tooth-meshing fundamental, and higher tooth-meshing harmonics, than those shown in Figure 5.1.

References

Beckenbach, E.F. and Bellman, R. (1965) *Inequalities*, Springer-Verlag, Berlin.

Drago, R.J. (1988) *Fundamentals of Gear Design*, Butterworths, Boston, MA.

Dudley, D.W. (1984) *Handbook of Practical Gear Design*, Revised edn, McGraw-Hill, New York.

Korn, G.A. and Korn, T.M. (1961) *Mathematical Handbook for Scientists and Engineers*, McGraw-Hill, New York. Republished by Dover, Mineola, NY.

Mark, W.D. (1978) Analysis of the vibratory excitation of gear systems: basic theory. *Journal of the Acoustical Society of America*, **63**, 1409–1430.

Mark, W.D. (1992a) Contributions to the vibratory excitation of gear systems from periodic undulations on tooth running surfaces. *Journal of the Acoustical Society of America*, **91**, 166–186.

Mark, W.D. and Reagor, C.P. (2001) *Performance-Based Gear-Error Inspection, Specification, and Manufacturing-Source Diagnostics*, AGMA Technical Paper 01FTM6, American Gear Manufacturing Association, Alexandria, Virginia.

Merritt, H.E. (1971) *Gear Engineering*, John Wiley & Sons, Inc., New York.

6

Diagnosing Manufacturing-Deviation Contributions to Transmission-Error Spectra

Although the central theme of this book is development and explanation of a method to enable computation of the transmission-error contributions from measured working-surface deviations made on a single gear, if that gear were to be meshed with a mating gear, the methodology required to carry out this task provides a great deal of information useful for diagnosing the manufacturing-deviation sources of the various harmonic contributions observed in meshing-gear-pair vibration spectra. Aspects of this diagnostic information that can be ascertained from general features of this methodology are described in the present chapter.

6.1 Main Features of Transmission-Error Spectra

Effects of Transmission Path

Unless stated otherwise, it is assumed that a meshing-gear-pair is operating at constant speed and transmitting constant torque. Such a gear-pair causes a vibration excitation (static transmission error) that is "filtered" by a structure and possible acoustic medium, and received by a transducer, for example, accelerometer or microphone. If the locations of the gear-pair and transducer are fixed, the structural/acoustic path between the gear-pair and transducer is normally well modeled as a linear time-invariant system. Let $H(f)$ denote the complex frequency response function (Newland, 1989, p. 4) of such a system (transducer/structural/acoustic

Performance-Based Gear Metrology: Kinematic-Transmission-Error Computation and Diagnosis,
First Edition. William D. Mark.
© 2013 John Wiley & Sons, Ltd. Published 2013 by John Wiley & Sons, Ltd.

path), and $\alpha_y(f)$ and $\alpha_\zeta(f)$ the complex Fourier series coefficients of the transducer response and static-transmission-error vibration excitation, respectively. Then, the response and excitation complex Fourier series coefficients are related by

$$\alpha_y(f) = H(f)\alpha_\zeta(f). \tag{6.1}$$

Consequently, the frequencies of the individual transmission-error excitation harmonics are unaffected by the transmission path, but the amplitudes and phases of these harmonics *are* affected. Nevertheless, because the frequency values of the individual gear-meshing harmonics can be identified in the transducer response, considerable diagnostic information can be obtained from the absolute value $|\alpha_y(f)|$ of the response spectra,

$$\left|\alpha_y(f)\right| = |H(f)| \left|\alpha_\zeta(f)\right|. \tag{6.2}$$

The contribution from working-surface deviations on a single gear of a meshing pair to the excitation $|\alpha_\zeta(f_n)|$ at rotational harmonic n is $[G_\zeta(n)]^{1/2}$, the square-root of Equation (5.18).

Harmonic Contributions from a Meshing-Gear-Pair

Because of the way in which a pair of gears mesh, the tooth-meshing periods are the same for both gears of the pair, and therefore, the tooth-meshing fundamental frequencies of the two gears coincide. Consider one of the two gears with N teeth. Let P be the rotation period of that gear. Therefore, the tooth-meshing period is P/N, since there are N teeth, and the tooth-meshing fundamental frequency is N/P cycles per second (Hertz). But the gear rotational fundamental frequency is $1/P$. Therefore, the tooth-meshing fundamental frequency is the Nth rotational harmonic frequency of the gear with N teeth, and there are, in principle, $N-1$ equispaced rotational harmonic frequencies between zero frequency and the tooth-meshing fundamental frequency N/P.

Figure 6.1 is an idealized sketch of the vibration-excitation transmission-error spectrum contributions from a single gear of a meshing pair. The horizontal axis labels the rotational harmonic numbers $n = 1, 2, \ldots, N, \ldots, 2N$, where the heavy harmonic lines $n = N$ and $n = 2N$ are, respectively, the tooth-meshing fundamental frequency and its first multiple. The lighter lines illustrate rotational harmonic amplitudes. In experimentally obtained vibration spectra, one normally observes several strong low-order rotational harmonic amplitudes $n = 1, 2, \ldots$ and also several so-called "sideband harmonics" located around each tooth-meshing harmonic $n = N$, $n = 2N, \ldots$ as sketched in Figure 6.1. In some rotational harmonic spectra, one or more isolated rotational harmonics labeled in the figure by D also are observed. Information will be provided in this chapter to aid in diagnosing the manufacturing-error sources of the rotational harmonics illustrated in Figure 6.1.

In most applications, the number of teeth on each gear of a meshing pair will differ. Because the locations of the tooth-meshing harmonics in measured vibration spectra arising from two meshing gears must coincide, as explained above, the locations of

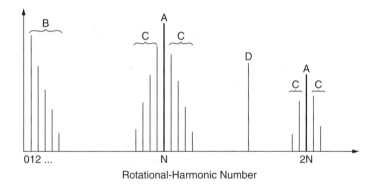

Figure 6.1   Sketch of dominant rotational harmonics caused by a single gear of a meshing pair operating at constant speed and transmitting constant loading. Abscissa labels rotational-harmonic numbers $n = 1, 2, \ldots$ The period of rotational harmonic $n = 1$ is the rotation period of the gear. All harmonics are integer multiples of $n = 1$. Rotational harmonic $n = N$ is the tooth-meshing fundamental harmonic with period equal to the gear rotation period divided by the number of teeth, N. Low-order rotational harmonics B and ''sideband'' harmonics C typically are strong. ''Ghost tone,'' when present, is labeled D (Adapted from Mark (1991))

the remaining rotational harmonics from the two gears generally will differ. In the case of "hunting-tooth" gear pairs (numbers of teeth have no common integer divisor except unity), the frequency locations of all rotational harmonic contributions, other than the tooth-meshing harmonics, will differ. For nonhunting-tooth gear pairs some of the rotational-harmonic frequency locations from each of the two meshing gears will coincide.

Linear System Model of Mesh Attenuation

In order to enable relatively straightforward implementation of the methodology developed herein, it has been necessary to require the user to assume a rectangular region of tooth-pair contact designated by the axial facewidth F, and roll-angle limits ϵ_t and ϵ_r utilized in Equation (3.7). Moreover, the tooth-pair stiffness per unit length of line of contact has been assumed to be a constant value, but this constant value is not required, as shown by Equation (5.13).

Real meshing gears only *approximately* satisfy these two assumptions. However, the general form of the linear-system model described by Equation (5.16) does not, in principle, require either of the above-two simplifying assumptions to be made. See, for example, Mark (1978, Figure 6 and Equation (40)) and Mark (1979, Equation (61)). The *general form* of this model described by Equation (5.16) does, in fact, sufficiently accurately represent the effects of meshing attenuation on the tooth-to-tooth variations $\epsilon_{Cj}(y, z)$, Equation (4.16), that are the source of the non-tooth-meshing rotational harmonics. This is especially true when the effects of loading onset and termination on the individual teeth are included in the mesh-attenuation functions

$\hat{\phi}_{k\ell}(n/N)$. As a consequence of this general linear system model, Equation (5.16), which describes the effects of gear-pair meshing action as *attenuation* of the harmonics contained in the spectra $B_{k\ell}(n)$, Equations (4.21) and (4.30), we can expect all such rotational harmonics n of $B_{k\ell}(n)$ to be present in the transmission error and transducer responses of real systems, but with attenuation governed by $\hat{\phi}_{k\ell}(n/N)$.

Sources of Tooth-Meshing Harmonics

Our principal goal here is diagnosing the manufacturing-deviation sources causing observed features in gear-vibration spectra measured under *constant* transmitted loading and *constant* rotational speed. Ordinarily, the strongest vibration-spectrum harmonics from a meshing-gear-pair are the tooth-meshing harmonics, $n = N, 2N, 3N, \ldots$, which are illustrated by the heavy lines shown in Figure 6.1. There are two dominant sources causing the tooth-meshing harmonics: tooth (and gear body) elastic deformations, and the average tooth-deviation surfaces from the two meshing gears, as explained below. Contributions to the tooth-meshing harmonics arise from both gears of a meshing pair. The simplest way to understand the vibration-excitation (static transmission error) source of the tooth-meshing harmonics is to imagine the *combined deviation* from perfect involute surfaces of the average *loaded* working-surfaces of the two meshing gears, which is the source of the tooth-meshing harmonics. Normally, the consistency of the geometric dimensions of gear teeth is such that the stiffness properties of every pair of meshing teeth on the two gears is the same, that is, there is negligible variation of tooth-pair stiffness properties among the various pairs of meshing teeth on a given gear-pair. However, because there are (generally small) geometric deviations on the working-surfaces among the differing teeth on each gear, it is necessary to consider the average working-surface from each of the two meshing gears, as defined by the middle term in Equation (4.10) and Figure 3.4. *The geometric deviation of this unloaded average working-surface from each of the two meshing gears combined with the elastic deformations of these average working-surfaces is the source of the tooth-meshing harmonics of a meshing-gear-pair.* Notice from Equation (4.12) that the DFT, Equation (4.5), evaluated at the tooth-meshing harmonic locations $n = pN$, $p = 0, \pm1, \pm2, \ldots$, is the average working-surface geometric deviation at each working-surface location y, z.

Gear-tooth-working-surfaces normally are intentionally modified from perfect involute surfaces by providing "relief" (removal of material) at locations of tooth-pair contact initiation and termination in such a manner that tooth-pair loading gradually increases from zero at contact initiation to a maximum then gradually decreases back to zero at contact termination. Hence, contributions to tooth-meshing harmonics of intentional tooth modifications and elastic deformations typically are of the same order of magnitude. (It is possible, in principle, to modify the working-surfaces to exactly compensate for elastic deformations at one nominal gear-pair loading.) The analytical construction of the average deviation surface $\bar{\eta}_C(y, z)$ given by Equation (3.47) can provide the design engineer with a very accurate representation of the *achieved* tooth modification surface to compare with his or her *desired* modification surface.

Sources of Remaining Rotational Harmonics

As mentioned above, each gear of a meshing pair will be a source of a generally distinct set of rotational harmonics, as illustrated in Figure 6.1. *If the gear-pair is operating at a constant rotational speed and transmitting constant loading, the rotational harmonic contributions to the transmission error (other than the tooth-meshing harmonics) are caused by the geometric deviations* $\epsilon_{Cj}(y, z)$ *of the individual working-surfaces* $j = 0, 1, 2, \ldots, N-1$ *from the mean deviation surface* $\bar{\eta}_C(y, z)$, *as defined by Equation (4.16).*

It follows from the DFT, Equation (4.5) applied to Equation (4.19), that the static transmission error rotational-harmonic contributions, other than the tooth-meshing harmonics, are caused by the deviations $\epsilon_{Cj}(y, z)$ of the individual working-surfaces from the mean working-surface of the gear. Notice that the lower limit in the summation in Equation (4.19) is $n = 1$, not zero, and from Equation (4.13) that the DFT, Equation (4.5) is periodic with period N. These deviations $\epsilon_{Cj}(y, z)$ of the individual working-surfaces from the mean working-surface are the source of "sidebands" on the tooth-meshing harmonics, "ghost tones," and low-order rational harmonics. It further follows from Equations (4.16), (4.19), and (4.22a) that all non-integer-multiple of N rotational harmonics n of the DFT spectra $B_{k\ell}(n)$, Equation (4.21a), $n \neq pN$, $p = 0, \pm 1, \pm 2, \ldots$ are caused by the tooth-to-tooth variations $\epsilon_{Cj}(y, z)$ of the individual working-surfaces from the mean working-surface. This rotational-harmonic behavior will be discussed in more detail in Section 6.2.

Effects of Torque and Speed Modulations

Because gear-pair-loading-caused elastic deformations provide significant contributions primarily to the tooth-meshing harmonics, any periodic gear-pair loading or torque variations will cause additional sideband contributions to the tooth-meshing harmonics in gear-vibration spectra. The resultant sideband frequencies will be displaced from the tooth-meshing harmonic locations by the frequencies of the torque variations. Because the earlier described rotational harmonics caused by tooth-to-tooth variations in the working-surfaces are substantially unaffected by loading variations, torque modulations will have only a secondary effect on these rotational harmonics.

Rotational speed variations will cause sidebands to all tooth-meshing and rotational harmonics, displaced from the original harmonic locations by multiples of the frequencies of the speed variations. Such speed variations will cause reductions in the original amplitudes of both tooth-meshing and rotational harmonics. If a speed variation frequency is a low-order multiple of the gear rotation frequency, tooth-meshing harmonic amplitudes can be reduced and rotational harmonic amplitudes increased.

6.2 Approximate Formulation for Generic Manufacturing Deviations

The transmission-error contribution $\zeta^{(\cdot)}(x)$ of gear (.), expressed as a function of roll-distance x, is given in terms of its complex Fourier series coefficients by

Equation (5.15). The Fourier series coefficients of working-surface-deviation contributions to $\zeta^{(\cdot)}(x)$ are provided by Equation (5.16) as functions of the mesh-attenuation functions $\hat{\phi}_{k\ell}(n/N)$ and the DFT, $B_{k\ell}(n)$, Equation (4.21), of the Legendre expansion coefficients, Equation (3.20), of the working-surface deviations of tooth j. The mesh-attenuation functions $\hat{\phi}_{k\ell}(n/N)$ are the remaining quantity to be determined.

Mesh-Attenuation Functions

For the case described by Equation (5.6) where the local tooth-pair stiffness per unit length of line of contact is assumed to be constant, it was shown that the transmission error contribution from working-surface deviations is independent of tooth stiffness, and is given by Equation (5.13), thereby negating the need to deal with tooth-stiffness properties. For this practically important case, it is shown by Equation (7.129), that a good approximation to the mesh-attenuation function $\hat{\phi}_{k\ell}(n/N)$ is

$$\hat{\phi}_{k\ell}(n/N) \approx (-i)^{k+\ell} \left[(2k+1)(2\ell+1)\right]^{1/2} \{ j_k (\pi Q_a n/N) j_\ell (\pi Q_t n/N)$$

$$- \sum_{\substack{p'=-\infty \\ \text{except} \\ p'=0}}^{\infty} j_0 (\pi Q_a p') j_0 (\pi Q_t p') j_k \left[\pi Q_a \left(\frac{n}{N} - p'\right)\right] j_\ell \left[\pi Q_t \left(\frac{n}{N} - p'\right)\right] \}, \qquad (6.3)$$

where the functions $j_m(\xi)$ denote (Antosiewicz, 1964, p. 437) spherical Bessel functions of the first kind of order m.

Because the form of each two-dimensional Legendre polynomial term $\psi_{yk}(y)\psi_{z\ell}(z)$ in the expansion of the working-surface deviation $\eta_{Cj}(y,z)$, Equation (3.18), differs for each k,ℓ pair, the mesh-attenuation function $\hat{\phi}_{k\ell}(n/N)$, Equation (6.3), is a function of the Legendre indices k and ℓ. As stated earlier, apart from its dependence on k, ℓ, and harmonic number ratio n/N, where N is the number of teeth, $\hat{\phi}_{k\ell}(n/N)$ is dependent only on the axial Q_a and transverse Q_t contact ratios which are determined by the assumed rectangular contact region.

Formulas for Contact Ratios

The axial contact ratio Q_a is defined by Equation (2.10), where F is the axial face width of the assumed contact region and Δ_a is the axial pitch, both illustrated in Figure 2.7b. It follows directly from the figure that Δ_a and the base pitch Δ are related by

$$\Delta_a = \Delta/\tan\psi_b \qquad (6.4)$$

where ψ_b is the base helix angle. Hence, from Equations (2.10) and (6.4), there follows

$$Q_a = F \tan\psi_b/\Delta \qquad (6.5a)$$

$$= F \cos\phi \tan\psi/\Delta, \qquad (6.5b)$$

by using Equation (2.8b). Equation (6.5b) expresses the axial contact ratio as a function of axial face width F of the assumed contact region, transverse pressure angle ϕ, pitch-cylinder helix angle ψ, and base pitch Δ.

The transverse contact ratio Q_t is defined by Equation (2.9), where L is the length of the path of contact in the transverse plane illustrated in Figures 2.6 and 2.7, and Δ is the base pitch. Hence, L is related to the roll angles ϵ_t and ϵ_r located at the tip and root of the defined contact region by

$$L = R_b \left(\epsilon_t - \epsilon_r \right),\tag{6.6}$$

where R_b is base-circle radius and the roll angles are measured in radians. See Equation (3.7) and note that D and L are related by

$$D = L \sin \phi,\tag{6.7}$$

which follows directly from Equation (3.3), because D is measured in units of z. Hence, the transverse contact ratio can be expressed from Equations (2.9) and (6.6) as

$$Q_t = R_b \left(\epsilon_t - \epsilon_r \right) / \Delta\tag{6.8a}$$

$$= N \left(\epsilon_t - \epsilon_r \right) / (2\pi)\tag{6.8b}$$

by using $N\Delta = 2\pi R_b$, the base-circle circumference, where N is the number of teeth. Equivalently (Lynwander, 1983, p. 42),

$$Q_t = N \times (\text{roll distance in degrees})/360°\tag{6.9}$$

where "roll distance in degrees" determines the length of the defined contact region in the transverse plane.

Total Contact Ratio

The roll-distance span over which a single helical tooth remains in contact with a mating tooth can be obtained with the aid of Figure 2.6. Consider the solid rectangular zone of contact shown in the upper portion of the figure. As roll distance x increases, a fictitious dashed line of contact first enters the zone of contact at the bottom right corner of the zone, moves across a distance L, then as x increases, this line of contact traverses up the left side of the zone, then exits. The roll distance x traversed across the bottom of the zone, is, simply, L, and the span in roll distance x traversed up the left side of the zone is $F \tan \psi_b$, as can be seen from Figure 2.6. But according to Equation (2.9), $L = Q_t \Delta$, and according to Equation (6.5a), $F \tan \psi_b = Q_a \Delta$. Therefore, *the total roll distance traveled while a single tooth is in contact is* $(Q_a + Q_t)\Delta$. The quantity $Q_a + Q_t$ is referred to as the "total contact ratio." For spur gears, the helix angle is zero, and therefore, $Q_a = 0$ for spur gears, which follows from Equation (6.5a). The above-determined single-tooth roll-distance span of $(Q_a + Q_t)\Delta$ will be seen to be important in understanding the sideband contributions of (accumulated) tooth-spacing errors.

Spherical Bessel Functions

The mesh-attenuation function approximation, Equation (6.3), is expressed in terms of spherical Bessel functions $j_m(\xi)$. These functions arise from the fact that they are, essentially, complex Fourier integral transforms of normalized Legendre polynomials (Bateman, 1954, p. 122; Antosiewicz, 1964, pp. 437, 438). They are defined (Antosiewicz, 1964, pp. 437, 438) in terms of Bessel functions of the first kind of order $m + 1/2$ by

$$j_m(\xi) \triangleq [\pi/(2\xi)]^{1/2} J_{m+\frac{1}{2}}(\xi). \tag{6.10}$$

The first four spherical Bessel functions are (Antosiewicz, 1964, pp. 437, 438)

$$j_0(\xi) = \frac{\sin \xi}{\xi} \tag{6.11a}$$

$$j_1(\xi) = \frac{\sin \xi}{\xi^2} - \frac{\cos \xi}{\xi} \tag{6.11b}$$

$$j_2(\xi) = \left(\frac{3}{\xi^3} - \frac{1}{\xi}\right) \sin \xi - \frac{3}{\xi^2} \cos \xi \tag{6.11c}$$

$$j_3(\xi) = \left(\frac{15}{\xi^4} - \frac{6}{\xi^2}\right) \sin \xi - \left(\frac{15}{\xi^3} - \frac{1}{\xi}\right) \cos \xi. \tag{6.11d}$$

A plot of these low-order spherical Bessel functions is shown in Figure 6.2. The asymptotic form of $j_m(\xi)$ is

$$j_m(\xi) = \xi^{-1} \cos [\xi - (m+1)\pi/2] + O\left(1/\xi^2\right), \quad \xi \to \infty \tag{6.12}$$

which is consistent with each member of Equation (6.11). This asymptotic form can be obtained from Equation (6.10) and Equation (37.4) of Sneddon (1961, p. 139) or Relton (1946, p. 183). It can be seen from Figure 6.2 and Equations (6.11) and (6.12) that the spherical Bessel functions are oscillatory functions with amplitudes that asymptotically decay with increasing ξ.

Formulation for Sideband Amplitudes

Apart from the low-order rotational harmonics, $n = \pm 1, \pm 2, \ldots$ (Figure 6.1), to be discussed in Sections 6.4 and 6.5, and the tooth-meshing harmonics $n = \pm N$, $\pm 2N, \ldots$, where N is the number of teeth, the remaining dominant harmonics in experimentally determined gear-meshing spectra are the so-called "sideband" harmonics in the immediate neighborhoods of the tooth-meshing harmonics, and in some spectra, so-called "ghost tones," which will be discussed in Section 6.6. Tooth-meshing harmonics $p = 1, 2, \ldots$ are rotational harmonics $n = pN$. If q denotes the rotational harmonic sideband order about tooth-meshing harmonic p, the rotational harmonic number of sideband q is $n = pN \pm q$. Hence, it follows from Equation (5.16)

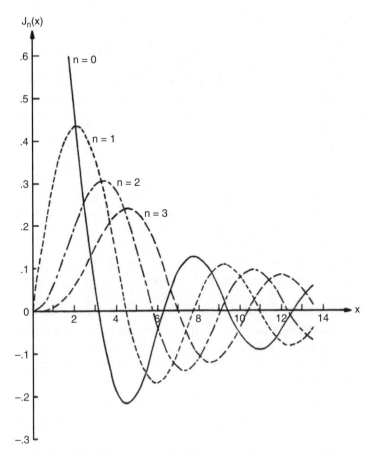

Figure 6.2   The first four spherical Bessel functions $j_n(x)$. Note that $j_0(0) = 1$ by Equation (6.11a) (From Antosiewicz (1964, p. 438), Reproduced with permission from the National Bureau of Standards)

that the complex Fourier series coefficients of sidebands q can be expressed as

$$\alpha_{pN\pm q}^{(\cdot)} = \sum_{k=0}^{\infty}\sum_{\ell=0}^{\infty} B_{k\ell}\,(pN\pm q)\,\hat{\phi}_{k\ell}\,(p\pm q/N) \qquad (6.13)$$

where, from Equation (6.3), with $n = pN \pm q$,

$$\hat{\phi}_{k\ell}\,(p\pm q/N) \approx (-i)^{k+\ell}\,[(2k+1)\,(2\ell+1)]^{1/2}\,\{j_k\,[\pi Q_a\,(p\pm q/N)]\,j_\ell\,[\pi Q_t\,(p\pm q/N)]$$

$$-\sum_{\substack{p'=-\infty \\ except \\ p'=0}}^{\infty} j_0\,(\pi Q_a p')\,j_0\,(\pi Q_t p')\,j_k\,[\pi Q_a(p-p'\pm q/N)]\,j_\ell\,[\pi Q_t\,(p-p'\pm q/N)]\}. \qquad (6.14)$$

Fundamental Source of Sideband Amplitudes

As described earlier, the quantity $B_{k\ell}(pN \pm q)$ in Equation (6.13) characterizes, in the frequency domain, the (complex) amplitudes of the working-surface-deviation contribution of Legendre term $k\ell$ before gear-meshing attenuation takes place, and $\hat{\phi}_{k\ell}(p \pm q/N)$ characterizes the meshing-attenuation effects on these deviations. Consequently, the *fundamental source* of the rotational-harmonic sidebands is $B_{k\ell}(pN \pm q)$. But as shown by Equation (4.21), with $n = \pm q = \pm 1$ *or* ± 2 *or* ± 3, say, the periodic property of $B_{k\ell}(n)$ yields

$$B_{k\ell}(pN \pm q) = B_{k\ell}(\pm q). \qquad (6.15)$$

Hence, as a consequence of the periodic nature of gear teeth, the low-order rotational harmonics $n = \pm q$ are again manifested in transmission-error rotational harmonic spectra as "sideband harmonics" located around the tooth-meshing harmonics $p = 1, 2, \ldots$ but with attenuation determined by the mesh-attenuation functions $\hat{\phi}_{k\ell}(p \pm q/N)$. As can be seen from Equations (4.22a,b), (4.19), (4.16), and (4.15), these non-tooth-meshing rotational harmonics are caused by deviations of the individual tooth-working-surfaces from the mean working-surface, that is, by tooth-to-tooth variations in the working-surfaces. As indicated by Equation (6.15), whenever these working-surface deviations vary slowly with tooth number, and therefore show up as low-order rotational harmonic contributions $n = q = 1, 2, \ldots$ in the DFT spectra of Equation (6.15), they again are found in the neighborhood of each tooth-meshing harmonic p.

Effects of Mesh-Stiffness Fluctuations

The first line in Equation (6.14) expresses the direct attenuating effect on Legendre term $k\ell$ of the tooth-working-surface deviations caused by averaging of these deviations over the lines of contact of all teeth in contact, as indicated by the integrations in Equations (5.11) and (5.12) and the summation over all teeth j in contact indicated in Equation (5.2). There remains in Equation (5.2) the effect on the transmission error $\zeta^{(.)}(x)$ caused by division of the summed integrated deviations by the total mesh stiffness $K_M(x)$. The effect of this division is expressed in the mesh-attenuation function by the second line in Equation (6.14). By utilizing the assumption, Equation (5.6), that the local tooth-pair stiffness per unit length of line of contact is a constant K, it was shown in Equation (5.13) that the working-surface-deviation contribution to the transmission error is independent of this local stiffness K, and the division by $K_M(x)$ becomes division by the total length $\ell(x)$ of all lines of tooth contact at roll-distance value x, as described by Equations (5.8) and (5.10).

The product form of the right-hand side of Equation (5.13) shows the tooth-deviation contribution multiplied by $1/\ell(x)$; that is, the tooth-deviation contribution is "amplitude modulated" by $1/\ell(x)$. The fundamental period of $\ell(x)$, and its reciprocal, is the tooth-meshing fundamental period. The strengths of the harmonics of $\ell^{-1}(x)$ are (approximately) the term $j_0(\pi Q_a p')j_0(\pi Q_t p')$, $p' = \pm 1, \pm 2, \cdots$ and the harmonics created by this modulation are these strengths multiplied by the terms $j_k[\ldots]j_\ell[\ldots]$

of the second line in Equation (6.14), which are harmonic-number-shifted versions of the "direct" terms in the first line, as is always observed in such amplitude modulation cases.

However, to understand the generation of rotational harmonic ("sideband", "ghost tone", etc.) contributions in measured vibration spectra, two facts pertaining to the above discussion are particularly important. First, for every term p' in the second line in Equation (6.14), there exists a term $p = p''$ in the first line where $p'' = p - p'$ of the second line. Therefore, the second line does not create any new harmonic terms, its role is to *modify* the terms in the first line, thereby modifying the harmonic *attenuating behavior*, but not generating new harmonic contributions. Second, rotational harmonic contributions can exist only at nonzero values of the discrete spectra $B_{k\ell}(pN \pm q)$ as is shown by the product relation $B_{k\ell}(pN \pm q)\, \hat{\phi}_{k\ell}(p \pm q/N)$ in Equation (6.13).

The particular case of Equation (6.14) where either Q_a or Q_t (or both) is an integer is of special interest. If Q is an integer, from Equation (6.11a) there follows for any integer $p' = \pm 1, \pm 2, \ldots$

$$j_0\left(\pi Q p'\right) = \frac{\sin\left(\pi Q p'\right)}{\pi Q p'} = 0, \quad Q = integer, \quad p' = \pm 1, \pm 2, \cdots \tag{6.16}$$

Therefore, it follows from Equation (6.16) that if either Q_a or Q_t is integer, the second line in Equation (6.14) provides no contribution predicted by variation in the total length of lines of contact $\ell(x)$ in Equation (5.13). This observation is consistent with Figure 2.7, which shows that if either Q_a or Q_t is an integer, no such variation in the total length $\ell(x)$ will take place. Hence, if either Q_a or Q_t is an integer, the (approximate) amplitude modulating effect arising from the division by $\ell(x)$ in Equation (5.13) is predicted to be zero.

6.3 Reduction of Results for Spur Gears

As is readily seen from Figure 2.6, helical gear results reduce to those for spur gears by setting the base helix angle ψ_b equal to zero. From Equation (6.5a), we therefore have for spur gears

$$Q_a = 0 \quad \text{for spur gears.} \tag{6.17}$$

Furthermore, for small ξ, one has (Antosiewicz, 1964, p. 437)

$$j_m(\xi) = \frac{\xi^m}{1 \times 3 \times 5 \ldots (2m + 1)}\left\{1 - \frac{\xi^2/2}{(2m + 3)} + \cdots\right\}, \tag{6.18}$$

and therefore,

$$j_0(0) = 1; \quad j_m(0) = 0, \quad m = 1, 2, 3, \ldots \tag{6.19a,b}$$

which is consistent with the behavior observed in Figure 6.2. Applying Equations (6.17) and (6.14) for spur gears, and then applying Equations (6.19a,b) for $m = k$, there follows *for spur gears*

$$\hat{\phi}_{0\ell}(p \pm q/N)$$

$$\approx (-i)^{\ell} (2\ell + 1)^{1/2} \left\{ j_{\ell} \left[\pi Q_t (p \pm q/N)\right] - \sum_{\substack{p'=-\infty \\ except \\ p'=0}}^{\infty} j_0(\pi Q_t p') j_{\ell} \left[\pi Q_t (p - p' \pm q/N)\right] \right\}$$

(6.20a)

$$\hat{\phi}_{k\ell}(p \pm q/N) = 0, \quad k = 1, 2, 3, \ldots .$$

(6.20b)

It follows, therefore, from Equation (6.20b) that the only nonzero contributions from the summation over k in Equation (6.13) are those from the term $k = 0$, that is, *for spur gears,*

$$\alpha_{pN \pm q}^{(.)} = \sum_{\ell=0}^{\infty} B_{0\ell}(pN \pm q)\hat{\phi}_{0\ell}(p \pm q/N),$$

(6.21)

which describes the transmission-error manufacturing-deviation complex Fourier series coefficients for spur gears.

Discussion of Spur Gear Results

From Equation (4.21a) there follows with $k = 0$,

$$B_{0\ell}(n) = \frac{1}{N} \sum_{j=0}^{N-1} c_{j,0\ell} \, exp(-i2\pi nj/N).$$

(6.22)

Moreover, from Equations (3.13) and (3.8), we have

$$\psi_{y0}(y) = 1,$$

(6.23)

and therefore, from Equation (3.20),

$$c_{j,0\ell} = \frac{1}{D} \int_{-D/2}^{D/2} \left[\frac{1}{F} \int_{-F/2}^{F/2} \eta_{Cj}(y, z)dy \right] \psi_{z\ell}(z)dz,$$

(6.24)

which provides the Legendre expansion coefficients $k = 0$, $\ell = 0, 1, 2, \ldots$ required in Equation (6.22).

As can be seen from Figure 2.6, when the base helix angle ψ_b is 0, the line of contact between meshing spur gear teeth lies in a direction parallel to the gear axes.

Consequently, the tooth deviation contribution to the transmission error at each radial location z of this line of contact is the average deviation of the working-surface taken along the line of contact. This average deviation will vary with radial location z on the working-surface. At each radial location z, this average deviation is expressed by the quantity within the brackets in Equation (6.24). Hence, only the one $k=0$ term is required in the Legendre expansion coefficient $c_{j,k\ell}$ for each tooth j, as shown by Equation (6.24). As a consequence of this simplification for spur gears, the double summation in Equation (6.13) for helical gears is reduced to the single summation in Equation (6.21) for spur gears.

Because the only axial expansion coefficient required for spur gears is the term $k=0$ given by Equation (6.24), it was suggested in Section 3.6 that the primary set of line-scanning tooth measurements for spur gears should ordinarily be chosen to be profile (radial line-scanning) measurements.

6.4 Rotational-Harmonic Contributions from Accumulated Tooth-Spacing Errors

Accumulated tooth-spacing errors, also called index errors, are errors in the *absolute* location of a tooth-working-surface on the base circle, relative to the working-surface location of some arbitrarily chosen tooth. Within the context of the Legendre-polynominal method of describing working-surface deviations, a tooth-spacing error is described by the two-dimensional Legendre term $k=0$, $\ell=0$, which describes the location on the base circle of the spatial average position of the surface, as described by Equation (3.24). Dependence on the arbitrary choice of a particular reference tooth can be removed, as shown by Equations (3.24–3.26). It is shown below that the transmission-error rotational-harmonic contributions from tooth-spacing errors differs from the contributions of all other error types.

Transmission-Error Contribution of a Slowly-Varying Accumulated Spacing Error

Consider a tooth-spacing error that varies only very slowly with tooth number on a gear with a large number of teeth. Such errors are often observed in gear measurements when the gear rotation axis used in measurement is accidently offset slightly from the base-cylinder axis used in tooth generation, leading to a one-cycle sinusoidal accumulated tooth-spacing-error contribution to the measurements. Assume the axis of rotation of this gear in operation coincides with the measurement axis, so that in meshing with a mating gear the once-per-revolution accumulated spacing error is experienced in the meshing action. Because such one-cycle accumulated spacing errors vary by only a minute amount over the roll-distance span $(Q_a + Q_t)\Delta$ where a single tooth on this gear continuously mates with each tooth during meshing with a mating gear, the tooth-deviation contribution $\eta_j(x)$ of this one-cycle spacing error experiences only negligible variation in x over this roll-distance span $(Q_a + Q_t)\Delta$.

Consider, now, the transmission-error contribution from this slowly varying accumulated spacing error. From Equations (5.12) and (5.8), there follows for this slowly-varying error $\tilde{\eta}_j(x)$

$$\tilde{\eta}_j(x) \approx \eta_j(x) \int_{y_{Aj}(x)}^{y_{Bj}(x)} d\ell = \eta_j(x)\ell_j(x), \tag{6.25}$$

because the error is essentially constant over the contact span in x where it meshes with a mating tooth. Moreover, because the spacing error $\eta_j(x)$ also is essentially constant over the roll-distance span in x of $(Q_a + Q_t)\Delta$ where it meshes with all teeth *in simultaneous contact* on the mating gear within the zone of contact (Figure 2.6), there follows from Equation (6.25),

$$\sum_j \tilde{\eta}_j(x) \approx \eta_j(x) \sum_j \ell_j(x) \tag{6.26a}$$

$$= \eta_j(x)\ell(x) \tag{6.26b}$$

from Equation (5.10). Hence, from Equations (6.26b) and (5.13), it follows that whenever the accumulated tooth-spacing error on a gear experiences negligible variation over roll-distance spans in x of $(Q_a + Q_t)\Delta$, it experiences *negligible attenuation* from the gear meshing action, and provides a direct unattenuated contribution,

$$\zeta^{(\cdot)}(x) \approx \eta_j(x) \tag{6.27}$$

to the transmission error (a perhaps obvious physical fact).

Approximate Mesh-Attenuation Function for Tooth-Spacing Errors

Let us now relate the above conclusion to the mesh-attenuation function ($k=0$, $\ell=0$) for accumulated tooth-spacing errors provided by Equation (6.14). First consider the $k=0$, $\ell=0$ version of Equation (6.14) for tooth-meshing harmonic contributions $p=0, \pm1, \pm2, \ldots$ only, that is, $q=0$:

$$\hat{\phi}_{00}(p) \approx j_0(\pi Q_a p) j_0(\pi Q_t p)$$

$$- \sum_{\substack{p'=-\infty \\ \text{except} \\ p'=0}}^{\infty} j_0(\pi Q_a p') j_0(\pi Q_t p') j_0\left[\pi Q_a \left(p - p'\right)\right] j_0\left[\pi Q_t \left(p - p'\right)\right] \tag{6.28}$$

where we recall from Equation (6.11a) that

$$j_0(\pi Q p) = \frac{\sin(\pi Q p)}{\pi Q p} \tag{6.29}$$

which exhibits its maximum value at $p=0$, where it is unity. Consequently, the dominant term p' in the summation in Equation (6.28) is the term $p'=p$. Retaining only this dominant term $p'=p$ in Equation (6.28) yields

$$\hat{\phi}_{00}(p) \approx j_0(\pi Q_a p) j_0(\pi Q_t p) - j_0(\pi Q_a p) j_0(\pi Q_t p) j_0(0) j_0(0) \qquad (6.30a)$$

$$= 0, \quad p = \pm 1, \pm 2, \ldots \qquad (6.30b)$$

because

$$j_0(0) = 1 \qquad (6.31)$$

as mentioned above. On the other hand, because the term $p'=0$ is excluded from the summation in Equation (6.28), for $p=0$ only the initial term in the right-hand side is retained yielding

$$\hat{\phi}_{00}(0) \approx j_0(0) j_0(0) = 1 \qquad (6.32)$$

according to Equation (6.31). The tooth-meshing harmonic $p=0$ of accumulated tooth-spacing errors represents a rigid-body rotation of a gear, which cannot be attenuated by meshing with a mating gear; hence, $\hat{\phi}_{00}(0) = 1$.

The purpose of this exercise was to show that by retaining only the dominant term $p'=p$ in the summation in Equation (6.28), we obtained exactly the same result described earlier in connection with Equations (6.25–6.27). The above considerations strongly suggest that reasonably accurate estimates of the *rotational harmonic* contributions from *tooth-spacing errors* should be obtained by using the mesh-attenuation function, Equation (6.14), for $k=0$, $\ell=0$, and retaining only the single dominant term $p'=p$ in the summation over p', that is, for $p=\pm1,\pm2,\ldots$

$$\hat{\phi}_{00}(p \pm q/N) \approx j_0 \left[\pi Q_a (p \pm q/N) \right] j_0 \left[\pi Q_t (p \pm q/N) \right]$$

$$- j_0(\pi Q_a p) j_0(\pi Q_t p) j_0(\pm \pi Q_a q/N) j_0(\pm \pi Q_t q/N),$$

$$p = \pm 1, \pm 2, \ldots \qquad (6.33a)$$

and for $p=0$,

$$\hat{\phi}_{00}(\pm q/N) \approx j_0(\pm \pi Q_a q/N) j_0(\pm \pi Q_t q/N),$$

$$p = 0 \qquad (6.33b)$$

since the term $p'=0$ is excluded from the summation in Equation (6.14). For sideband order $q=0$, Equation (6.33a) reduces exactly to Equation (6.30a), and Equation (6.33b) reduces exactly to Equation (6.32). Thus, the above-described approximation leading to Equations (6.33a,b) is based on (assumed) continuity of the mesh-attenuation function $\hat{\phi}_{00}(p \pm q/N)$ for small to modest sideband orders q about the tooth-meshing harmonics $p=0,\pm1,\pm2,\ldots$

Low-Order Rotational Harmonic and "Sideband" Contributions from Tooth-Spacing Errors

First consider the low-order rotational harmonics $n = \pm 1, \pm 2, \ldots$ arising from accumulated tooth-spacing errors. These low-order rotational harmonics are the "sidebands" around "tooth-meshing harmonic" $p = 0$, which represents a rigid-body rotational of the gear. Therefore, for this special case $q = n$ we should re-write the mesh-attenuation function, Equation (6.33b), using $q = n$:

$$\hat{\phi}_{00}(n/N) \approx j_0(\pi n Q_a/N) j_0(\pi n Q_t/N), \quad n = 0, \pm 1, \pm 2, \ldots \qquad (6.34)$$

where we have used the fact that $j_0(\xi)$ is an even function of ξ, that is,

$$j_0(-\xi) = j_0(\xi), \qquad (6.35)$$

as is readily seen from Equation (6.11a).

Unfortunately, the value of $j_0(\xi)$ at $\xi = 0$ is not shown in Figure 6.2; it is $j_0(0) = 1$. The behavior of $j_0(\xi)$ near $\xi = 0$ is an upside-down parabola, which can be seen from the first few terms of Maclaurin expansion of $j_0(\xi)$:

$$j_0(\xi) = \frac{\sin \xi}{\xi} = \frac{1}{\xi}\left(\xi - \frac{\xi^3}{3!} + \frac{\xi^5}{5!} - \cdots\right) \qquad (6.36a)$$

$$\approx 1 - \frac{\xi^2}{6} + \frac{\xi^4}{120}, \quad \xi < 1.5 \qquad (6.36b)$$

beyond which the behavior of $j_0(\xi)$ can be seen from Figure 6.2. It follows from the behavior of $j_0(\xi)$ for small $|\xi|$, Equations (6.36), that for typical values of axial Q_a and transverse Q_t contact ratios, and typical numbers of teeth N, the values of $j_0(\pi n Q_a/N)$ and $j_0(\pi n Q_t/N)$ do not differ significantly from unity for the first few rotational harmonics $n = \pm 1, \pm 2, \ldots$, and therefore, the mesh-attenuation function $\hat{\phi}_{00}(n/N)$ given by Equation (6.34) also does not differ significantly from unity for these low-order rotational harmonics. Hence, in this low-order rotational harmonic region $n = \pm 1, \pm 2, \ldots$, the DFT contributions from accumulated tooth-spacing errors, $B_{00}(n)$, Equation (4.21a), remain essentially unattenuated and provide direct contributions to the Fourier series coefficients $\alpha_n^{(.)}$ of the transmission error, as can be seen from Equation (5.16). This observation is particularly true in the case of spur gears, for which $Q_a = 0$, and therefore $j_0(\pi n Q_a/N) = 1$.

Consider, now, the contrasting case of "sideband harmonics" $q = 1, 2, \ldots$ located in the immediate neighborhoods of the tooth-meshing harmonics $p = \pm 1, \pm 2, \ldots$ The approximate mesh-attenuation function for accumulated tooth-spacing errors for these sideband locations $q = 1, 2, \ldots$ is given by Equation (6.33a). For $q = 1, 2, \ldots$ the last two terms of $j_0(\pm \pi Q_a q/N) j_0(\pm \pi Q_t q/N)$ in Equation (6.33a) are identical to the right-hand side in Equation (6.34) for $q = n$, and we have just shown that for typical values of Q_a, Q_t, and N, these terms do not differ significantly from unity.

Next, consider the term $j_0[\pi Q(p \pm q/N)]$, $Q = Q_a$ or Q_t, in Equation (6.33a), which, according to Equation (6.11a) is

$$j_0 \left[\pi Q(p \pm q/N) \right] = \frac{\sin\left[\pi Q(p \pm q/N) \right]}{\pi Q(p \pm q/N)} \tag{6.37a}$$

$$= \frac{\sin(\pi Qp)\cos(\pi qQ/N) \pm \cos(\pi Qp)\sin(\pi qQ/N)}{\pi Q(p \pm q/N)}, \tag{6.37b}$$

by using an elementary trigonometric identity. The smallest tooth-meshing harmonic applicable in Equation (6.37) is $p = \pm 1$. Therefore, at least for the first few sideband harmonics $q = 1, 2, \ldots$, the quantity q/N is small in comparison with p, and for these few sideband harmonics the denominator in Equation (6.37) does not differ appreciably from πQp. This approximation improves with increasing values of the number of teeth N. Next consider the second term in the numerator of Equation (6.37b). The absolute value of $\cos (\pi Qp)$ is bounded by unity. Furthermore, for relatively small contact ratio Q and large number of teeth N, for the first few sideband harmonics $q = 1, 2, \ldots$ one has $\sin(\pi qQ/N) \ll 1$. Turning to the first term in the numerator of Equation (6.37b), for small q, small contact ratio, and large number of teeth N, $\cos(\pi qQ/N) \approx 1$. Hence, for these approximate values, $j_0[\pi Q(p \pm q/N)]$ does not differ appreciably from $[\sin(\pi Qp)/(\pi Qp)] = j_0(\pi Qp)$. Comparing Equation (6.33a) with Equation (6.30a), it follows from the above considerations that for $(q/N) \ll 1$, $\pi qQ_a/N \ll 1$, and $\pi qQ_t/N \ll 1$, the term preceded by a negative sign in the right-hand side of Equation (6.33a) will tend to cancel the first term in that side, just as was observed in Equation (6.30a). Consequently, for $\pi qQ_a/N \ll 1$, and $\pi qQ_t/N \ll 1$, the mesh-attenuation function $\hat{\phi}_{00}(p \pm q/N)$ given by Equation (6.33a) will strongly attenuate the accumulated tooth-spacing-error sideband harmonics about tooth-meshing harmonics $p = \pm 1, \pm 2, \ldots$ that arise from the DFT $B_{00}(\pm q) = B_{00}(pN \pm q)$ of accumulated tooth-spacing errors, Equation (4.21) with $n = q$.

Role of Contact Ratios in Attenuating Tooth-Spacing-Error Low-Order Rotational and ''Sideband'' Harmonics

In the case of Equation (6.34), we have seen from the behavior of $j_0(\xi)$, Equation (6.36), that the very low-order rotational harmonics $n = \pm 1, \pm 2, \ldots$ arising from accumulated tooth-spacing errors are *negligibly attenuated* by the meshing action with a mating gear whenever $(\pi nQ_a/N) \ll 1$, and $(\pi nQ_t/N) \ll 1$. Furthermore, we have just seen above that sideband orders $q = \pm 1, \pm 2, \ldots$ arising from accumulated tooth-spacing errors are *strongly attenuated* whenever $\pi qQ_a/N \ll 1$, and $\pi qQ_t/N \ll 1$.

Consider the *nth* rotational harmonic of an accumulated tooth-spacing error. The wavelength (period) of this *nth* harmonic, measured on the base circle, is $N\Delta/n$. One-third of this wavelength is about $N\Delta/(\pi n)$. Any single tooth on the subject gear is in continuous contact with teeth on a mating gear for a roll-distance interval of $(Q_a + Q_t)\Delta$, where $(Q_a + Q_t)$ is the "total contact ratio". Therefore, for the *nth* rotational harmonic of an accumulated tooth-spacing error to be unchanging (approximately constant) over the roll-distance span for which a single tooth is in continuous contact with teeth on a mating gear, we require the above-described approximately one-third wavelength to be large in comparison with the single-tooth contact span of $(Q_a + Q_t)\Delta$, that is,

$$\frac{N\Delta}{\pi n} \gg (Q_a + Q_t)\Delta \tag{6.38}$$

or

$$\frac{\pi n(Q_a + Q_t)}{N} \ll 1 \tag{6.39}$$

which, with $q = n = 1, 2, \ldots$ is consistent with the criteria obtained above for strong attenuation of tooth-spacing error "sidebands." Note from Equation (6.37) that $q = 1$, $2, \ldots$ denote "*sideband*" *order and not* (absolute) rotational harmonic number. More-over, the criterion of Equation (6.39) is identical with the physical argument of Equations (6.25–6.27) showing that "large wavelength" accumulated tooth-spacing errors cannot be attenuated by the meshing action with a mating gear. *To summarize: The transmission-error contributions of low-order rotational harmonics n = 1, 2, … of accumulated tooth-spacing errors k = 0, ℓ = 0 that satisfy Equation (6.39) experience negligible attenuation from the meshing action with a mating gear, and provide negligible sideband contributions in the neighborhoods of the tooth-meshing harmonics. (For spur gears, Q_a = 0.)*

The rotational harmonic spectrum results displayed in Figures 4.1 and 5.1, computed by the "exact" algorithms delineated in Chapters 7 and 8, are consistent with the above-described rotational harmonic behavior of accumulated tooth-spacing errors. In the discussion of Figure 4.1, it was pointed out that the dominant source of rotational harmonic $n = 1$, shown in Figure 4.1, is accumulated tooth-spacing errors. In Figure 4.1, we observe that the (unattenuated) rms amplitude of $n = 1$ is again found in "sidebands" $q = \pm 1$ about the tooth-meshing harmonic locations $p = 1$, which is rotational harmonic $n = N = 38$, and $p = 2$ which is $n = 2N = 76$, with the same amplitude as $n = 1$. However, in the computed mesh-attenuated rms spectrum shown in Figure 5.1b, rotational harmonic $n = 1$ exhibits essentially the same unattenuated strength shown in Figure 4.1, but the "sideband" harmonics $q = \pm 1$ about tooth-meshing harmonic locations $n = N = 38$ and $n = 2N = 76$ are very strongly attenuated in comparison with the strength of $n = 1$. In fact, these "sideband" strengths at $q = \pm 1$ about $n = 38$ and $n = 76$ almost surely arise from the linear "wobble" of the tooth-surfaces displayed in Figure 4.2. The meshing attenuation of such non-tooth-spacing tooth-to-tooth errors is considered next.

6.5 Rotational-Harmonic Contributions from Tooth-to-Tooth Variations Other Than Tooth-Spacing Errors

As we have described in Section 3.4, if the axis of rotation used in measuring gear teeth does not coincide with the base-cylinder axis used in tooth generation, but the two axes are parallel, then a once-per-revolution small linear profile error will be found in the working-surface measurements. If the two axes are not parallel, then a once-per-revolution linear lead error also will be found in the measurements. If the axis of rotation of the gear in operation coincides with the measurement axis, these linear errors, which vary from one tooth to another, provide contributions to the transmission error.

The DFT, Equation (4.21), resulting from such errors will exhibit a strong once-per-revolution rotational harmonic $n = 1$. But, because the DFT, Equation (4.21), is periodic, that is, $B_{k\ell}(n) = B_{k\ell}(pN + n)$, $p = 0, \pm1, \pm2, \ldots$ working-surface errors that vary slowly with tooth number, thereby generating low-order rotational harmonics $n = \pm1, \pm2, \ldots$ in $B_{k\ell}(n)$, also provide equal contributions to the $B_{k\ell}(n)$ about harmonic locations $n = pN$, which are the tooth-meshing-harmonic locations. These rotational-harmonic locations in the neighborhoods of the tooth-meshing-harmonic locations are the so-called "sideband" locations, denoted by $n = q$ in Equation (6.13). The mesh-attenuation functions $\hat{\phi}_{k\ell}(p \pm q/N)$ in Equation (6.13) govern the attenuation of the periodic DFTs, $B_{k\ell}(pN + q)$. Equation (6.14) provides an approximation to these mesh-attenuation functions.

Low-Order Rotational-Harmonic Contributions from Errors Other Than Tooth-Spacing Errors

First, consider attenuation of the low-order rotational harmonics $n = \pm1, \pm2, \ldots$ For this case, $p = 0$ and "sideband" q is rotational-harmonic number $q = n$. For these low-order rotational harmonics, Equation (6.14) becomes

$$\hat{\phi}_{k\ell}(\pm n/N) \approx (-i)^{k+\ell} [(2k + 1)(2\ell + 1)]^{1/2} \{ j_k (\pm\pi Q_a n/N) j_\ell (\pm\pi Q_t n/N)$$

$$- \sum_{\substack{p'=-\infty \\ except \\ p'=0}}^{\infty} j_0 (\pi Q_a p') j_0 (\pi Q_t p') j_k [\pi Q_a(-p' \pm n/N)] j_\ell [\pi Q_t(-p' \pm n/N)] \}. \qquad (6.40)$$

Consider the terms in the summation over p' in Equation (6.40). Because the term $p' = 0$ is excluded from the summation, each of the terms in the summation is in the range where the asymptotic approximation, Equation (6.12), is applicable. Especially for helical gears, $Q_a > 0$, the product of the magnitude of the denominators, ξ^{-1} in Equation (6.12), in the summation is at least $\pi^4 Q_a^2 Q_t^2$ leading to an exceedingly small contribution of each term in the summation relative to unity. These terms arise from the "modulating effect" on the mesh-attenuation functions caused by the variation in total contact line length $\ell(x)$ in Equation (5.13). For large contact ratios, Q_a and Q_t, the fractional fluctuation of the total length $\ell(x)$ of tooth contact lines decreases, and this behavior is manifested by the denominator behavior $\pi^4 Q_a^2 Q_t^2$. Moreover, if either Q_a or Q_t is an integer (or both integers) there is no fluctuation in total contact line length $\ell(x)$, as can be seen from the $j_0(\pi Q p')$ terms in the second line in Equations (6.40) and (6.11a) and Figure 2.7, leaving the first line in Equation (6.40) as the only contribution. However, for low-order rotational harmonics $n = 1, 2, \ldots$, large numbers of teeth N, and small to moderate contact ratios Q_a or Q_t, the arguments $\pi Q n/N$ in the first line in Equation (6.40) are very small, yielding very small contributions of

$j_k(\pm \pi Q_a\, n/N) j_\ell(\pm \pi Q_t\, n/N)$ for the first few rotational harmonics, as can be seen from Figure 6.2. For spur gears, the only nonzero k term is $k=0$, as shown by Equation (6.20). Consequently, for the first few rotational harmonics $n = 1, 2, \ldots$, for k and ℓ not both zero, Equation (6.40) normally predicts that

$$\left| \hat{\phi}_{k\ell}\,(\pm n/N)\right| \ll 1, \quad n = 1, 2, \ldots, \quad k \text{ and } \ell \text{ not both zero.} \tag{6.41}$$

The discrete error spectra $B_{k\ell}(n)$ caused by manufacturing-error contributions that vary slowly with tooth number, such as linear lead or linear profile errors, will have strong low-order rotational harmonic contributions $n = 1, 2, \ldots$. However, as indicated by Equation (6.41), for manufacturing errors other than accumulated tooth-spacing errors, $k=0, \ell=0$, such strong low-order contributions from the error spectra $B_{k\ell}(n)$ will normally be strongly attenuated by the meshing action with a mating gear. In contrast to this behavior, for $k=0$ and $\ell=0$ describing accumulated tooth-spacing errors, we observed from Equations (6.34) to (6.36) that

$$\hat{\phi}_{00}(\pm n/N) \approx 1, \quad n = 1, 2, \ldots \tag{6.42}$$

indicating negligible attenuation arising from the meshing action with a mating gear. *Therefore, the low-order rotational-harmonic contributions $n = 1, 2, \ldots$ observed in transmission-error spectra and structural responses that are caused by manufacturing deviations on gear teeth are almost always caused by accumulated tooth-spacing-error contributions with small to negligible contributions from other forms of tooth-to-tooth variations in working-surface deviations.*

Working-Surface-Deviation Contributions to ''Sideband'' Rotational Harmonics

Let us next consider the complex Fourier series coefficients of the working-surface deviation contribution to the tooth-meshing harmonics $n = pN, p = \pm 1, \pm 2, \ldots$ which are given by Equation (6.13) with sideband order $q=0$:

$$\alpha_{pN}^{(.)} = \sum_{k=0}^{\infty} \sum_{\ell=0}^{\infty} B_{k\ell}(pN)\hat{\phi}_{k\ell}(p). \tag{6.43}$$

From Equation (4.21a), we observe with $n=pN$ that, for the exponential term in that equation, we have

$$\exp(-i2\pi nj/N) = \exp(-i2\pi pj) = 1 \tag{6.44}$$

because both p and j are integers. Hence, from Equations (4.21a) and (6.44), there follows

$$B_{k\ell}(pN) = \frac{1}{N} \sum_{j=0}^{N-1} c_{j,k\ell} = \bar{c}_{k\ell} \tag{6.45}$$

which is the average value of Legendre coefficient $k\ell$, averaged over all of the teeth. This tooth-average Legendre coefficient $\bar{c}_{k\ell}$ is the expansion coefficient of the average working-surface deviation, as can be seen by Equations (3.45–3.47), where we have dropped the unnecessary prime displayed there. Consequently, DFT spectrum contributions $B_{k\ell}(n)$, $n \neq pN$, represent rotational-harmonic contributions of the individual working-surfaces j that differ from the average working-surface, as can be seen from Equation (4.22a) for $n \neq pN$.

But because of the periodic behavior of the DFT spectra $B_{k\ell}(n)$, Equation (4.21b), the first few low-order rotational harmonics $n = \pm 1, \pm 2, \ldots$ again occur with equal amplitude as $B_{k\ell}(pN \pm 1)$, $B_{k\ell}(pN \pm 2) \ldots$, which are sideband harmonics around tooth-meshing harmonic p. The equispaced teeth on a gear effectively "sample" the slowly varying geometric deviations of the working-surfaces from the mean working-surface in exactly the same way that equispaced sampling of slowly varying signals create "aliasing" harmonics. Because the teeth are truly discrete, the low-order harmonics $n = \pm 1, \pm 2, \ldots$ again show-up as sidebands around the tooth-meshing harmonics $n = pN$.

Let us now relate the rotational-harmonic contributions of tooth-to-tooth variations of working-surface deviations to the tooth-meshing-harmonic contributions $B_{k\ell}(pN)$ of the average working-surface, Equation (6.45). The Legendre expansion coefficients of the working-surface *variations* are $\delta c_{j,k\ell} \triangleq c_{j,k\ell} - \bar{c}_{k\ell}$, where $\bar{c}_{k\ell}$ is given by Equation (6.45). The DFT, Equation (4.21a), of the Legendre expansion coefficient of the varying working-surface contribution is

$$B'_{k\ell}(n) \triangleq \frac{1}{N} \sum_{j=0}^{N-1} \left(c_{j,k\ell} - \bar{c}_{k\ell} \right) \exp\left(-i2\pi nj/N \right) \tag{6.46a}$$

$$= B_{k\ell}(n) - \bar{c}_{k\ell} \frac{1}{N} \sum_{j=0}^{N-1} \exp\left(-i2\pi nj/N \right) \tag{6.46b}$$

$$= B_{k\ell}(n), \quad n \neq pN, \quad p = 0, \pm 1, \pm 2, \ldots \tag{6.46c}$$

because the summation in the right-hand side of Equation (6.46b) is identically zero, as shown in Appendix 6.A at the end of this chapter. In other words, as mentioned above, the DFT spectra $B_{k\ell}(n)$, $n \neq pN$, $p = 0, \pm 1, \pm 2, \ldots$ describe the contributions of working-surface deviations that differ from the average working-surface.

Now consider a single *rotational harmonic* contribution of Legendre component $k\ell$, which can be expressed using the inverse DFT, Equation (4.4), as

$$\delta_{j,k\ell} = A\bar{c}_{k\ell} \exp(i2\pi nj/N). \tag{6.47}$$

With $A = 1$, the rotational-harmonic amplitude of $\delta_{j,k\ell}$ is equal to the amplitude $\bar{c}_{k\ell}$ of the Legendre coefficient of the average working-surface deviation, Equation (3.47).

Forming the DFT, Equation (4.21a) of $\delta_{j,k\ell}$ above, we obtain

$$\delta B_{k\ell}(n) = A\bar{c}_{k\ell}\frac{1}{N}\sum_{j=0}^{N-1}\exp\left(i2\pi nj/N\right)\exp\left(-i2\pi nj/N\right)$$

$$= A\bar{c}_{k\ell}\frac{1}{N}\sum_{j=0}^{N-1}1$$

$$= A\bar{c}_{k\ell},\tag{6.48}$$

and furthermore, from Equation (4.21b),

$$\delta B_{k\ell}(n) = \delta B_{k\ell}(n+pN),\tag{6.49}$$

because Equation (6.48) utilized the DFT, Equation (4.21a).

In particular, if the working-surface variation, Equation (6.47), is utilized to describe a low-order rotational harmonic, $n=\pm1,\pm2,\ldots$ then from Equation (6.49), that harmonic amplitude will again show up as sidebands around the tooth-meshing harmonics $p=1,2,\ldots$. If the amplitude $A\bar{c}_{k\ell}$ of that low-order rotational harmonic contribution is comparable to the amplitude $\bar{c}_{k\ell}$ of the average working-surface contribution, then it follows from Equation (6.45) that the DFT sideband contribution will be comparable to the DFT tooth-meshing harmonic contribution.

The transmission-error complex Fourier series contributions from the individual Legendre components $k\ell$ are obtained by multiplying the above-described DFT spectrum contributions by the mesh-attenuation functions $\hat{\phi}_{k\ell}\left(p\pm q/N\right)$, Equation (6.14), as indicated by Equation (6.13). In the neighborhoods of the tooth-meshing harmonics $p=1,2,\ldots$, the arguments of the functions $j_k[\ldots]$ and $j_\ell[\ldots]$ in Equation (6.14) fall in the asymptotic range where Equation (6.12) is valid, except for the negligible term $p'=p$ in the second line, k and ℓ not both zero (Figure 6.2). Hence, from Equations (6.12) and (6.14), it is readily seen that the value of the mesh-attenuation function of Equation (6.14) at sideband locations $q=1,2,\ldots$ does not differ significantly from its value at the associated tooth-meshing harmonic p, especially for large numbers of teeth N and low-order sidebands $q=1,2,\ldots$. *Therefore, except for accumulated tooth-spacing errors, $k=0$, $\ell=0$, the above analysis shows that if amplitudes $A\bar{c}_{k\ell}$ of the low-order rotational harmonic contributions $n=1,2,\ldots$ of Legendre components $k\ell$ of tooth-to-tooth variations in working-surface deviations are comparable to the amplitudes $\bar{c}_{k\ell}$ of the average working-surface-deviation components, then the kinematic transmission-error sideband contributions from the tooth-to-tooth variation Legendre components $k\ell$ will be comparable to the tooth-meshing-harmonic contributions from these same Legendre components.*

To summarize the results of Sections 6.4 and 6.5, the above analysis has shown that low-order rotational-harmonic variations in accumulated tooth-spacing errors will provide strong non-attenuated contributions to the low-order rotational harmonics of transmission-error spectra, but negligible contributions to the sidebands of tooth-meshing harmonics.

Conversely, low-order rotational-harmonic variations of working-surface deviations other than accumulated tooth-spacing errors will provide significant sideband contributions in the neighborhoods of the tooth-meshing harmonics, but negligible contributions to the low-order rotational harmonics of transmission-error spectra.

Log-log plots of the approximation to the mesh-attenuation functions, Equation (6.3), are provided in Figure 6.3. All plots were computed using transverse contact ratio $Q_t = 1.819$, axial contact ratio $Q_a = 3.19$. The absolute values $|\hat{\phi}_{k\ell}(n/N)|$ are shown.

Figure 6.3a shows $|\hat{\phi}_{00}(n/N)|$, which describes the mesh-attenuation function for accumulated tooth-spacing errors, $k = 0$, $\ell = 0$. In the low-order rotational-harmonic region, the value of $|\hat{\phi}_{00}(n/N)|$ is unity, indicating negligible attenuation, as described above. All of the remaining plots, Figure 6.3b–f, for $k\ell$ pairs of 0,1; 1,0; 0,2; 1,1; and 2,0 show substantial attenuation in the low-order rotational-harmonic region, as explained above. In Figure 6.3a, showing the results for accumulated tooth-spacing errors, $k = 0$, $\ell = 0$, nulls at integer multiples of the number of teeth, $(n/N) = 1$, 2, . . . , are shown, illustrating the above-described attenuation of accumulated tooth-spacing sideband contributions. None of the other five plots show such attenuation, as explained above.

Justification of the Term ''Mesh-Attenuation Function''

Equation (5.21) was given to justify use of the term "mesh-attenuation function", under the assumption that the local tooth-pair stiffness $K_{Tj}(x, y)$ and total mesh stiffness $K_M(x)$ both are constant. Fluctuations in roll-distance x of total mesh stiffness $K_M(x)$ are represented in the approximate mesh-attenuation function, Equation (6.3), by the summation over p'. Ignoring this term, there follows from Equation (6.3)

$$\sum_{k=0}^{\infty}\sum_{\ell=0}^{\infty}|\hat{\phi}_{k\ell}(n/N)|^2 \approx \sum_{k=0}^{\infty}(2k+1)j_k^2(\pi Q_a n/N)$$

$$\times \sum_{\ell=0}^{\infty}(2\ell+1)j_\ell^2(\pi Q_t n/N). \tag{6.50a}$$

Each of the summations in the right-hand side of Equation (6.50a) is unity (Antosiewicz, 1964, p. 440, Equation (10.1.50)), thereby yielding

$$\sum_{k=0}^{\infty}\sum_{\ell=0}^{\infty}|\hat{\phi}_{k\ell}(n/N)|^2 \approx 1, \tag{6.50b}$$

the result described by Equation (5.21).

6.6 Rotational-Harmonic Contributions from Undulation Errors

In the general analysis method presented up to this juncture, tooth-working-surface deviations were expanded in two-dimensional normalized Legendre polynomials,

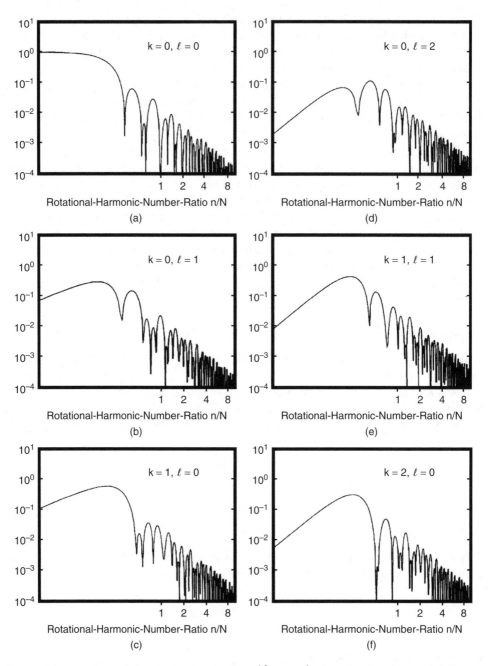

Figure 6.3 Log-log plots of absolute values $\left|\hat{\phi}_{k\ell}(n/N)\right|$ of approximate mesh-attenuation functions, Equation (6.3), for k,ℓ pairs of 0,0; 0,1; 1,0; 0,2; 1,1; and 2,0. All plots computed for transverse contact ratio $Q_t = 1.819$ and axial contact ratio $Q_a = 3.19$. $\hat{\phi}_{k\ell}(n/N)$ is dimensionless. (Adapted from Mark (1979, Figures 22–27))

as described by Equation (3.18), where each term $\psi_{yk}(y)\psi_{z\ell}(z)$ was regarded as an "elementary error." This method is capable of describing *any* working-surface deviations. However, it is known that "generating type" manufacturing operations (Baxter, 1962, pp. 1–19) can produce periodic errors on tooth-working-surfaces commonly called "undulation errors" (Merritt, 1971, p. 82; Cluff, 1992, pp. 7, 38, 91–97). In the present section, we postulate a generic form of sinusoidal undulation error and determine the requirements for this error to be nonattenuated by the meshing action with a mating gear. Such undulation errors are responsible for "ghost tones" in gear transmission-error spectra.

We shall continue to use the notation established earlier, but with added subscript u to denote specific application to the single generic undulation waveform denoted by $k\ell$. From Equation (4.30b), the rotational-harmonic nonattenuated spectrum contribution from this single undulation waveform $k\ell$ is

$$G_\eta(n)_u = 2\left|B_{k\ell}(n)_u\right|^2, \tag{6.51}$$

and from Equation (5.18b) its mesh-attenuated rotational-harmonic contribution is

$$G_\zeta(n)_u = 2\left|B_{k\ell}(n)_u\hat{\phi}_{k\ell}(n/N)_u\right|^2 \tag{6.52a}$$

$$= 2\left|B_{k\ell}(n)_u\right|^2\left|\hat{\phi}_{k\ell}(n/N)_u\right|^2. \tag{6.52b}$$

Thus, if this single undulation error experiences no attenuation during meshing action, its mesh-attenuation function must satisfy

$$\left|\hat{\phi}_{k\ell}(n/N)_u\right| = 1, \tag{6.53}$$

thereby yielding $G_\zeta(n)_u = G_\eta(n)_u$ for rotational-harmonic n generated by the undulation error.

Rotational harmonic $n = 20$ in Figures 4.1 and 5.1 exhibits essentially this behavior, having an rms value of 100 µin. ($2\frac{1}{2}$ µm) in Figure 4.1 and 80 µin. (2 µm) in Figure 5.1b. The manufacturing deviation causing rotational harmonic $n = 20$ is shown in Figure 4.5 on two adjacent teeth. It is approximately sinusoidal with a period of 1.9 tooth spacings, as described in Chapter 4. It is a sinusoidal undulation error.

A detailed analysis of such undulation errors has been carried out (Mark, 1992a) in the manner described by Equations (6.51–6.53). But it is very important to distinguish the above-described approach outlined by Equations (6.51–6.53) from the general treatment contained in the remainder of this book. In the general treatment, no assumptions are made pertaining to the working-surface deviations on the teeth; they are measured, and the working-surface non-mesh-attenuated and mesh-attenuated rotational-harmonic spectra are computed and diagnosed from these measurements. In the presently discussed treatment of undulation errors, we postulate a general form of working-surface sinusoidal undulation error, and seek to determine under what conditions its rotational-harmonic components will remain unattenuated by the meshing action with a mating gear, with mesh-attenuation function thereby satisfying Equation (6.53).

Representation of Undulation Errors

The treatment below follows Mark (1992a). Any generic deviation on the rectangular working-surface of tooth j, Figure 3.4, can be represented by the two-dimensional Fourier integral

$$\eta_{Cj}(y,z) = \int\limits_{-\infty}^{\infty} \int\limits_{-\infty}^{\infty} \hat{\eta}_{Cj}(\mu,v) \exp\left[i2\pi(\mu y + vz)\right] d\mu dv \qquad (6.54)$$

where

$$\hat{\eta}_{Cj}(\mu,v) = \int\limits_{-D/2}^{D/2} \int\limits_{-F/2}^{F/2} \eta_{Cj}(y,z) \exp\left[-i2\pi(\mu y + vz)\right] dy dz. \qquad (6.55)$$

The representation of $\eta_{Cj}(y,z)$ by Equation (6.54) is a representation of the surface deviation by a continuum of complex sinusoids of infinitesimal amplitude. Undulation errors are periodic errors with finite amplitudes. Any number of undulation errors on tooth j can be represented as in Equation (6.54) by replacing the integrals by summations; that is,

$$\eta_{Cj}(y,z) = \sum_{k}\sum_{\ell} c_j(k,\ell) \exp\left[i2\pi(\mu_k y + v_\ell z)\right], \qquad (6.56)$$

where discrete wavenumbers (spatial frequencies) μ_k and v_ℓ in Equation (6.56) have replaced their continuous counterparts in Equation (6.54).

Working-surface deviations are real, not complex, quantities. To obtain the corresponding representation by real sinusoids, each wavenumber term k,ℓ in Equation (6.56) must be accompanied by a negative term $-k,-\ell$ with negative wavenumbers $\mu_{-k} = -\mu_k$ and $v_{-\ell} = -v_\ell$, and with complex amplitude $c_j(-k,-\ell)$ equal to the complex conjugate of its complement $c_j(k,\ell)$; that is,

$$c_j(-k,-\ell) = c_j^*(k,\ell). \qquad (6.57)$$

The real undulation harmonic $u_j(k,\ell)$ with y and z direction wavenumbers μ_k and v_ℓ, respectively, is represented by the sum of the k,ℓ term and its complement $-k,-\ell$; that is,

$$\begin{aligned} u_j(k,\ell) &= c_j(k,\ell) \exp\left[i2\pi\left(\mu_k y + v_\ell z\right)\right] \\ &\quad + c_j^*(k,\ell) \exp\left[-i2\pi\left(\mu_k y + v_\ell z\right)\right] \\ &= 2R_e\left\{c_j(k,\ell) \exp\left[i2\pi\left(\mu_k y + v_\ell z\right)\right]\right\}, \end{aligned} \qquad (6.58)$$

where $R_e\{\ldots\}$ denotes the real part of the complex quantity within the braces. Using straightforward complex algebra, the real undulation harmonic $u_j(k,\ell)$ can be

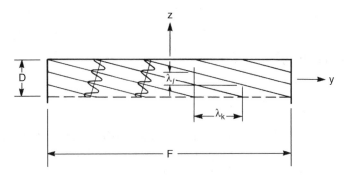

Figure 6.4 "Pure" undulation error on tooth-working-surface illustrating axial-wavelength λ_k and radial-wavelength λ_ℓ

expressed (Mark, 1992a) in terms of the absolute value of $c_j(k,\ell)$ and its phase ϕ_j as

$$u_j(k,\ell) = 2\left|c_j(k,\ell)\right| \cos\left[2\pi(\mu_k y + v_\ell z) + \phi_j\right],\tag{6.59}$$

where the phase ϕ_j is related to the real and imaginary parts of $c_j(k,\ell)$ by

$$c_j(k,\ell) = \left|c_j(k,\ell)\right|(\cos\phi_j + i\sin\phi_j)$$

$$= \left|c_j(k,\ell)\right|\exp(i\phi_j).\tag{6.60}$$

The real undulation harmonic $u_j(k,\ell)$, Equation (6.59), represents a generic two-dimensional sinusoidal wave pattern on the working-surface of tooth j, as illustrated in Figure 6.4. The axial wavelength is $\lambda_k = 1/\mu_k$ and the radial wavelength is $\lambda_\ell = 1/v_\ell$. The location of any spatial shift in this wave pattern is controlled by the phase ϕ_j in Equation (6.59).

Mesh-Attenuation Function for Undulation Errors

The approximate mesh-attenuation function $\hat{\phi}_{k\ell}(n/N)_u$ for the sinusoidal undulation $u_j(k,\ell)$ is (Mark, 1992a, Equation (54))

$$\hat{\phi}_{k\ell}(n/N)_u = j_0\left\{\pi Q_a\left[(n/N) - q_k\right]\right\} j_0\left\{\pi Q_t\left[(n/N) - \rho_{k\ell}q_k\right]\right\}$$

$$- \sum_{\substack{p'=-\infty \\ except \\ p'=0}}^{\infty} j_0\left(\pi Q_a p'\right) j_0\left(\pi Q_t p'\right) \times$$

$$\times j_0\left\{\pi Q_a\left[(n/N) - q_k - p'\right]\right\} j_0\left\{\pi Q_t\left[(n/N) - \rho_{k\ell}q_k - p'\right]\right\}\tag{6.61}$$

where we have added a subscript u on $\hat{\phi}_{k\ell}(n/N)_u$ to indicate that it applies *only* to the sinusoidal undulation error, Equation (6.58). The new parameters in Equation (6.61) are

$$q_k \triangleq \Delta_a / \lambda_k \qquad (6.62)$$

where Δ_a is axial pitch, Figure 2.7, and

$$\rho_{k\ell} \triangleq \frac{D/A}{\lambda_\ell / \lambda_k} \qquad (6.63)$$

where D and A are both illustrated in Figure 3.2, and axial λ_k and radial λ_ℓ wavelengths are illustrated in Figure 6.4. (Symbol q_k is completely different from sideband-order q introduced earlier in this chapter.)

Requirement for Zero Attenuation of Undulation Errors

Just as in the mesh-attenuation function, Equation (6.3), the first terms on the right-hand side of Equation (6.61) represent the attenuation provided by the integrations, Equation (5.12), of the working-surface deviations over the lines of tooth contact, whereas the summation over p' in Equation (6.61) accounts for the modulating effect on the mesh-attenuation function arising from the fluctuation in the total length $\ell(x)$ of these lines of contact in Equation (5.13). *Consequently, for there to exist no attenuation of the working-surface deviations, at each roll-distance location x the manufacturing deviation must be exactly the same value at all points on each line of contact. But this value of working-surface deviation can vary for differing values of roll-distance x.* (The integrations in Equation (5.12) and summation in Equation (5.13), followed by division by $\ell(x)$ in Equation (5.13), form the average of the working-surface deviations over all lines of contact at each particular roll-distance location x.)

The terms $j_0 \{\ldots\}$ in Equation (6.61) are the zero-order spherical Bessel function, Equation (6.11a), $j_0(\xi) = \sin \xi / \xi$, which is unity only at $\xi = 0$. Therefore, from Equations (6.61) to (6.63), the requirement for zero attenuation provided by the undulation-error mesh-attenuation function is

$$\rho_{k\ell} \triangleq \frac{D/A}{\lambda_\ell / \lambda_k} = 1 \qquad (6.64a)$$

or

$$\frac{D}{A} = \frac{\lambda_\ell}{\lambda_k} = \frac{\mu_k}{\nu_\ell}, \qquad (6.64b)$$

since wavenumbers and wavelengths are reciprocals, and

$$q_k \triangleq \frac{\Delta_a}{\lambda_k} = \frac{n}{N} \qquad (6.65a)$$

or

$$n\lambda_k = N\Delta_a. \qquad (6.65b)$$

The two requirements, Equations (6.64b) and (6.65b), for zero attenuation by the undulation-error mesh-attenuation function are easy to interpret. Consider Equation (6.64b) first. From Figure 3.2, we observe that the ratio D/A is the slope of a line of contact in the y, z tooth coordinate system. From Figure 6.4, the ratio λ_ℓ/λ_k is the slope of an undulation-error line of constant phase in the same y, z tooth coordinate system. Therefore, the condition (6.64b) requires that the phase of an undulation error must be constant along lines of tooth contact on tooth-working-surfaces, which, if satisfied, means that the undulation error will be constant along *each* line of contact.

Now consider the second requirement (Equation (6.65b)). Because Δ_a is the axial pitch, Equation (2.13) and Figure 2.7, it follows that $N\Delta_a$ is the combined axial spacing of all of the N working-surfaces in exactly one gear rotation (Figure 2.6), which is the period of rotational harmonic $n = 1$. Consequently, because λ_k is the axial-direction wavelength of the undulation rotational harmonic n, the requirement (Equation (6.65b)) is that an integer multiple $n\lambda_k$ of undulation-error cycles exist on the working-surfaces in one gear rotation. The requirement (Equation (6.65b)) is entirely equivalent to its radial analog,

$$n(\lambda_\ell/\sin\phi) = N\Delta, \tag{6.66}$$

where λ_ℓ is undulation radial wavelength; hence, $\lambda_\ell/\sin\phi$ is the projection of the radial wavelength onto the plane of contact (Figure 3.1), and therefore, onto the base circle of circumference $N\Delta$. If the equivalent condition (6.65b) or (6.66) and that of (6.64b) are met, then no attenuation of the undulation error will occur in the meshing action with a mating gear.

The condition (6.64b) places a requirement on the ratio of wavenumbers μ_k/ν_ℓ in the undulation model of Equation (6.59), and the equivalent conditions (6.65b), or (6.66) place requirements on the actual wavenumber values $\mu_k = \lambda_k^{-1}$ or $\nu_\ell = \lambda_\ell^{-1}$. There remains in the undulation waveform model of Equation (6.59), the dependences of the amplitude $2\left|c_j(k, \ell)\right|$ and the phase ϕ_j on tooth number $j = 0, 1, \ldots, N-1$. The variations of these two quantities with tooth number j control the behavior of the spectrum contribution $\left|B_{k\ell}(n)_u\right|$ in Equation (6.52), where for the undulation waveform model of Equations (6.58–6.60), (Mark, 1992a, Equation (67)),

$$B_{k\ell}(n)_u = \frac{1}{N}\sum_{j=0}^{N-1} c_j(k, \ell)\exp(-i2\pi nj/N) \tag{6.67a}$$

$$= \frac{1}{N}\sum_{j=0}^{N-1}\left|c_j(k, \ell)\right|\exp\left\{-i\left[(2\pi nj/N) - \phi_j\right]\right\}, \tag{6.67b}$$

where Equation (6.60) was used in going to the second line, and subscript u has been added to indicate that Equation (6.67b) applies only to the undulation model of Equations (6.58–6.60). Hence, even if the requirements of Equations (6.64b) and (6.65b) for zero attenuation, Equation (6.53), are satisfied, it is possible that only one rotational harmonic, or alternatively, many rotational harmonics, will appear in the spectrum $\left|B_{k\ell}(n)_u\right|$ and in $G_\zeta(n)_u$ in Equation (6.52).

In Mark (1992a), a reasonably general phase model is introduced that is tailored to working-surface errors likely to be produced by gearing imperfections in gear manufacturing machines, as is illustrated by the following examples.

Classic Manufacturing Source of "Ghost Tones"

When gears are cut or finished, they often are mounted fixed on a rotary table. The rotational position of the table therefore controls the position of the workpiece gear relative to the fixed location of the cutter or finisher. Consequently, errors in table rotational position will result in gear-tooth working-surface errors. In some gear cutting and finishing operations, at each fixed rotational position of the worktable, the gear cutter or finisher traverses along a path identical to the path of a line of tooth contact, for example, Drago (1988, p. 106). In such operations, any error in table rotational position becomes an error on the tooth being generated, and this error is constant along the path that coincides with a line of contact, therefore satisfying Equation (6.64a,b).

In the classical ghost-tone case, the rotational position of the worktable is controlled by a worm-wormwheel drive. In the case of a single-threaded worm (Faires and Keown, 1960, p. 245), an imperfection in the worm-wormwheel drive will create a periodic variation in the rotational position of the worktable with the number of cycles of this periodic variation equal to the number of teeth on the wormwheel. This periodic variation in worktable rotational position, relative to the position of the fixed cutter, causes a comparable periodic error on the workpiece teeth, thereby satisfying Equations (6.65a,b), where n is the number of teeth on the wormwheel (or a multiple of the number of teeth). This is the "classical" ghost-tone manufacturing source, (Cluff 1992, pp. 7, 38, 91–97). Because Equations (6.64a) and (6.65a) both are satisfied by such kinematically generated errors, no attenuation resulting from the meshing action with a mating gear will take place. In such cases, a "ghost" tone will appear in the transmission-error spectrum with rotational harmonic number n equal to the number of teeth on the worktable-drive wormwheel, and possibly also, additional rotational harmonics may appear with rotational harmonic numbers equal to integer multiples of the number of teeth on the wormwheel.

Unintended Contribution to Tooth-Meshing Harmonics

A special case of the above-described behavior can occur if the number of teeth N_w on the wormwheel is an integer multiple p of the number of teeth N on the workpiece gear,

$$N_w = pN. \tag{6.68}$$

Because a nonattenuated rotational harmonic number $n = N_w$ transmission-error contribution is generated by worktable rotational-position errors, as described above, when the numbers of teeth N_w and N satisfy Equation (6.68), we have

$$\frac{N_w}{N} = \frac{n}{N} = p \tag{6.69}$$

which is tooth-meshing harmonic number p. In this case, the nonattenuated "ghost" tone contribution is superimposed on tooth-meshing harmonic number p.

Ordinarily, the combined effect of intentional tooth modifications and tooth elastic deformations are strongly attenuated by the meshing action with a mating gear. In contrast to this strong attenuation, a "ghost" tone contribution to a tooth-meshing harmonic p arising from the condition, Equation (6.68), will experience no attenuation, as described above. It therefore can be a dominant contribution to this tooth-meshing harmonic.

Non-Integer Number of Undulation Cycles

Suppose the number of teeth on the rotary-table wormwheel is an odd number, and the worm is double threaded (Faires and Keown, 1960, p. 245). Moreover, suppose the table rotational-position errors are caused by an imperfect worm (Cluff, 1992, p. 96). Then, after a single rotation of the worktable, the phase of the periodic error imposed on the teeth of the workpiece gear would not match the phase generated at the start of the rotation, but would be out of match by one-half wavelength because the number of teeth on the wormwheel is odd. In this case, instead of Equation (6.65b), one would have

$$\left(n \pm \frac{1}{2}\right)\lambda_k = N\Delta_a, \tag{6.70}$$

depending on how the final half-cycle is counted.

Because the workpiece gear is circular, its transmission-error contributions can only occur at rotational harmonic values that are integer multiples of the fundamental rotational harmonic $n=1$. Therefore, as one might guess from Equation (6.70), two strong "ghost" tone transmission-error rotational harmonics would be generated in this case at rotational harmonic values

$$n = \frac{N\Delta_a}{\lambda_k} \pm \frac{1}{2}. \tag{6.71}$$

Moreover, a large number of much weaker rotational harmonics also would be generated (Mark, 1992a, Case III).

Parametric Sensitivity of Undulation Mesh-Attenuation Function

As described above, the two critical parameters of the undulation mesh-attenuation function, Equation (6.61), are q_k and $\rho_{k\ell}$, Equations (6.62) and (6.63), respectively. For the term $j_0\{\pi Q_a[(n/N) - q_k]\}$ to be unity (no attenuation), Equation (6.65b) must be satisfied. This condition is *least* satisfied when an integer number plus one-half of undulation wavelengths occur in one workpiece gear rotation, as described by Equations (6.70) and (6.71). In this case, there follows from Equations (6.62) and (6.71),

$$\frac{n}{N} - q_k = \frac{n}{N} - \frac{\Delta_a}{\lambda_k} = \pm\frac{1}{2N}, \tag{6.72}$$

and for this case

$$j_0\{\pi Q_a[(n/N) - q_k]\} = j_0\left[\pm(Q_a/N)\pi/2\right]. \tag{6.73}$$

But $j_0(\xi) = \sin\xi/\xi$ only begins to decay significantly from unity at $\xi = \pi/2$ where $j_0(\pi/2) = 2/\pi = 0.637$. Because gears always must satisfy

$$Q_a \ll N, \tag{6.74}$$

no significant undulation attenuation is predicted by the behavior of q_k in Equation (6.73).

The principal attenuation of undulation errors therefore is controlled by the parameter $\rho_{k\ell}$, Equation (6.63), which describes the ratio of the slope D/A of the lines of tooth contact (Figure 3.2) to the slope λ_ℓ/λ_k of the lines of constant undulation-error phase (Figure 6.4). When this ratio is significantly different from unity, the lines of tooth contact will average over several to many undulation-error cycles, and significant attenuation will be achieved. However, if this ratio, Equation (6.63), is unity, as in gear-tooth generating operations, for example, Drago (1988, p. 106), then no significant attenuation of periodic undulation errors will take place when the subject gear is meshed with a mating gear. The rotational harmonic frequency spectrum of the resultant "ghost tones" (one tone or many tones) will be controlled (Mark, 1992a) by the phase ϕ_j and amplitude $\left|c_j(k, \ell)\right|$ behavior, $j = 0, 1, \ldots, N - 1$ of the undulation errors, Equation (6.59), on the individual working-surfaces j, as indicated by Equations (6.52) and (6.67b).

Diagnosing ''Ghost-Tone'' Manufacturing Sources

Any kinematic (or dynamic) source in gear manufacturing machines causing a periodic error in tooth-working-surfaces is a potential source of ghost tones. Once a ghost-tone rotational harmonic is identified, the working-surface deviations causing this harmonic can be computed from detailed working-surface measurements, as described in Equation (4.42) or (4.45). The working-surface error pattern on one or more teeth provided by such a computation should enable the user to diagnose the kinematic manufacturing-machine source causing this harmonic. The rotational harmonic number of the ghost tone often is the best clue.

As described in Section 5.4, rotational harmonic $n = 20$ in Figure 5.1 is a "ghost tone", having experienced very little attenuation from its rms value shown in the error spectrum in Figure 5.1a. Its periodic behavior on two adjacent teeth was shown in Figure 4.5. Because $n = 20$ is approximately one-half the rotational harmonic number $n = N = 38$ of the tooth-meshing fundamental of that particular gear, the resultant undulation error displayed in Figure 4.5 has a relatively long wavelength.

Second ''Ghost Tone'' Example

A second helical gear having a known "ghost tone" at rotational harmonic $n = 144$ was measured in detail by the method described in Chapter 3. This gear has $N = 59$ teeth. Therefore, in contrast to the above-mentioned ghost tone at rotational harmonic $n = 20$ which occurs only about midway to the tooth-meshing fundamental at $n = N = 38$, the ghost-tone harmonic $n = 144$ of the 59 tooth gear falls between the second $n = 2N = 118$ and third $n = 3N = 177$ tooth-meshing harmonics. Hence, the undulation error causing rotational harmonic $n = 144$ will be seen to have a correspondingly shorter wavelength than that seen in Figure 4.5.

Because the amplitudes of undulation errors causing ghost tones are known to often be very small, it was decided to measure the subject gear by the method described in Chapter 3 twice, once using profile measurements as the primary set, and the second time using lead measurements as the primary set. If these two independent sets of measurements yielded essentially the same predictions of transmission-error ghost-tone rms amplitude of rotational harmonic $n = 144$, then we could be confident of the results of the measurements and computations.

The axial Q_a and the transverse Q_t contact ratios of the subject gear analysis region were computed by Equations (6.5) and (6.8), respectively, to be

$$Q_a = 1.1645 \qquad Q_t = 1.6094. \qquad (6.75a,b)$$

Using $n = 144$, $N = 59$, and the above values of Q_a and Q_t, it was shown in Appendix 3.B that for interpolation across profile measurements in the axial direction, 14 profile measurements would be required, Equation (3.B.12), and for interpolation across lead measurements in the radial direction 17 or 18 lead measurements would be required, Equation (3.B.14). Thus, in the measurement set using the 14 profile measurements as the primary set, 5 lead measurements as the secondary set were used, and in the second independent set of measurements using lead measurements as the primary set, 18 lead measurements were used, and 5 profile measurements as the secondary set were used. All of these scanning measurements were located at the zeros of scaled Legendre polynomials of order equal to the number of scanning measurements as described in Chapter 3.

The rms tooth-deviation spectrum $[G_n(n)]^{1/2}$, Equation (4.30b), computed from the measurement set consisting of the 14 profile measurements and 5 lead measurements, is shown in Figure 6.5 using a logarithmic amplitude scale. As noted by Equation (4.31), it is periodic with period in n equal to $n = N = 59$, the location of the tooth-meshing fundamental $p = 1$. The rms amplitude of rotational harmonic $n = 144$ is seen from Figure 6.5 to be about 1.3×10^{-4} mm $= 0.13$ μm (5.1 μin.). The rms tooth-deviation spectrum $[G_n(n)]^{1/2}$, Equation (4.30b), computed from the measurement set consisting of 18 lead measurements and 5 profile measurements is shown in Figure 6.6, also using a logarithmic amplitude scale. The rms amplitude of rotational harmonic $n = 144$ is seen from Figure 6.6 to be about 1.6×10^{-4} mm $= 0.16$ μm (6.3 μin.). There is nothing to suggest from these two spectra that rotational harmonic $n = 144$ would produce a ghost tone.

The low-order rotational harmonics in Figure 6.6, especially $n = 1$, are much stronger than those in Figure 6.5. These strong low-order harmonics are believed to have been caused by an arbor problem when the measurements were taken. Because of the periodic character of the spectra, these strong low-order harmonics again show up as strong sidebands about the locations $n = 59$, 118, and 177 of the tooth-meshing harmonics. But any arbor-problem effect on rotational harmonic $n = 144$ was minimal, since its value in Figure 6.6 is only slightly larger than in Figure 6.5.

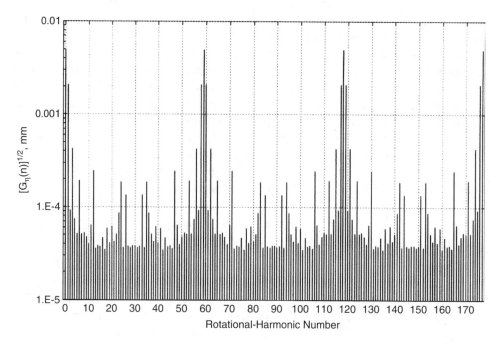

Figure 6.5 Rotational harmonic rms spectrum $(G_\eta(n))^{1/2}$, square-root of Equation (4.30b), of working-surface deviations of $N = 59$ tooth helical gear computed from 14 profile measurements and 5 lead measurements made on every tooth. Logarithmic amplitude scale in millimeters

The rms transmission-error spectra $[G_\zeta(n)]^{1/2}$, Equation (5.18b), $n = 1, 2, \ldots,$ computed from the working-surface measurements used in Figures 6.5 and 6.6, respectively, are shown in Figures 6.7 and 6.8, using logarithmic amplitude scales. In both of these latter two figures, the rms amplitude of rotational harmonic number $n = 144$ is almost exactly 10^{-4} mm $= 0.10\,\mu$m ($4\,\mu$in.). Notice also very good agreement in Figures 6.7 and 6.8, of the rms values of the tooth-deviation contribution to the tooth-meshing harmonics $p = 1$ at $n = N = 59$ and $p = 2$ at $n = N = 118$.

In contrast to the behavior of $n = 144$ in the tooth-deviation rms spectra of Figures 6.5 and 6.6, the rms amplitude of rotational harmonic $n = 144$ in the computed rms transmission-error spectra of Figures 6.7 and 6.8 is about a factor of 10 larger than its rotational harmonic neighbors. (This particular gear was known to have a ghost-tone problem at rotational harmonic number $n = 144$.)

The tooth-working-surface deviation contribution of rotational harmonic n is given by Equation (4.42), where $B_{k\ell}(n)_e$ and $B_{k\ell}(n)_o$ are given by Equations (4.36) and (4.37), respectively, $c_{j,k\ell}$ is given by Equation (3.20), and $\psi_{yk}(y)$ and $\psi_{z\ell}(z)$ are given by Equations (3.13) and (3.14), respectively. For the measurement set of Figures 6.5 and 6.7 that utilized the 14 profile measurements, the working-surface deviation

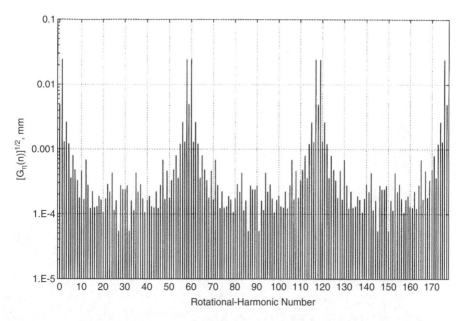

Figure 6.6 Rotational harmonic rms spectrum $(G_\eta(n))^{1/2}$, square-root of Equation (4.30b), of working-surface deviations of same $N = 59$ tooth helical gear computed from 18 lead measurements and 5 profile measurements made on every tooth. Logarithmic amplitude scale in millimeters

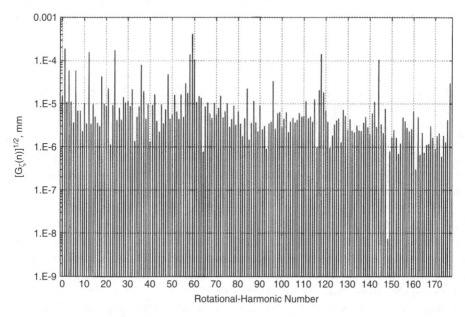

Figure 6.7 Rotational harmonic rms spectrum $(G_\zeta(n))^{1/2}$, square-root of Equation (5.18b), of mesh-attenuated working-surface deviations of $N = 59$ tooth helical gear computed from same working-surface measurements as Figure 6.5. Logarithmic amplitude scale in millimeters

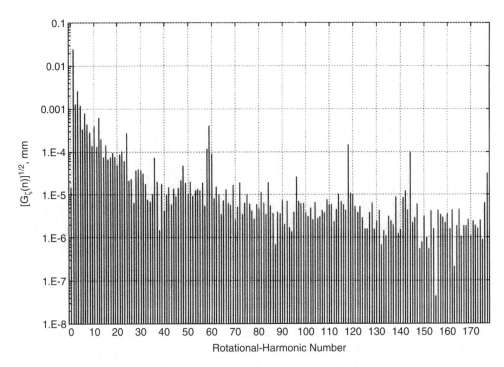

Figure 6.8 Rotational harmonic rms spectrum $(G_\zeta(n))^{1/2}$, square-root of Equation (5.18b), of mesh-attenuated working-surface deviations of $N=59$ tooth helical gear computed from same working-surface measurements as Figure 6.6. Logarithmic amplitude scale in millimeters

contribution of rotational harmonic $n=144$ is shown in Figure 6.9, and for the measurement set of Figures 6.6 and 6.8 that utilized the 18 lead measurements, the working-surface deviation contribution of $n=144$ is shown in Figure 6.10. Both Figures 6.9 and 6.10 show the result of Equation (4.42) for a typical tooth number $j=5$. Both of these figures display a sinusoidal-like undulation with six and a fraction cycles. Even though Figure 6.9 was generated using profile measurements as the primary set and Figure 6.10 was generated using lead measurements as the primary set, both figures display essentially the same behavior and the same sinusoidal phase.

In Section 4.4, use of the DFT in the generation of spectra from equispaced sampling of continuous waveforms was discussed. In that application of the DFT to sampled continuous waveforms, frequencies generated beyond $n=N/2$ are "aliased" frequencies. In the present application, $N=59$ teeth; therefore, rotational harmonic $n=144$ is 4.88 times $59/2$. In contrast to use of the DFT in the spectrum analysis of sampled continuous waveforms, because gear teeth are discrete, the DFT is the exact mathematical tool required in the present analysis, not an approximation as in its application to sampled continuous waveforms.

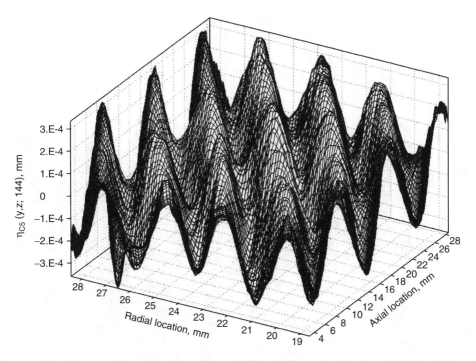

Figure 6.9 Computation by Equation (4.42) of the contribution from rotational harmonic $n = 144$ to the working-surface deviations of tooth $j = 5$ using same working-surface measurements as in Figures 6.5 and 6.7. All dimensions in millimeters

Understanding Rotational Harmonic Error-Pattern Generation, Equation (4.42)

Let us revisit what Equation (4.42) tells us, and does not tell us, about working-surface error patterns. The mesh-attenuation function for undulation errors, Equations (6.61–6.63), will be an aid in this understanding. Here is what Equation (4.42) does: Pick a single tooth j, and pick a single point y, z on that tooth-working-surface, Figure 3.4. Pick a rotational harmonic n. Then, from the working-surface deviation at location y, z on all N teeth of the gear, the computation of $\eta_{Cj}\, (y, z; n)$ by Equation (4.42) yields the unique contribution of that rotational harmonic n to working-surface j at location y, z. *But Equation (4.42) treats each location y,z independently of all other locations on the working-surfaces. Whatever error pattern we observe in our computations is a consequence of the behavior of the contribution of that rotational harmonic n from all of the N teeth on the gear.*

The contrast between Figure 4.3, generated for rotational harmonic $n = 19$, and Figure 4.5, generated for rotational harmonic $n = 20$, both by Equation (4.42), illustrates the above comment very nicely. The contributions of the error patterns on all of the teeth from these particular rotational harmonics $n = 19$ and $n = 20$, are responsible for the very different error patterns observed in Figures 4.3 and 4.5.

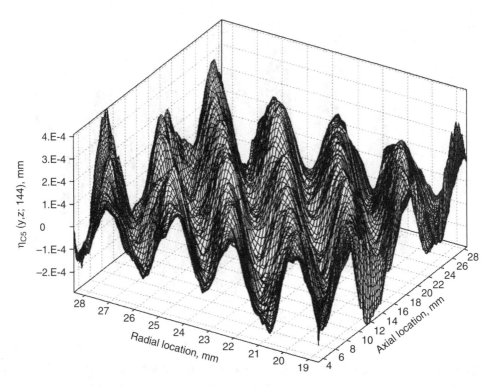

Figure 6.10 Computation by Equation (4.42) of the contribution from rotational harmonic $n = 144$ to the working-surface deviations of tooth $j = 5$ using same working-surface measurements as in Figures 6.6 and 6.8. All dimensions in millimeters

In the case of rotational harmonic $n = 20$, very little attenuation was observed between the nonattenuated and attenuated spectra of Figures 5.1a and 5.1b. A similar lack of attenuation of rotational harmonic $n = 144$ was observed between the nonattenuated and attenuated spectra of Figures 6.5 versus 6.7 and 6.6 versus 6.8. In all three of these cases, the undulation-error requirement for zero attenuation, Equation (6.65) or (6.66), is automatically satisfied by Equation (4.42), and because we observe nearly perfect sinusoidal behavior in Figures 4.5, 6.9, and 6.10, the additional undulation-error requirement for nonattenuation, Equation (6.64), is very nearly satisfied in all three of these undulation-error examples.

Because the zero-attenuation requirement, Equation (6.65) or (6.66), is automatically satisfied by Equation (4.42), an improvement for *interpretation* of the results of application of Equation (4.42), *for undulation errors and ghost tones*, can be obtained by modifying the result of the computation of Equation (4.42) so as to approximately satisfy the requirement, Equation (6.64), which is the requirement that the error be the same constant value along each possible line-of-contact location. Hence, the *approximate* contribution to the transmission error of the surface generated by Equation (4.42) can be obtained, at every line of contact location, by forming the *average* of the surface along a line that would coincide with the line of contact. At each

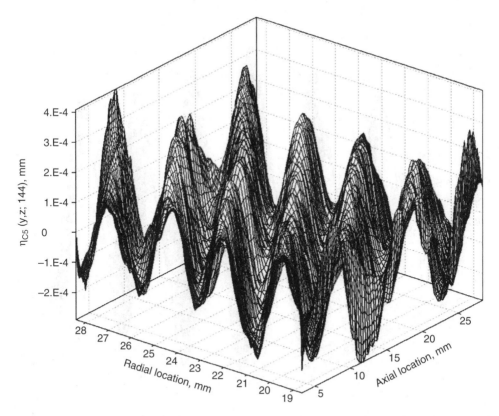

Figure 6.11   Same exact data as Figure 6.10, but different viewing angle. All dimensions in millimeters

line of contact location, this is a "line integral" integration along the line of contact divided by the length of that line of contact within the rectangular zone of contact used in making the tooth-working-surface measurements (see Equations (5.8–5.13)).

The result of such an averaging is illustrated by Figures 6.11 and 6.12. Figure 6.11 displays the result of the same implementation of Equation (4.42) shown in Figure 6.10, but at a slightly different viewing angle. Figure 6.12 shows the result of forming the above-described "line-integral average" of Figure 6.11 along all possible line of contact locations, plotted as a constant value along each line of contact location.

This computation was carried out for tooth number $j = 5$. The average single-sided amplitude of the near-perfect sinusoidal undulation shown in Figure 6.12 is about 1.5×10^{-4} mm, which is very close to the value of $\sqrt{2} \times 10^{-4}$ mm that would be predicted from the rms amplitude of 10^{-4} mm shown in Figure 6.8.

Contribution of "Random" Working-Surface Errors

The reader might be familiar with "random noise" which is uncorrelated from one instant to the next, and that produces "white noise" which describes a constant

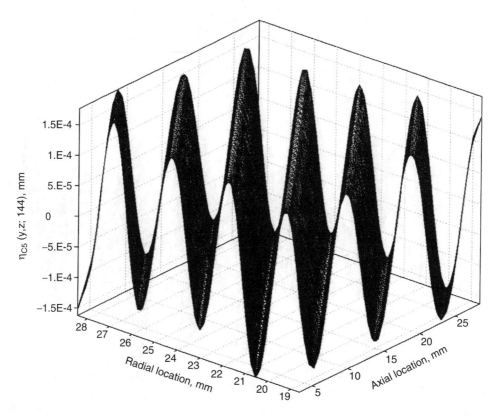

Figure 6.12 Computed from data shown in Figure 6.11 by forming the line-integral average value within the rectangular analysis region along all line-of-contact locations within the analysis region, plotted as a constant value at each line-of-contact location. All dimensions in millimeters

frequency spectrum. Random errors on the working-surfaces of gear teeth that are uncorrelated among the teeth behave in a similar manner, potentially producing a contribution to the amplitude of every rotational harmonic n in the spectrum $[G_\zeta(n)]^{1/2}$, Equation (4.30b). Consequently, when Equation (4.42) is used to generate the contribution of rotational harmonic n to the working-surface deviation of tooth j, it includes the contribution from any such random deviations, in addition to the contribution from systematic deviations such as we have seen in Figures 4.3, 4.5, and 6.9–6.11. The fact that the fractional attenuation was small in each of the undulation-error examples of Figures 4.5 and 6.9–6.11, in going from the rms tooth-deviation spectrum to the rms transmission-error spectrum, was an indication that the contribution from random-deviation errors was small in each of these examples.

The ghost-tone example shown next illustrates a case in which such random tooth deviations are not small, also illustrating why the interpretation provided by the "line-integral averaging" shown in Figure 6.12 can be useful.

Third Ghost-Tone Example

In this third example, we were provided a helical gear with $N = 51$ teeth. We were told this gear was the source of a ghost tone somewhere between the tooth-meshing fundamental harmonic at $n = N = 51$ and the second tooth-meshing harmonic at $n = 2N = 102$, but no other information pertaining to the location of the ghost tone was provided. In this case, as in that described above, we decided to make two predictions using independent sets of measurements. In one set, 17 scanning profile measurements and seven scanning lead measurements were used; in the second set, 17 scanning lead measurements and seven scanning profile measurements were used. (Better choices would have been to use more than 17 in the primary sets and fewer than seven in the secondary sets.)

Normally, choice of the number of scanning measurements to use in the primary set should be based on the known ghost-tone rotational harmonic number (or the largest rotational harmonic number of interest). In our predictions we learned that $n = 72$ was the dominant ghost-tone harmonic of interest. The transverse and axial contact ratios of the rectangular measurement region (assumed zone of contact) of this gear were $Q_t = 2.024$ and $Q_a = 2.534$. Using $N = 51$, $n = 72$, and $Q_t = 2.024$, the number of radial cycles of the undulation harmonic $n = 72$ is determined from Equation (3.B.6) to be 2.857; and using $Q_a = 2.534$, the number of axial cycles is determined from Equation (3.B.3) to be 3.577. Rounding 2.857 up to $m = 3$, Table 3.B.1 gives $k = 5$, and rounding 2.534 up to $m = 3$, also gives $k = 5$. Because interpolation across lead measurements interpolate in the radial direction, use the Equation (3.B.7) yields for the required number of primary lead measurements

$$n' = 2.857\pi + 5 = 13.98 \tag{6.76}$$

and because interpolation across profile measurements interpolate in the axial direction, use of Equation (3.B.7) yields for the required number of primary profile measurements,

$$n' = 3.577\pi + 5 = 16.24. \tag{6.77}$$

Because the number of 17 primary measurements was used in both sets of independent measurements, both measurement sets were adequate to accurately interpolate undulation errors giving rise to rotational harmonic $n = 72$.

The rms tooth-deviation spectrum $[G_\eta(n)]^{1/2}$, computed by Equation (4.30b) from the 17 lead measurements and seven profile measurements, is shown in Figure 6.13, and the rms transmission-error spectrum $[G_\zeta(n)]^{1/2}$ of Equation (5.18b), $n = 1, 2, \ldots$, computed from the same measurement set, is shown in Figure 6.14. Both spectra are plotted on a logarithmic amplitude scale. In Figure 6.14, the tooth-deviation contributions to the transmission-error tooth-meshing harmonics at $n = N = 51$, $n = 2N = 102$, and $n = 3N = 153$ clearly are significant. In addition, a strong rotational harmonic at $n = 72$ is clearly present in the transmission-error spectrum in Figure 6.14, together with a number of strong harmonics immediately adjacent to $n = 72$. The rms amplitude of $n = 72$ shown in Figure 6.14 is almost exactly 10^{-4} mm $= 0.1$ μm(4 μin.),

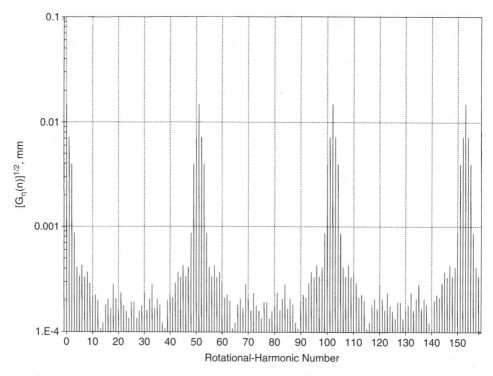

Figure 6.13   Rotational harmonic rms spectrum $(G_\eta(n))^{1/2}$, square-root of Equation (4.30b), of working-surface deviations of $N = 51$ tooth helical gear computed from 17 lead measurements and 7 profile measurements made on every tooth. Logarithmic amplitude scale in millimeters

which coincidentally, is the same rms amplitude of ghost-tone $n = 144$ shown in the transmission-error spectra of Figures 6.7 and 6.8. The rms transmission-error spectrum computed from the alternative measurement set utilizing 17 profile measurements and seven lead measurements showed the same behavior as Figure 6.14, but the rms amplitude of $n = 72$ was computed to be slightly higher, with a value of 1.2×10^{-4} mm $= 0.12\,\mu$m$(5\,\mu$in.), also surrounded by several strong adjacent harmonics, just as in Figure 6.14. The consistency of these two transmission-error predictions convinced us of the existence of a significant ghost tone at rotational harmonic number $n = 72$. However, as we have shown earlier, one would not have guessed from the tooth-deviation rms spectrum shown in Figure 6.13 that rotational harmonic $n = 72$ would have provided a significant contribution to the transmission-error spectrum.

In the cases of the previously-discussed ghost-tone rms spectra of Figures 5.1a and 5.1b, 6.5 and 6.7, and 6.6 and 6.8, the fractional reduction in rms amplitude in going from the tooth-deviation rms spectrum, $[G_\eta(n)]^{1/2}$, to the transmission-error rms spectrum, $[G_\zeta(n)]^{1/2}$, was significantly less than 50%. However, the rms amplitude of $[G_\eta(n)]^{1/2}$, for $n = 72$ in Figure 6.13 is about 2.3×10^{-4} mm, whereas the rms amplitude of, $[G_\zeta(n)]^{1/2}$, for $n = 72$ in Figure 6.14 is about 1.0×10^{-4} mm, significantly more than

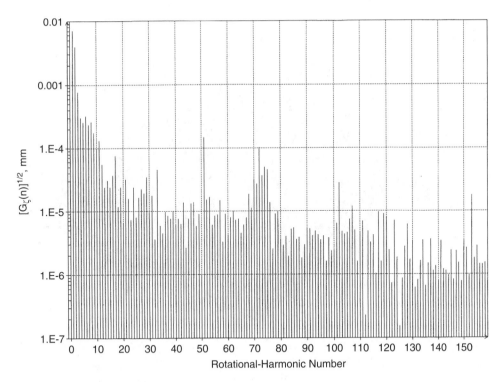

Figure 6.14 Rotational harmonic rms spectrum $(G_\zeta(n))^{1/2}$, square-root of Equation (5.18b), of mesh-attenuated working-surface deviations of $N=51$ tooth helical gear computed from same working-surface measurements as Figure 6.13. Logarithmic amplitude scale in millimeters

a 50% reduction. This larger reduction in rms amplitude suggests that some of the attenuated fraction of $n=72$ might be attributable to "random" working-surface errors, as described above. The results shown below indicate that this explanation is likely correct.

Figures 6.15–6.17 show the contribution of rotational harmonic $n=72$ to working-surface deviations computed by Equation (4.42). Figures 6.15 and 6.16 show the results of this computation on two consecutive teeth, numbers 15 and 16, computed from the measurement set that utilized 17 lead measurements and seven profile measurements on every tooth. These two figures show a clear periodic undulation error on a portion of the working-surfaces, together with randomly appearing deviations. There is a phase shift of the undulation between Figures 6.15 and 6.16.

Figure 6.17 shows the contribution of rotational harmonic $n=72$ to tooth number 16, obtained from the measurement set that utilized 17 profile measurements and seven lead measurements, to be compared with Figure 6.16 for the same tooth obtained from the other measurement set. Although the detailed behavior of the "random" deviations differs between these two plots, the behavior of the undulation

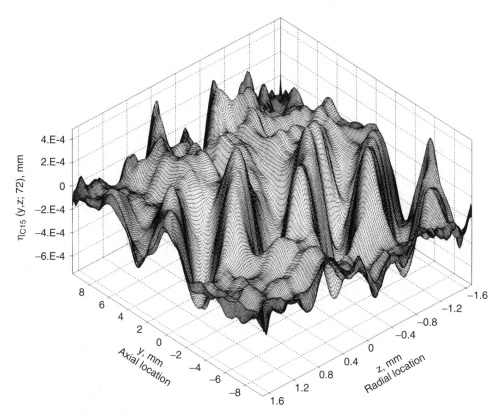

Figure 6.15 Computation by Equation (4.42) of the contribution from rotational harmonic $n = 72$ to the working-surface deviations of tooth $j = 15$ of the $N = 51$ tooth helical gear using same working-surface measurements as in Figures 6.13 and 6.14. All dimensions in millimeters

contribution, especially its phase, is very nearly the same, verifying the consistency between the results of the two independent sets of measurements.

The peak single-sided amplitude of the undulation error in these three figures is about 5×10^{-4} mm $= 0.5$ μm (20 μin.). But the undulation is present on only a fraction of the working-surfaces. The transmission error arises from the averaging action of tooth deviations along lines of contract, as illustrated by Figures 6.11 and 6.12. Therefore, the above-cited single-sided amplitude of about 0.5 μm is consistent with the rms transmission-error contribution of 0.1 μm. Forming the average value of Figures 6.15–6.17 along lines of contact, as illustrated by Figures 6.11 and 6.12, would have helped to clarify the transmission-error contribution provided by the working-surface deviations shown in Figures 6.15–6.17.

The results of two additional computations for this particular gear are of interest. Figure 6.18 shows the computation of the working-surface deviation from a perfect involute surface of tooth number 51. Some ripple of the surface is clearly present. This computation was carried out using Equation (3.18), where the expansion

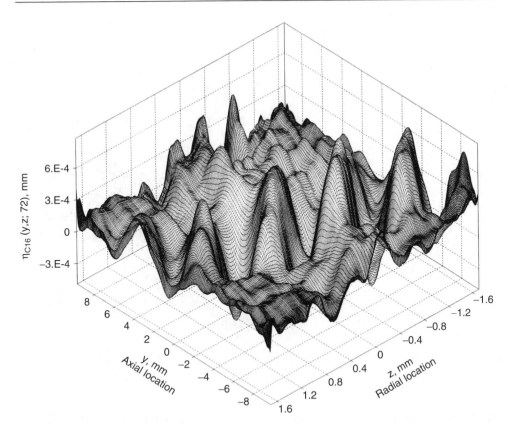

Figure 6.16 Computation by Equation (4.42) of the contribution from rotational harmonic $n = 72$ to the working-surface deviations of adjacent tooth $j = 16$ of the $N = 51$ tooth helical gear using same working-surface measurements as in Figures 6.13–6.15. All dimensions in millimeters

coefficients, Equation (3.20), were evaluated from the 17 scanning lead measurements by Equations (3.32) and (3.37), and the seven scanning profile measurements by Equations (3.41) and (3.43).

Figure 6.19 shows the computation of the average deviation surface, obtained by forming the average over all 51 of the individual working-surfaces. In forming this average of all 51 surfaces, the ripples clearly shown in Figure 6.18, typical of the individual working-surface deviations on this gear, have been completely removed by the averaging over all 51 surfaces. The computation of Figure 6.19 was carried out by using Equations (3.46) and (3.47). A comparable computation of the average deviation surface was carried out using the other measurement set consisting of 17 scanning profile and seven scanning lead measurements. That computation created a surface indistinguishable from that shown in Figure 6.19. The maximum variation (modification from the involute) of the surfaces shown in Figures 6.18 and 6.19 is about 0.025 mm = 25 μm, or slightly more, a value consistent with the elastic deformation of steel gear teeth under "full" loading.

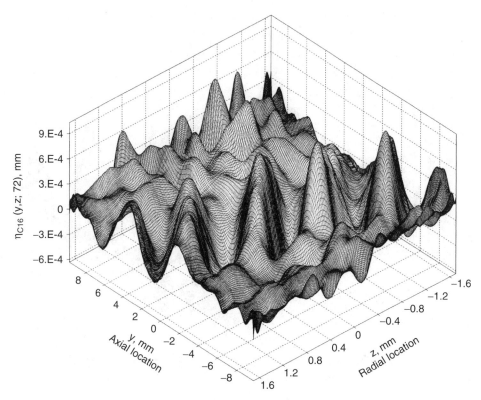

Figure 6.17 Computation by Equation (4.42) of the contribution from rotational harmonic $n = 72$ to the working-surface deviations of tooth $j = 16$ of the $N = 51$ tooth helical gear computed from 17 profile measurements and 7 lead measurements made on every tooth. All dimensions in millimeters

Comparison with Sound-Spectrum Measurement

After the above-described computations for this third ghost-tone example were completed, and the final report was submitted, we were provided with a high-resolution sound-spectrum measurement that had been carried out earlier using this same gear. This high-resolution spectrum is the continuous-line spectrum shown in Figure 6.20. The strongest harmonic shown there is rotational harmonic $n = 72$. In the immediate vicinity of $n = 72$ is a group of weaker harmonics at rotational harmonic numbers $n = 71$, 70, 69, 68 below $n = 72$, and at harmonic numbers $n = 73$, 74, 75, and 76 above $n = 72$. The amplitude scale of the spectrum in Figure 6.20 is linear, not logarithmic as in the transmission-error rotational harmonic spectrum shown in Figure 6.14. We were told that when the spectrum shown in Figure 6.20 was obtained, the gear rotational speed was varied so that each rotational harmonic shown there occurred at exactly the same frequency as the others, thereby eliminating any differing transmission-path resonance effects for the different measured rotational harmonic amplitudes.

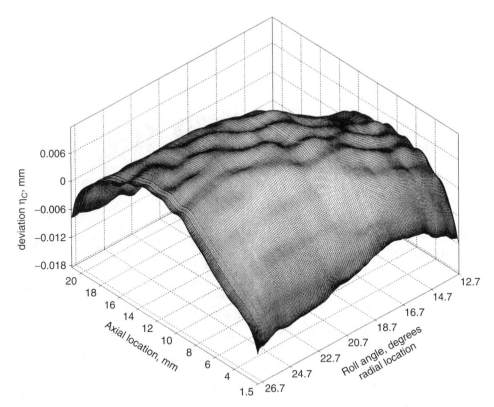

Figure 6.18 Computation by Equation (3.18) of the working-surface deviation of tooth $j=51$ of the $N=51$ tooth helical gear generated from 17 lead measurements and 7 profile measurements made on that tooth. Deviation sign convention reversed to yield convex appearance. Amplitude dimension in millimeters

After we received the sound spectrum shown in Figure 6.20, we replotted our computations of the transmission-error rotational harmonic amplitudes on a linear-amplitude scale. Because the continuous spectrum shown in Figure 6.20 is a sound spectrum and our computations are transmission-error spectra, there is an unknown multiplicative factor (transfer function) unavailable for comparison of the two spectra. Therefore, we adjusted the amplitude scale factor of our computations of the transmission-error spectra so that the amplitude of rotational harmonic $n=72$ in our two computations was in agreement with the amplitude of $n=72$ of the sound spectrum measurement. We then replotted, on the sound spectrum, the amplitudes of the eight neighboring rotational harmonics from each of our two computations, as can be seen in Figure 6.20.

The rms amplitude of our computation of rotational harmonic $n=72$ was about 10^{-4} mm $=0.1$ μm (4 μin.) (Figure 6.14). Hence, the rotational harmonics in the neighborhood of $n=72$ have computed amplitudes of about 1–2 μin. Given the values of these amplitudes, the agreement with the sound-spectrum amplitudes is

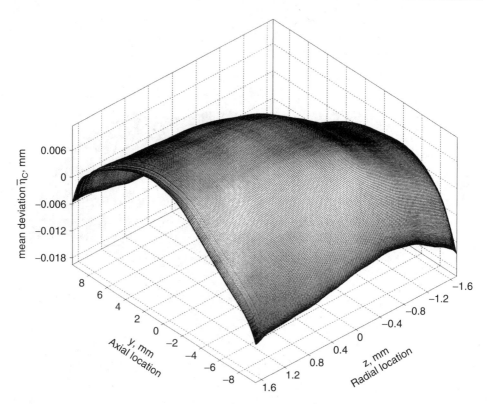

Figure 6.19 Computation by Equations (3.46) and (3.47) of the average deviation surface of all $N = 51$ teeth of the 51 tooth helical gear generated from 17 lead and 7 profile measurements made on all 51 teeth. Deviation sign convention reversed to yield convex appearance. All dimensions in millimeters

remarkable, although imperfect. An explanation of how such "sideband harmonics" of a dominant ghost tone can be generated is found in Mark (1992a, Case III p. 175).

Independent Verification of Second Ghost-Tone Example

In order to verify the validity of the predicted results of the earlier-described second ghost-tone example, and of operation of the software, the same helical gear utilized in those measurements was again repeatably measured on *another* (Gleason M&M) machine, and the software was used to compute the rms transmission-error rotational harmonic spectrum (using a different computer). In two of these additional tests, 18 scanning lead measurements were used as the primary measurement set and five scanning profile measurements were used as the secondary measurement set (measuring all teeth as is required). Complete measurement sets were taken twice, in the second complete measurement set, starting with measuring the first tooth

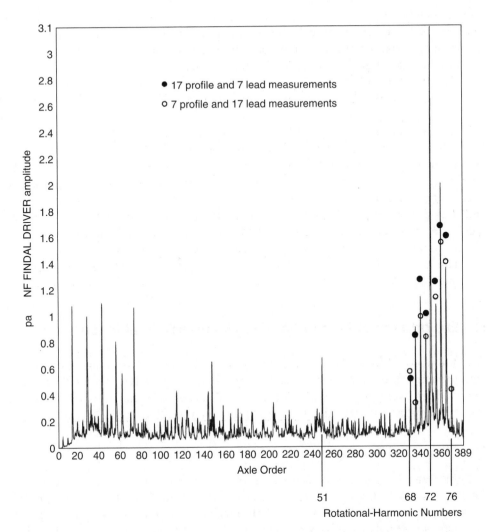

Figure 6.20 High-resolution sound spectrum generated by a 51 tooth helical gear meshing with a higher-quality gear compared with two kinematic-transmission-error spectrum computations. Both kinematic-transmission-error predictions were forced to agree with strongest acoustic-harmonic amplitude at $n = 72$, thereby showing agreement of predicted rotational harmonic amplitudes at $n = 68-76$ relative to measured acoustic harmonic amplitudes. Linear-amplitude ordinate. Solid-point amplitudes computed from 17 profile and 7 lead measurements made on every tooth; hollow-point amplitudes computed from 7 profile and 17 lead measurements

approximately $90°$ from where the first tooth was measured in the earlier set. The third complete measurement set was made using 18 scanning profile measurements as the primary measurement set and five scanning lead measurements as the secondary measurement set. As shown earlier for this particular gear by Equations (3.B.12) and (3.B.14), 18 primary scanning lead or profile measurements were sufficient to accurately interpolate the undulation error causing rotational harmonic number $n = 144$ for this particular gear.

From each of these three additional independent sets of measurements, the transmission-error rotational harmonic rms spectrum was computed. Agreement among all three of these computed spectra was excellent. In particular, the rms amplitude of rotational harmonic number $n = 144$ in each of the transmission-error spectrum predictions was almost exactly 10^{-4} mm, the same value computed earlier and shown in Figures 6.7 and 6.8, with no more variation in this amplitude among these three new computations than is seen between the amplitudes of rotational harmonic number $n = 144$ between Figures 6.7 and 6.8. Consistency of transmission-error spectrum predictions made from gear measurements taken on a second measurement machine therefore was verified.

6.7 Explanation of Factors Enabling Successful Predictions

"Common sense" suggests that what clearly has been achieved in the above-described measurements and example computations should not be possible. Outlined below are some factors likely responsible for enabling the success of these measurements and computations, and for the success of any future computations using this technology.

How Small Is 0.1 μm (4 μin.)?

A tightly packed package of 500 sheets of copier paper is 2 in. thick; therefore, 1000 sheets is 4 in. thick, and one sheet is 4×10^{-3} in. thick, or 4000 μin. thick. The wavelength of the visible spectrum of light ranges from 16 to 28 μin. Hence, 4 μin. is 1/1000th of the thickness of a sheet of copier paper and about 1/5 of the wavelength of the center of the visible spectrum. Thus, the 4 μin. rms amplitude of the transmission-error contribution of rotational harmonic number $n = 144$ in the above-described second ghost-tone example, Figures 6.7 and 6.8, and of rotational harmonic number $n = 72$ of the third ghost-tone example, Figure 6.14, is about 1/1000th of the thickness of a sheet of copier paper and about 1/5 of the wavelength of light, an apparently very small dimension. Because the surface finish of gear teeth ranges from about 6 to 36 μin., depending on manufacturing and finishing processes (Dudley, 1984, p. 5.71), the above value of 4 μin. rms amplitude of ghost-tone transmission-error contributions is smaller than typical surface roughness values. Yet, it is consistent with what can be regarded as precision machining (Nakazawa, 1994, p. 12), and therefore not unreasonable. Moreover, it has been shown earlier, for example, Mark (1998), that dedicated gear metrology equipment is capable of making measurements of comparable precision.

Mathematical Model and Computational Accuracy

It was shown by Equation (3.2) that the transmission error of a meshing gear-pair can be decomposed into additive contributions from each of the two meshing gears. By making the assumption that the tooth-pair stiffness per unit length of line of contact is a constant value (not needed), it was shown in Equations (5.2), (5.9), and (5.13) that the transmission-error contributions arising from tooth-working-surface deviations from equispaced perfect involute surfaces of each of two meshing gears is independent of total mesh stiffness $K_M(x)$, thereby allowing each of these transmission-error contributions to be considered independently from that of the mating gear. The remaining necessary assumption pertains to the assumed zone of contact of the mating gear teeth, which has been assumed to be a rectangular region in the y, z Cartesian coordinate system illustrated in Figures 3.2 and 3.4. At each instantaneous rotational position of the gear-pair, the transmission-error contribution of a single gear of the meshing pair arising from these tooth-working-surface deviations is the average value of the deviations, averaged along all lines of contact (Figure 3.2) within the zone of contact illustrated in Figure 2.6, as symbolically illustrated by Equations (5.8–5.13). Within the context of the above-described assumptions, the computational algorithms developed in the following chapters are exact. Moreover, there is no real limitation imposed on the computational-accuracy capabilities of present-day computers. The remaining considerations pertain to measurement of the working-surface deviations of all N teeth on a gear within the above-described rectangular zone of tooth contact, and to the accuracy of such measurements.

Interpolation of Line-Scanning Tooth Measurements

The axial face-width F and tooth-depth D, Figure 3.4, define the assumed zone of tooth contact and axial Q_a and transverse Q_t contact ratios of the working-surface analysis region. Ordinarily, the primary set of line-scanning measurements will be chosen from the smaller of Q_a or Q_t, as described in Appendix 3.B. Choice of this number of line-scanning measurements is dependent on the largest rotational harmonic of the transmission error of interest. If that number of line-scanning measurements is determined by Equation (3.B.7) and Table 3.B.1, then the resultant number of line-scanning measurements of the primary set of measurements (either lead or profile) will accurately interpolate working-surface rotational harmonic contributions at least up to the chosen rotational harmonic number, as illustrated by Figure 3.B.2. (However, some higher harmonic contributions can contribute to the Legendre-polynomial version of "aliasing.") Interpolation of lead measurements is carried out by Equations (3.32) and (3.37), and of profile measurements by Equations (3.41) and (3.43). The working-surface deviations of each tooth $j = 0, 1, \ldots, N-1$ are thus described by a two-dimensional sequence of Legendre expansion coefficients, $c'_{j,k\ell}, k = 0, 1, 2, \ldots, \ell = 0, 1, 2, \ldots$. If the above-described rules are followed, these expansion coefficients will provide a very accurate representation of the deviations of each working-surface $j = 0, 1, \ldots, N-1$ up to and including the rotational harmonic number used in determining the number of line-scanning measurements of the primary set, as further expanded on below.

Effects of Consistent Linear-Axis Errors and Measurement-Probe Bias Errors

The typical dedicated gear-measurement machine has three linear axes, and a rotary-axis on which the gear is fixed in place. Motion of these axes is controlled in a coordinated fashion so that the base of the measurement probe generates a perfect involute helicoid (perfect involute helical tooth), or a perfect involute in the case of spur gears. The probe measures the deviation from its base motion of the perfect involute helicoids, which is measurement of the deviation of the surface from a perfect involute surface. All such base motions contain errors, however small. But what is exceedingly important here is where these base-motion errors, and any probe errors, show up in rotational harmonic spectrum and transmission error computations.

Equation (4.16) describes the deviation $\epsilon_{Cj}(y, z)$ of tooth j from the mean working-surface deviation $\bar{\eta}_C(y, z)$, Equation (4.15); and Equation (4.19) shows that only the non-tooth-meshing rotational harmonics, $n \neq pN$, contribute to the deviation $\epsilon_{Cj}(y, z)$ of each tooth from the mean-deviation surface. But when gear teeth are measured, at *each* location y, z on *all* tooth-working-surfaces, the linear axes are in exactly the same position. Therefore, systematic (consistent) linear-axis errors contribute only to prediction of the mean deviation surface, Equation (4.15) – for example, Figures 3.14 and 6.19. Prediction by Equation (4.16) of the deviation $\epsilon_{Cj}(y, z)$ of each tooth j from the mean is not significantly affected by these consistent linear-axis errors; and according to Equation (4.19), predictions of non-tooth-meshing rotational harmonic contributions, $n \neq pN$, also are not substantially affected by consistent linear-axis errors.

Differential errors in rotary-axis positions within roll-angle spans of *individual* involute tooth measurements will show up in computed deviations $\epsilon_{Cj}(y, z)$ from the mean, and in non-tooth-meshing rotational harmonic computations, but such *differential* rotary-axis errors over small rotary-axis spans can be very small (Mark, 1998). Moreover, in the case of helical gears, these rotary-axis differential errors will be averaged along lines of tooth contact (Figure 3.2), significantly reducing their contributions in helical-gear transmission-error computations. This averaging effect is likely to have been at least partially instrumental in the successful computation of the very-small-amplitude ghost-tone harmonic contributions shown in Figures 6.7, 6.8, and 6.14.

Therefore, if linear-axis position errors are *consistent* and rotary-axis *differential* errors over involute roll-angle spans are small or negligible, then prediction of non-tooth-meshing rotational harmonic amplitudes and contributions $\epsilon_{Cj}(y, z)$ will be unaffected by these axis errors.

Probe consistent bias errors affect measurement values in the same manner as consistent linear-axis errors – that is, affect predicted tooth-meshing harmonic amplitudes and computations of mean tooth-surface deviations only. Therefore, for prediction of non-tooth-meshing rotational harmonics, the accuracy requirement for probes is consistency of bias errors and differential accuracy only (measurement of differences). Moreover, in computing rotational harmonic amplitudes, a great deal of "statistical averaging" takes place, which tends to minimize the effects of any short-span rotary-axis errors, and of "random" errors in both linear-axis positions and probe differential displacements.

Effects of Measurement Errors on Legendre Expansion Coefficients and on their DFTs $B_{k\ell}(n)$

It follows from Equations (3.45) to (3.47) that the average value $\bar{c}_{k\ell}$, Equation (3.46), of the Legendre expansion coefficients $k = 0, 1, 2, \ldots; \ell = 0, 1, 2, \ldots$ characterizes the mean working-surface deviation, and therefore, the $\bar{c}_{k\ell}$ contain any measurement errors of the mean tooth deviations. Moreover, it follows directly from Equation (6.46) that the DFT's, $B_{k\ell}(n), n \neq pN$, of the Legendre expansion coefficients are independent of the Legendre expansion coefficients $\bar{c}_{k\ell}$ of the mean working-surface deviation $\bar{\eta}_C(y, z)$, Equation (3.45). Consequently, the contribution $\eta_{Cj}(y, z;n)$, Equation (4.42), of each rotational-harmonic n, $n \neq pN$, to each working-surface j, and the contributions to the rotational-harmonic spectra $G_\zeta(n)$, Equation (4.30b), and transmission-error spectra $G_\zeta(n)$, Equation (5.18), $n \neq pN$, all are independent of the measured mean working-surface deviation $\bar{\eta}_C(y, z)$, and therefore are unaffected by consistent linear-axis position errors and measurement-probe consistent bias errors.

The Legendre expansion term $k = 0$, $\ell = 0$ is a special case. This Legendre term describes the absolute rotational-position error of tooth j. Absolute rotary-axis errors will directly contribute to errors in $c_{j,00}$.

Effects of Surface Roughness and Statistical Averaging

Forming the integrations of lead measurements in Equation (3.32), and of profile measurements in Equation (3.41), effectively smooths and negates the effects of short-wavelength surface roughness.

Many of the operations in generating the rotational harmonic spectra require averages and summations involving measured surface characteristics, for example, Equations (4.21a), (4.30b), and (5.18b). The effect of these averages and summations is to partially average and eliminate "random" measurement errors. Generation of the rotational harmonic contributions to the working-surface deviations by Equation (4.42) also involves such effective averaging (summation) operations, tending to eliminate "random" measurement errors.

Appendix 6.A. Validation of Equation (6.46)

Define

$$S \triangleq \sum_{j=0}^{N-1} \exp(-i2\pi nj/N) \qquad (6.A.1)$$

$$= \sum_{j=1}^{N} \exp(-i2\pi nj/N) \qquad (6.A.2)$$

because

$$\exp(-i2\pi n) = \cos(2\pi n) - i\sin(2\pi n) = 1$$
$$= \exp(0) \tag{6.A.3}$$

since n is an integer. Equation (6.A.2) is a finite geometric series (Hildebrand, 1952, p. 259),

$$S = \sum_{j=1}^{N} r^j = \frac{r^{N+1} - r}{r - 1} \ , \quad r \neq 1 \tag{6.A.4}$$

with

$$r = \exp(-i2\pi n/N) \tag{6.A.5}$$

yielding (Hildebrand, 1952, p. 260),

$$S = \frac{\sin(\pi n)}{\sin(\pi n/N)} \exp\left[-i\pi n(N+1)/N\right] , \quad \frac{n}{N} \neq 0, \pm 1, \pm 2, \ldots$$
$$= 0 \ , \qquad\qquad\qquad\qquad \frac{n}{N} \neq 0, \pm 1, \pm 2, \ldots , \tag{6.A.6}$$

because $n = \pm 1, \pm 2$ always is an integer.

References

Antosiewicz, H.A. (1964) Bessel functions of fractional order, in *Handbook of Mathematical Functions With Formulas, Graphs, and Mathematical Tables*, Chapter 10 (eds M. Abramowitz and I.A. Stegun), U.S. Government Printing Office, Washington, DC. Republished by Dover, Mineola, NY, pp. 435-478.

Bateman, H. (1954) in *Tables of Integral Transforms*, vol. **1** (ed. A. Erdelyi), McGraw-Hill, New York.

Baxter, M.L. (1962) Basic theory of gear-tooth action and generation, in *Gear Handbook*, Chapter 1 (ed. D.W. Dudley), 1st edn, McGraw-Hill, New York, pp. 1-1, 1-21.

Cluff, B.W. (ed.) (1992) *Gear Process Dynamics*, 7th edn, American Pfauter Limited Partnership, Loves Park, IL.

Drago, R.J. (1988) *Fundamentals of Gear Design*, Butterworths, Boston, MA.

Dudley, D.W. (1984) *Handbook of Practical Gear Design*, Revised edn, McGraw-Hill, New York.

Faires, V.M. and Keown, R.M. (1960) *Mechanism*, 5th edn, McGraw-Hill, New York. Republished by Robert, E. Krieger, Huntington, NY.

Hildebrand, F.B. (1952) *Methods of Applied Mathematics*, 1st edn, Prentice-Hall, Englewood Cliffs, NJ.

Lynwander, P. (1983) *Gear Drive Systems: Design and Application*, Marcel Dekker, New York.

Mark, W.D. (1978) Analysis of the vibratory excitation of gear systems: basic theory. *Journal of the Acoustical Society of America*, **63**, 1409–1430.

Mark, W.D. (1979) Analysis of the vibratory excitation of gear systems. II: tooth error representations, approximations, and application. *Journal of the Acoustical Society of America*, **66**, 1758–1787.

Mark, W.D. (1991) Gear noise, in *Handbook of Acoustical Measurements and Noise Control*, Chapter 36 (ed. C.M. Harris), 3rd edn, McGraw-Hill, New York.

Mark, W.D. (1992a) Contributions to the vibratory excitation of gear systems from periodic undulations on tooth running surfaces. *Journal of the Acoustical Society of America*, **91**, 166–186.

Mark, W.D. (1998) Method for precision calibration of rotary scale errors and precision determination of gear tooth index errors. *Mechanical Systems and Signal Processing*, **12**, 723–752.

Merritt, H.E. (1971) *Gear Engineering*, John Wiley & Sons, Inc., New York.

Nakazawa, H. (1994) *Principles of Precision Engineering*, Oxford University Press, Oxford.

Newland, D.E. (1989) *Mechanical Vibration Analysis and Computation*, Longman Scientific & Technical, Essex. Republished by Dover, Mineola, NY.

Relton, F.E. (1946) *Applied Bessel Functions*, Blackie and Son, London. Republished by Dover, New York.

Sneddon, I.N. (1961) *Special Functions of Mathematical Physics and Chemistry*, Oliver and Boyd, Edinburgh.

7

Transmission-Error Decomposition and Fourier Series Representation

Once the highest rotational harmonic of interest of the transmission-error contribution from a gear to be measured is specified, it was shown in Chapter 3 that an adequate representation of the working-surface-deviations of all teeth on that gear can be obtained from line-scanning lead and profile measurements made by currently available dedicated gear metrology equipment. The resultant representation of the working-surface-deviation of each tooth is given by the expansion, Equation (3.18), in two-dimensional normalized Legendre polynomials, where the expansion coefficients are evaluated, utilizing Gaussian quadrature, by Equations (3.37) and (3.43). This procedure yields a measured mathematical representation of the deviation of every point y, z on the working-surface of every tooth $j = 0, 1, \ldots, N-1$.

Because the working-surfaces are a collection of N equispaced surfaces, it was explained in Chapter 4 that the required rotational harmonic representation of the tooth deviations at each point y, z on the working-surfaces, Figure 3.4, is the discrete Fourier transform, Equation (4.5) with respect to tooth number j. When combined with the Legendre representation of Equation (3.18), this lead to the DFT representation, Equation (4.22), where $B_{k\ell}(n)$ is the DFT, Equation (4.21), of the Legendre expansion coefficients.

The transmission error contribution from working-surface-deviations is described by Equation (5.2), which is a function of the integrals, Equation (5.5), of stiffness weighted deviations of the individual teeth, integrated over all lines of tooth contact. If the local tooth-pair stiffness per unit length of line of contact is assumed to be constant, it was then shown symbolically by Equation (5.13) that the transmission-error contribution from tooth deviations is independent of this local stiffness, and is

Performance-Based Gear Metrology: Kinematic-Transmission-Error Computation and Diagnosis,
First Edition. William D. Mark.

the average value of the tooth deviations, averaged over all lines of contact at that particular rotational position of the gear. This average value of tooth deviations is a generally fluctuating quantity, periodically fluctuating with the period equal to the period of rotation of the gear, which is the period of rotational harmonic $n = 1$.

The mathematical operations on the working-surface-deviations, symbolically illustrated by Equations (5.2–5.13), are linear operations, and consequently are amenable to a transfer-function approach. The resultant complex Fourier series coefficients of the transmission-error contribution from working-surface-deviations was expressed by Equation (5.16), where the mesh-attenuation function $\hat{\phi}_{k\ell}(n/N)$ characterizes in the frequency domain the instantaneous averaging of the working-surface-deviations over all lines of tooth contact illustrated in Figures 2.6 and 3.2. It is instructive to compare Equation (5.16) with Equation (4.22), where $\hat{\phi}_{k\ell}(n/N)$ represents the averaging action of the two-dimensional Legendre term $k\ell$ in Equation (4.22a).

The goal in organizing the book has been to present as much of it as could be developed using relatively straightforward mathematical tools. Once the requirement of forming the DFT, Equation (4.5), of the working-surface-deviations is accepted (this requirement is not a priori obvious), it was possible to fully develop the mean-square rotational harmonic spectrum, Equation (4.30), of working-surface-deviations, and in particular, the expression Equation (4.42) for the working-surface-deviations on teeth $j = 0, 1, 2, \ldots, N - 1$ causing any prescribed rotational harmonic contribution n to the transmission error, an important tool for diagnosing the manufacturing sources of ghost tones (or any other tones).

However, to compute the transmission-error contributions from measured working-surface-deviations, the mathematical form of the mesh-attenuation function $\hat{\phi}_{k\ell}(n/N)$ in Equation (5.16) is required. In this chapter, the approximation to $\hat{\phi}_{k\ell}(n/N)$ given by Equation (6.3) is derived. In addition, equations and algorithms for "exact" computation of the $\hat{\phi}_{k\ell}(n/N)$ mesh-attenuation functions are derived. In particular, it is rigorously shown that the required frequency-domain characterization of the working-surface-deviations is the DFT, Equation (4.21), of the Legendre expansion coefficients, which is a direct consequence of the DFT, Equation (4.5), and the Legendre representation, Equation (3.18), of the working-surface-deviations.

To facilitate ease in computer programming for prediction of the transmission-error contributions from measured working-surface-deviations in both "time" and "frequency" domains, the final forms of the derived computational algorithms involve only real (not complex) quantities. The required derivations are less straight-forward than those found in the preceding chapters. The derivation begins with decomposition of the transmission error into its constituent components.

7.1 Decomposition of the Transmission Error into its Constituent Components

The transmission error of a meshing-gear-pair is the combined error in the rotational positions of the two meshing gears from the positions that would transmit an exactly constant speed ratio. For parallel-axis gears with nominally involute working-surfaces, the *lineal* transmission error of the gear-pair can be decomposed into the deviation of the position of each of the two meshing gears from the position occupied by its rigid perfect involute counterpart with equispaced teeth, as described by Equation (3.2). This form of the transmission error allows us to deal with transmission-error contributions arising from the geometric working-surface-deviations of the teeth of a single gear from their equispaced perfect involute counterparts – the subject of this book.

Contact of Involute Helical Teeth

Let us return to the geometric description of involute helical gears illustrated by Figures 2.3–2.6. Instead of regarding Figure 2.3a as the unwrapping of a cord from the base cylinders, let it illustrate the unwrapping of a belt of finite axial width from the base cylinders, and let Figure 2.3b illustrate the resultant belt drive operating between the two base cylinders. Draw a straight line on this belt at an angle ψ_b with a base-cylinder element (which is parallel to the base-cylinder axis). Let the contact point of the two involute curves in Figure 2.3b be on this straight line drawn on the belt. When this straight line is projected onto the upper portion of Figure 2.6, it is seen to be the line of contact between the mating involute helical gear teeth. Consequently, equispaced lines drawn parallel on the fictitious belt drive riding on the two base cylinders of a pair of involute helical gears, as illustrated in Figure 2.6, provide an *exact* representation of the lines of contact on the working-surfaces of parallel-axis helical gears with equispaced rigid perfect involute teeth. The transverse-plane distance between these lines of contact is the base pitch Δ, as illustrated in Figure 2.6. Because the tooth-working-surfaces of both meshing gears have finite curvature, without elastic deformations these contact lines have infinitesimal width.

To relate the lines of contact shown in Figure 2.6 to lines of contact on tooth-working-surfaces, it is necessary to define the coordinates x, y of lines of contact on the fictitious belt illustrated in the upper portion of Figure 2.6. Coordinate x designates the rotational positions of the two rigid perfect involute counterparts to the actual gears under consideration:

$$x \triangleq R_b^{(1)} \theta^{(1)} = R_b^{(2)} \theta^{(2)}, \tag{7.1}$$

where $\theta^{(1)}$ and $\theta^{(2)}$ designate the instantaneous rotational positions of the perfect involute counterparts to the gears under consideration, as illustrated in Figure 2.5. Angles $\theta^{(1)}$ and $\theta^{(2)}$ are "measured" in radians. As indicated in Figure 2.6, the origin $x = 0$ is defined to be the rotational positions of the two perfect counterparts to the actual gears when the line of contact of tooth-pair $j = 0$ is exactly centered in the zone of contact illustrated in the figure. Coordinate y designates axial location, with origin $y = 0$ axially centered in the zone of contact.

Consider a generic point P on a typical line of contact, as illustrated in Figure 2.6. Such a point P on the same line of contact on the tooth-working-surface is shown in

Figure 3.2. *It follows from Figures 2.6 and 3.2 that for a generic tooth-pair j, specification of coordinate values x, y designates a unique point of contact on the working-surfaces of the designated tooth-pair j.*

Expression for Lineal Transmission Error

The expression for the transmission error derived below assumes that the shaft centerlines of the two meshing gears remain parallel and fixed. (When the resultant expression for the transmission error is used as the vibratory excitation, inclusion of shaft-mounting elastic deformations can, and should be, accounted for as explained, for example, in Mark (1992b).) As described in Section 3.1, the (lineal) transmission error is defined as

$$\zeta(x) \triangleq R_b^{(1)} \delta\theta^{(1)}(x) - R_b^{(2)} \delta\theta^{(2)}(x) \tag{7.2}$$

where $\delta\theta^{(1)}(x)$ and $\delta\theta^{(2)}(x)$ denote the instantaneous rotational deviations of meshing gears (1) and (2) from the positions of their rigid perfect involute counterparts. The negative sign associated with $\delta\theta^{(2)}(x)$ in Equation (7.2) arises from the sign convention of $\theta^{(2)}$ illustrated in Figure 2.5. Hence, $\zeta(x)$, Equation (7.2), describes the lineal distance the two meshing gears approach one another in the plane of contact relative to the positions of their rigid perfect involute counterparts. Therefore, the transmission-error contribution from each of the two meshing gears is defined as positive when it results from the effective removal of material from the working-surfaces of their perfect involute counterparts.

Referring to Figure 2.6, if the shafts of the two meshing gears remain fixed and parallel, the same lineal approach must take place at each point on all lines of tooth contact shown in the figure. Therefore, if we denote local elastic deformation by u and local geometric working-surface-deviation by η, both "measured" in a direction defined by the intersection of the transverse plane and plane of contact, and both defined as positive for effective removal of material from equispaced perfect involute surfaces, then there follows

$$\zeta(x) = u_j^{(1)}(x,y) + \eta_j^{(1)}(x,y) + u_j^{(2)}(x,y) + \eta_j^{(2)}(x,y), \tag{7.3}$$

where superscripts (1) and (2) denote the contributions of the two meshing gears. Equation (7.3) must be satisfied at all axial locations y of every tooth pair j in contact at each instantaneous rotational position designated by x; it describes the transmission error as additive contributions from each of the two meshing gears, as in Equation (7.2). (The assumption that gear shafts remain parallel can be relaxed, leading to a three-component definition of transmission error (Mark, 1989, 1987) and substantial additional complexity.)

It is convenient to combine the elastic deformation contributions into one contribution, that is,

$$u_j(x,y) \triangleq u_j^{(1)}(x,y) + u_j^{(2)}(x,y); \tag{7.4}$$

hence, from Equation (7.3),

$$\zeta(x) = u_j(x,y) + \eta_j^{(1)}(x,y) + \eta_j^{(2)}(x,y) \tag{7.5}$$

which can be solved for $u_j(x,y)$,

$$u_j(x,y) = \zeta(x) - \eta_j^{(1)}(x,y) - \eta_j^{(2)}(x,y), \tag{7.6}$$

which always is non-negative within the zone of tooth contact.

For $\zeta(x)$ to be independent of axial location y and the individual tooth-pair numbers j within the zone of contact, the local elastic deformations $u_j(x,y)$ of the individual tooth pairs j must vary with y and j so that the force-deformation relationship of the entire gear-pair mesh is satisfied. Let $K_{Tj}(x,y)$ denote the local tooth-pair stiffness per unit length of line of contact as defined in Appendix 7.A, and $W_j(x)$ the total force transmitted by tooth pair j in the direction defined by the intersection of the plane of contact and the transverse plane. Let dl denote differential line of contact length, that is, from Figure 2.6,

$$d\ell = \sec \psi_b dy. \tag{7.7}$$

Then, from the definition of $K_{Tj}(x,y)$, there follows

$$W_j(x) = \int K_{Tj}(x,y)\, u_j(x,y)\, d\ell$$

$$= \sec \psi_b \int_{y_{Aj}(x)}^{y_{Bj}(x)} K_{Tj}(x,y)u_j(x,y)dy, \tag{7.8}$$

where $y_{Aj}(x)$ and $y_{Bj}(x)$ denote the endpoint locations of the individual lines of contact of tooth-pairs j that determine the zone of contact, as illustrated in Figure 2.6. Substituting Equation (7.6) into Equation (7.8), there follows

$$W_j(x) = \sec \psi_b \int_{y_{Aj}(x)}^{y_{Bj}(x)} K_{Tj}(x,y)\left[\zeta(x) - \eta_j^{(1)}(x,y) - \eta_j^{(2)}(x,y)\right] dy. \tag{7.9}$$

Let us define

$$\tilde{K}_{Tj}(x) \triangleq \sec \psi_b \int_{y_{Aj}(x)}^{y_{Bj}(x)} K_{Tj}(x,y)dy, \tag{7.10}$$

which is the total stiffness of tooth-pair j within the zone of contact determined by the limits of integration in Equation (7.10), and for gears (1) and (2),

$$\tilde{\eta}_{Kj}^{(\cdot)}(x) \triangleq \sec \psi_b \int_{y_{Aj}(x)}^{y_{Bj}(x)} K_{Tj}(x,y)\eta_j^{(\cdot)}(x,y)dy, \tag{7.11}$$

which is the local-stiffness-weighted geometric deviation of tooth j of gear $(\cdot)$ integrated along the line of contact located by coordinate x. Then, substituting Equations (7.10) and (7.11) into Equation (7.9), there follows for the force transmitted by tooth pair j,

$$W_j(x) = \zeta(x)\tilde{K}_{Tj}(x) - \tilde{\eta}_{Kj}^{(1)}(x) - \tilde{\eta}_{Kj}^{(2)}(x), \qquad (7.12)$$

because $\zeta(x)$ is independent of y.

The total force $W(x)$ transmitted by the entire gear mesh is the superposition of the forces $W_j(x)$, Equation (7.8), arising from the elastic deformations of the individual tooth pairs j in contact at instant x:

$$W(x) = \sum_j W_j(x)$$

$$= \zeta(x)\sum_j \tilde{K}_{Tj}(x) - \sum_j \left[\tilde{\eta}_{Kj}^{(1)}(x) + \tilde{\eta}_{Kj}^{(2)}(x) \right], \qquad (7.13)$$

upon substitution of Equation (7.12). Define the total mesh stiffness at instant x as

$$K_M(x) \triangleq \sum_j \tilde{K}_{Tj}(x), \qquad (7.14)$$

divide Equation (7.13) by $K_M(x)$, and solve for $\zeta(x)$:

$$\zeta(x) = \frac{W(x)}{K_M(x)} + \frac{\sum_j \tilde{\eta}_{Kj}^{(1)}(x)}{K_M(x)} + \frac{\sum_j \tilde{\eta}_{Kj}^{(2)}(x)}{K_M(x)}. \qquad (7.15)$$

Equation (7.15) expresses the instantaneous (fluctuating) transmission error in terms of three additive contributions, a contribution arising from elastic deformations of the teeth, and contributions arising from (stiffness weighted) geometric deviations of the tooth-working- surfaces from each of the two meshing gears. Because of our sign convention: positive transmission-error contributions arise from effective "removal" of material from the tooth-working-surfaces of rigid perfect involute gears, the three components in the right-hand side of Equation (7.15) are additive. Noting that only the first component on the right-hand side of Equation (7.15) is directly dependent on mesh loading, Equation (7.15) can be more concisely expressed as

$$\zeta(x) = \zeta_W(x) + \zeta^{(1)}(x) + \zeta^{(2)}(x), \qquad (7.16)$$

with obvious correspondences between the individual terms of Equations (7.15) and (7.16). The results given by Equations (7.15) and (7.16) are those cited earlier by Equations (5.1–5.5) in Chapter 5, where Equation (7.7) has been utilized.

The additive property of the three terms on the right-hand side of Equation (7.15) is the feature that allows us to justifiably define, describe, and analyze the transmission-error contributions arising from the tooth-working-surface-deviations of a single gear of a meshing pair, without regard to the working-surface-deviations on the mating gear and elastic deformations. The principal assumption is the requirement to specify the (rectangular) zone of tooth-pair contact, which for a specific gear is equivalent to specifying the operating axial Q_a and transverse Q_t contact ratios.

The derivation of Equation (7.15) follows along the lines of Remmers (1972), as modified by Dr. Ranganath Nayak in a Bolt Beranek and Newman internal memorandum, but without the definition and rigorous determination of local tooth-pair stiffness $K_{Tj}(x, y)$, (Mark, 1978), described in Appendix 7.A.

7.2 Transformation of Locations on Tooth Contact Lines to Working-Surface Coordinate System

Location-Coordinate Transformation

As shown by Equations (7.15) and (7.11), the contribution to the transmission error $\zeta(x)$ from the tooth-working-surface-deviation $\eta_j^{(\cdot)}(x, y)$ at any particular value of roll-distance x is dependent on the working-surface-deviations along the entire tooth contact line for that tooth j. For each tooth j and each value of roll-distance x, the location of the line of contact of tooth-pair j in the plane of contact is uniquely determined, as shown by the upper portion of Figure 2.6. Therefore, a unique point on this line of contact is determined by specifying the axial coordinate y, once j and x are specified, as illustrated by the point P' shown in Figure 2.6. This line of contact shown in Figure 2.6 coincides exactly with a line of contact on the working-surface of tooth j, as illustrated by Figure 3.2. Consequently, specification of j, x, and y designates a unique point on the working-surface of tooth j. Because tooth-working-surface-deviations are measured at working-surface coordinates determined by the measurement process, a coordinate transformation is required to transform the above-described x, y coordinates for a generic tooth j to the working-surface coordinates y, z of Figure 3.2 used in gear-tooth measurements.

First, define

$$s \triangleq x - j\Delta \tag{7.17}$$

which transforms roll distance x of tooth j to a roll-distance coordinate exactly centered in the zone of contact illustrated in the upper portion of Figure 2.6. Notice that for $j = 0$, $s = x$, and with $x = 0$ in the upper portion of Figure 2.6, the dashed line of contact is exactly centered in the zone of contact at $y = 0$. This same dashed line of contact, when projected onto the y, z coordinate system shown in Figure 3.2, passes through the origin $y = 0$, $z = 0$ of that Cartesian coordinate system. Thus, as can be seen from the upper portion of Figure 2.6, specifying values of s and y determines a unique point on the line of contact of a tooth, which, in turn, determines a unique point in the Cartesian y, z coordinate system illustrated in Figure 3.2. This coordinate transformation is shown in Appendix 7.B to be

$$y = y \tag{7.18}$$

$$z = \beta s + \gamma y \tag{7.19}$$

where

$$\beta = \pm \sin \phi \tag{7.20}$$

$$\gamma = \pm \sin \phi \cos \phi \tan \psi \tag{7.21a}$$

$$= \pm \sin \phi \tan \psi_b, \tag{7.21b}$$

where ϕ, ψ, and ψ_b are always taken as positive (non-negative) values. In gear tooth-working-surface measurements, the coordinate z is related to roll angle by Equation (3.6). The algebraic signs for β and γ are to be chosen from Table 7.1. Right-hand and left-hand refer to helices (Drago, 1988, pp. 483–484), and driver and follower refer to driving versus following gear of a meshing pair. See Equations (7.B.14–7.B.17) and Table 7.B.2.

Line-Integral Transformation

According to Equations (7.15) and (7.16), the transmission-error contributions from tooth-working-surface-deviations of a single gear ($\cdot$) are given by

$$\zeta^{(\cdot)}(x) = \frac{\sum_j \tilde{\eta}_{Kj}^{(\cdot)}(x)}{K_M(x)} \tag{7.22}$$

where contributions from individual teeth j to the numerator are given by Equation (7.11), and contributions to the denominator are given by Equations (7.14) and (7.10). Equations (7.10) and (7.11) are expressed as line integrals of functions

Table 7.1 Signs for β and γ

	Sign for β	Sign for γ
Right-hand driver	+	−
Left-hand driver	+	+
Right-hand follower	−	+
Left-hand follower	−	−

$K_{Tj}(x,y)$ and $K_{Tj}(x,y)\eta_j^{(\cdot)}(x,y)$ over the line of contact of a generic tooth j, Figure 3.2. Consequently, to compute the transmission-error contribution, Equation (7.22), it is necessary to transform the integrands $K_{Tj}(x,y)$ and $K_{Tj}(x,y)\eta_j^{(\cdot)}(x,y)$ to the Cartesian coordinate system y,z illustrated by Figure 3.2. Because the subject of this book is tooth-working-surface-deviation contributions from only a single gear, at this juncture we shall drop the use of $(\cdot)$ indicating a specific gear, as in Equations (5.2–5.13), unless confusion might arise.

Denote the integrand of either Equation (7.10) or (7.11) by $f_j(x,y)$, which is a function of tooth j, roll-distance x, and axial-location y on the line of contact of a generic tooth j in the x,y coordinate system illustrated in Figure 2.6. As shown by Equations (7.17–7.19), a point on the line of contact of tooth j at roll-distance x and axial location y is uniquely located in the s,y,z coordinate system by Equations (7.17–7.19), where $z = \beta s + \gamma y$. Denote the value of the function $f_j(x,y)$ for any j,x,y, but described in the Cartesian coordinate system y,z illustrated in Figure (3.2), by $f_{Cj}(y,z)$, C for Cartesian. Then, we have from Equations (7.17–7.19),

$$f_j(x,y) = f_j(s+j\Delta,y) \equiv f_{Cj}(y,\beta s + \gamma y). \tag{7.23}$$

By appropriately identifying the integrand in Equation (7.10) or (7.11) with $f_j(x,y)$, each of these integrals can be expressed upon substitution of Equation (7.23), as

$$\tilde{f}_j(x) \triangleq \sec \psi_b \int_{y_{Aj}(x)}^{y_{Bj}(x)} f_j(x,y)dy \tag{7.24a}$$

$$= \sec \psi_b \int_{y_A(s)}^{y_B(s)} f_{Cj}(y,\beta s + \gamma y)dy \tag{7.24b}$$

$$\equiv \underset{\sim}{f}_j(s) = \underset{\sim}{f}_j(x-j\Delta), \tag{7.24c}$$

where we again have used the tilde to denote integration over a line of contact, and in Equation (7.24c) the tilde is moved under f to indicate the shift from roll-distance x to shifted roll-distance $s = x - j\Delta$. When Equation (7.24a), which is a function of roll-distance x, is transformed to Equation (7.24b), a function of $s = x - j\Delta$, the limits of integration become independent of tooth number j, as indicated in Equation (7.24b).

Let us now apply the notation described by Equations (7.23) and (7.24) to the integration in Equation (7.10). When the local tooth-pair stiffness per unit length of line of contact $K_{Tj}(x,y)$ is expressed in the Cartesian y,z coordinate system of Figure 3.2, as in Equation (7.23), it is no longer dependent on tooth-pair number j, as can readily be understood from Figure 3.2. Let us denote this local tooth-pair stiffness, expressed in the Cartesian y,z coordinate system, by $K_{TC}(y,z)$. Then, using

Equation (7.19), application of Equation (7.23) to the local tooth-pair stiffness per unit length of line of contact gives

$$K_{Tj}(x,y) = K_{Tj}(s + j\Delta, y) \equiv K_{TC}(y, \beta s + \gamma y), \tag{7.25}$$

where, as noted, $K_{TC}(y,z)$ is independent of j. Using the tilde notation of Equation (7.24a) to denote integration along the line of tooth-pair contact, we then have from Equations (7.24a–c), for the integration described by Equation (7.10),

$$\tilde{K}_{Tj}(x) = \sec\psi_b \int_{y_A(s)}^{y_B(s)} K_{TC}(y, \beta s + \gamma y)dy \tag{7.26a}$$

$$\equiv \underset{\sim}{K}_T(s) = \underset{\sim}{K}_T(x - j\Delta), \tag{7.26b}$$

recognizing that when expressed as a function of shifted roll-distance location s, the above integrated tooth-pair local stiffness is independent of tooth-pair number j.

The integration described by Equation (7.11) is dealt with in the same manner as that of Equation (7.10), except that the final result remains dependent on tooth-number j. Denote the local-stiffness-weighted deviation of tooth j in the integrand of Equation (7.11) by

$$\eta_{Kj}(x,y) \triangleq K_{Tj}(x,y)\eta_j(x,y) \tag{7.27}$$

where the identification $(\cdot)$ of the specific gear has been dropped, as noted earlier. Denote this same locally-stiffness-weighted deviation described in the Cartesian y, z coordinate system of Figure 3.2 by

$$\eta_{KCj}(y,z) \triangleq K_{TC}(y,z)\eta_{Cj}(y,z), \tag{7.28}$$

where, as mentioned above, the local tooth-pair stiffness $K_{TC}(y,z)$ is independent of tooth-pair j when expressed in the Cartesian y, z coordinate system. Then applying the notation of Equation (7.23) to Equations (7.27) and (7.28), we have

$$\eta_{Kj}(x,y) = \eta_{Kj}(s + j\Delta, y) \equiv \eta_{KCj}(y, \beta s + \gamma y), \tag{7.29}$$

where subscript C again has been used to denote the Cartesian coordinate system. Applying the integration of Equations (7.24a–c) to the integration, Equation (7.11), and using Equations (7.27–7.29), there follows

$$\tilde{\eta}_{Kj}^{(\cdot)}(x) = \sec\psi_b \int_{y_A(s)}^{y_B(s)} \eta_{KCj}(y, \beta s + \gamma y)dy \tag{7.30a}$$

$$\equiv \underset{\sim}{\eta}_{Kj}^{(\cdot)}(s) = \underset{\sim}{\eta}_{Kj}^{(\cdot)}(x - j\Delta), \tag{7.30b}$$

as in the general case described by Equations (7.24a–c).

7.3 Fourier-Series Representation of Working-Surface-Deviation Transmission-Error Contribution

The rotational period of the gear $(\cdot)$ under consideration is the period of its fundamental rotational-harmonic contribution to the transmission error. The independent variable used to describe these transmission-error contributions is the roll-distance variable

$$x = R_b^{(\cdot)} \theta^{(\cdot)} \tag{7.31}$$

described by Equation (7.1). If N is the number of teeth on the gear of interest and Δ is the base pitch (tooth spacing in roll-distance x), then the complex Fourier-series representation (Mark, 1978) of the transmission-error contribution, Equation (7.22), from the tooth-working-surface-deviations $\eta_j(x, y)$ of Equation (7.27), or $\eta_{Cj}(y, z)$ of Equation (7.28), $j = 0, 1, \ldots, N-1$ can be expressed as

$$\zeta^{(\cdot)}(x) = \sum_{n=-\infty}^{\infty} \alpha_n^{(\cdot)} \exp\left[i2\pi nx/(N\Delta)\right] \tag{7.32}$$

where the complex Fourier series coefficients $\alpha_n^{(\cdot)}$ are given by

$$\alpha_n^{(\cdot)} = \frac{1}{N\Delta} \int_{-N\Delta/2}^{N\Delta/2} \zeta^{(\cdot)}(x) \exp\left[-i2\pi nx/(N\Delta)\right] dx. \tag{7.33}$$

As indicated by Equations (5.16) and (5.18), in order to be able to compute the mesh-attenuated mean-square rotational-harmonic spectrum, $G_\zeta(n)$, Equation (5.18), from the tooth-working-surface-deviations of the gear $(\cdot)$ of interest, a method to compute the Fourier series coefficients $\alpha_n^{(\cdot)}$ is required. Moreover, once these coefficients have been computed, the transmission-error contribution $\zeta^{(\cdot)}(x)$ can be computed as a function of roll-distance x by using Equation (7.32).

Expression for Fourier-Series Coefficients

Let index $l = 1, 2, \ldots$ count the revolutions of the subject gear $(\cdot)$. Consider the working-surface of a single tooth j. According to Equations (7.11) and (7.22), the working-surface-deviation contribution of that single tooth j to the transmission error is $\tilde{\eta}_{Kj}^{(\cdot)}(x)/K_M(x)$. Because the period of rotation of the gear in roll-distance x is $N\Delta$, the contribution of this single tooth j to $\zeta^{(\cdot)}(x)$ can be expressed for all x as

$$K_M^{-1}(x) \sum_{\ell=-\infty}^{\infty} \tilde{\eta}_{Kj}^{(\cdot)}(x - \ell N\Delta)$$

which is periodic with period $N\Delta$. Thus, the transmission-error contribution $\zeta^{(\cdot)}(x)$ from all N teeth $j = 0, 1, \ldots, N-1$ is

$$\zeta^{(\cdot)}(x) = K_M^{-1}(x) \sum_{j=0}^{N-1} \sum_{\ell=-\infty}^{\infty} \tilde{\eta}_{Kj}^{(\cdot)}(x - \ell N\Delta). \tag{7.34}$$

But, as illustrated by Figure 2.6, the individual working-surfaces are nominally equally spaced in x, where the spacing in x is the base pitch Δ. The transformation described by Equations (7.30a,b) can be used to re-locate the individual tooth-deviation terms $\tilde{\eta}_{Kj}^{(\cdot)}(x - \ell N\Delta)$ in Equation (7.34), in roll-distance, to the centers of the individual working-surfaces (Figure 3.4). Using Equations (7.30a,b) to accomplish this re-location, and recognizing that the (reciprocal) total mesh stiffness $K_M^{-1}(x)$ is periodic in x with period Δ, as indicated by Equations (7.14) and (7.26a,b), we have, from Equations (7.30a,b) and (7.34),

$$\zeta^{(\cdot)}(x) = \sum_{j=0}^{N-1} \sum_{\ell=-\infty}^{\infty} K_M^{-1}(x - j\Delta - \ell N\Delta)\eta_{\sim Kj}^{(\cdot)}(x - j\Delta - \ell N\Delta), \tag{7.35}$$

which has put Equation (7.34) into manageable form.

The summation over index l in Equation (7.35) is in the form of a "rep" function, Appendix 7.C, Equation (7.C.23), with repetition interval of $N\Delta$, the rotation period of the gear, the repeating function being $K_M^{-1}(x - j\Delta)\eta_{\sim Kj}^{(\cdot)}(x - j\Delta)$. Using the notation of Equation (7.C.23), Equation (7.35) can be further expressed as

$$\zeta^{(\cdot)}(x) = \sum_{j=0}^{N-1} \mathrm{rep}_{N\Delta} \left[K_M^{-1}(x - j\Delta)\eta_{\sim Kj}^{(\cdot)}(x - j\Delta) \right]. \tag{7.36}$$

Equation (7.36) expresses the transmission-error contribution $\zeta^{(\cdot)}(x)$ as a summation of contributions from the individual teeth $j = 0, 1, \ldots, N-1$. Each such individual tooth contribution repeats in x with a repetition interval of $N\Delta$, the base-circle circumference. Because Equation (7.36) expresses the transmission error $\zeta^{(\cdot)}(x)$ as a summation of contributions from the individual teeth, $j = 0, 1, \ldots, N-1$, we can form the Fourier series coefficients of the individual terms $\mathrm{rep}_{N\Delta}[\ldots]$ in Equation (7.36), and then sum these coefficients to obtain the Fourier series coefficients of $\zeta^{(\cdot)}(x)$. We require some additional notation to carry this out. Define the Fourier transform of a generic function $f(x)$ as

$$\mathcal{F}_x[f(x); g] \triangleq \int_{-\infty}^{\infty} f(x) \exp\left(-i2\pi gx\right) dx. \tag{7.37}$$

The individual terms in the summation over j in Equation (7.36) are rep functions with repetition interval of $N\Delta$. Using the notation of Equation (7.37), it follows from

Equations (7.C.23–7.C.27) that, with the repetition interval of the rep function in Equation (7.36) being $N\Delta$, the Fourier series coefficients α_n of $\zeta^{(\cdot)}(x)$, Equation (7.36), can be expressed as

$$\alpha_n^{(\cdot)} = \sum_{j=0}^{N-1} \frac{1}{N\Delta} \mathcal{F}_x \left[K_M^{-1}(x - j\Delta)\eta_{\underset{\sim}{Kj}}^{(\cdot)}(x - j\Delta); \frac{n}{N\Delta} \right], \tag{7.38}$$

where, as indicated by Equation (7.37), the Fourier transform in Equation (7.38) is to be taken with respect to the variable x.

Again, using the notation of Equation (7.37), for a generic function $f(x)$, substituting $s = x - j\Delta$, we have

$$\mathcal{F}_x \left[f(x - j\Delta); g \right] = \int_{-\infty}^{\infty} f(x - j\Delta) exp \left(-i2\pi gx \right) dx$$

$$= exp \left(-i2\pi gj\Delta \right) \int_{-\infty}^{\infty} f(s) \exp \left(-i2\pi gs \right) ds$$

$$= exp \left(-i2\pi gj\Delta \right) \mathcal{F}_s \left[f(s); g \right]. \tag{7.39}$$

Applying the generic result of Equation (7.39), with

$$f(x - j\Delta) = K_M^{-1}(x - j\Delta)\eta_{\underset{\sim}{Kj}}^{(\cdot)}(x - j\Delta), \tag{7.40}$$

to the Fourier transform within Equation (7.38), there follows directly with $g = n/(N\Delta)$,

$$\alpha_n^{(\cdot)} = \frac{1}{N} \sum_{j=0}^{N-1} \frac{1}{\Delta} \mathcal{F}_s \left[K_M^{-1}(s)\eta_{\underset{\sim}{Kj}}^{(\cdot)}(s); \frac{n}{N\Delta} \right] \exp(-i2\pi nj/N). \tag{7.41}$$

Let us now use the caret symbol to denote the Fourier transform of a generic function $f(s)$,

$$\hat{f}(g) \equiv \mathcal{F}_s \left[f(s); g \right] \equiv \int_{-\infty}^{\infty} f(s) \exp \left(-i2\pi gs \right) ds. \tag{7.42}$$

Applying the caret symbol to the Fourier transforms of $K_M^{-1}(s)$ and $\eta_{\underset{\sim}{Kj}}^{(\cdot)}(s)$ in Equation (7.41), there follows from the convolution theorem for Fourier transforms, Equation (7.C.14c),

$$\mathcal{F}_s \left[K_M^{-1}(s)\eta_{\underset{\sim}{Kj}}^{(\cdot)}(s); g \right] = \int_{-\infty}^{\infty} \hat{K}_M^{-1}(v)\hat{\eta}_{\underset{\sim}{Kj}}^{(\cdot)}(g - v)dv, \tag{7.43}$$

which expresses the Fourier transform of the product of $K_M^{-1}(s)$ and $\underset{\sim}{\eta}_{Kj}^{(\cdot)}(s)$ as the convolution of their Fourier transforms. But the mesh stiffness $K_M(s)$, and its reciprocal $K_M^{-1}(s)$ are periodic, with period Δ, the base pitch. Applying Equation (7.C.7), the Fourier transform $\hat{K}_M^{-1}(v)$ of the reciprocal mesh stiffness $K_M^{-1}(s)$ can be expressed in terms of its complex Fourier-series coefficients,

$$
\alpha_{(1/K)p} \triangleq \frac{1}{\Delta} \int_{-\Delta/2}^{\Delta/2} K_M^{-1}(s) \exp\left(-i2\pi ps/\Delta\right) ds, \tag{7.44}
$$

as

$$
\hat{K}_M^{-1}(v) = \sum_{p=-\infty}^{\infty} \alpha_{(1/K)p} \delta(v - p/\Delta) \tag{7.45}
$$

where $\delta(\cdot)$ is the Dirac delta function, Equations (7.C.6a, b). Substituting Equation (7.45) into Equation (7.43), there follows

$$
\mathcal{F}_s\left[K_M^{-1}(s)\underset{\sim}{\eta}_{Kj}^{(\cdot)}(s); g\right] = \int_{-\infty}^{\infty} \sum_{p=-\infty}^{\infty} \alpha_{(1/K)p} \delta(v - p/\Delta) \hat{\underset{\sim}{\eta}}_{Kj}^{(\cdot)}(g - v) dv
$$

$$
= \sum_{p=-\infty}^{\infty} \alpha_{(1/K)p} \hat{\underset{\sim}{\eta}}_{Kj}^{(\cdot)}\left(g - \frac{p}{\Delta}\right), \tag{7.46}
$$

where the property, Equation (7.C.6c), of the Dirac delta function has been used. Furthermore, substituting Equation (7.46) into Equation (7.41) with $g = n/(N\Delta)$, there follows directly

$$
\alpha_n^{(\cdot)} = \frac{1}{N} \sum_{j=0}^{N-1} \left[\frac{1}{\Delta} \sum_{p=-\infty}^{\infty} \alpha_{(1/K)p} \hat{\underset{\sim}{\eta}}_{Kj}^{(\cdot)}\left(\frac{n - Np}{N\Delta}\right)\right] \exp\left(-i2\pi nj/N\right). \tag{7.47}
$$

If we define

$$
F_n(j) \triangleq \frac{1}{\Delta} \sum_{p=-\infty}^{\infty} \alpha_{(1/K)p} \hat{\underset{\sim}{\eta}}_{Kj}^{(\cdot)}\left(\frac{n - Np}{N\Delta}\right), \tag{7.48}
$$

then Equation (7.47) can be expressed as

$$
\alpha_n^{(\cdot)} = \frac{1}{N} \sum_{j=0}^{N-1} F_n(j) \exp\left(-i2\pi nj/N\right), \tag{7.49}
$$

which is the form of a discrete Fourier transform, Equation (4.3), except that $F_n(j)$ is a function of harmonic number n as well as tooth number j.

Introduction of Working-Surface-Deviation Generic Expansion Functions

Let us suppose now that the stiffness-weighted working-surface-deviations of a generic tooth j, expressed as a function of the Cartesian y, z coordinate system as in Equation (7.28), can be represented by a linear superposition of generic (real) expansion functions $\psi_{KCm}(y, z)$, where index m denotes the individual expansion functions, that is,

$$\eta_{KCj}(y, z) = \sum_m c_{j,m} \psi_{KCm}(y, z),\tag{7.50}$$

where subscripts K and C denote stiffness weighting and the Cartesian coordinate system, as before, and where the same set of expansion functions $\psi_{KCm}(y, z)$ is used for every tooth $j = 0, 1, \ldots, N-1$. Then, carrying out the integration over the line of contact as in Equation (7.30), with $z = \beta s + \gamma y$ as in Equation (7.19), there follows from Equations (7.30a, b) and (7.50),

$$\underset{\sim}{\eta}_{Kj}^{(\cdot)}(s) = \sum_m c_{j,m} \sec \psi_b \int_{y_A(s)}^{y_B(s)} \psi_{KCm}(y, \beta s + \gamma y)dy.\tag{7.51}$$

Again, using the notational convention of Equations (7.24b, c) with $f_{Cj} = \psi_{KCm}$, we define

$$\underset{\sim}{\psi}_{Km}(s) \triangleq \sec \psi_b \int_{y_A(s)}^{y_B(s)} \psi_{KCm}(y, \beta s + \gamma y)dy,\tag{7.52}$$

which, when substituted into Equation (7.51), yields

$$\underset{\sim}{\eta}_{Kj}^{(\cdot)}(s) = \sum_m c_{j,m} \underset{\sim}{\psi}_{Km}(s).\tag{7.53}$$

As in Equation (7.42), define the Fourier transform of the line integral of the expansion function, Equation (7.52), as

$$\underset{\sim}{\hat{\psi}}_{Km}(g) \triangleq \int_{-\infty}^{\infty} \underset{\sim}{\psi}_{Km}(s) \exp(-i2\pi gs)ds.\tag{7.54}$$

Then, again as in Equation (7.42), the Fourier transform $\hat{\eta}_{Kj}^{(\cdot)}(g)$ of $\underset{\sim}{\eta}_{Kj}^{(\cdot)}(s)$, Equation (7.53), can be expressed using Equation (7.54) as

$$\underset{\sim}{\hat{\eta}}_{Kj}^{(\cdot)}(g) = \sum_m c_{j,m} \underset{\sim}{\hat{\psi}}_{Km}(g).\tag{7.55}$$

Let us now define for each of the expansion functions m,

$$\phi'_m(s) \triangleq \Delta^{-1} K_M^{-1}(s)\underset{\sim}{\psi}_{Km}(s),\tag{7.56}$$

where $K_M^{-1}(s)$ is the reciprocal mesh stiffness and $\psi_{Km}(s)$ is the line integral, Equation (7.52), of the expansion function $\psi_{KCm}(y,z)$ in Equation (7.50). The Fourier transform $\hat{\phi}'_m(g)$ of Equation (7.56) is expressed, using the notation of Equation (7.42), as

$$\hat{\phi}'_m(g) = \Delta^{-1}\mathcal{F}_s\left[K_M^{-1}(s)\psi_{Km}(s); g\right] \tag{7.57a}$$

$$= \frac{1}{\Delta}\sum_{p=-\infty}^{\infty}\alpha_{(1/K)p}\hat{\psi}_{Km}\left(g - \frac{p}{\Delta}\right), \tag{7.57b}$$

where we again have used the convolution theorem for Fourier transforms, as used in Equation (7.46), with the definitions of Equations (7.44) and (7.54). Setting $g = (n - Np)/(N\Delta)$ in Equation (7.55), and substituting the result into Equation (7.48) yields

$$F_n(j) = \frac{1}{\Delta}\sum_{p=-\infty}^{\infty}\alpha_{(1/K)p}\sum_m c_{j,m}\hat{\psi}_{Km}\left(\frac{n - Np}{N\Delta}\right)$$

$$= \sum_m c_{j,m}\frac{1}{\Delta}\sum_{p=-\infty}^{\infty}\alpha_{(1/K)p}\hat{\psi}_{Km}\left(\frac{n}{N\Delta} - \frac{p}{\Delta}\right)$$

$$= \sum_m c_{j,m}\hat{\phi}'_m\left(\frac{n}{N\Delta}\right), \tag{7.58}$$

where Equation (7.57) was used in the last line.

A prime has been added to $\phi'_m(s)$ and $\hat{\phi}'_m(g)$ in Equations (7.56) and (7.57) because, in the present derivation, $\hat{\phi}'_m(g)$ reduces to $\hat{\phi}'_m[n/(N\Delta)]$ in Equation (7.58). However, it is later shown in Equation (7.94) that the mesh-attenuation functions $\hat{\phi}'_{k\ell}[n/(N\Delta)]$ are, in fact, independent of base pitch Δ, as was indicated earlier, for example, in Equations (5.16) and (6.3). To distinguish this initial apparent dependence of $\hat{\phi}'_m[n/(N\Delta)]$ on Δ from its final independence of Δ, a prime has been added on $\phi'_m(s)$ and $\hat{\phi}'_m(g)$, as indicated above.

Finally, defining

$$B_m(n) \triangleq \frac{1}{N}\sum_{j=0}^{N-1}c_{j,m}\exp\left(-i2\pi nj/N\right), \tag{7.59}$$

we have by substituting Equation (7.58) into Equation (7.49),

$$\alpha_n^{(\cdot)} = \frac{1}{N}\sum_{j=0}^{N-1}\sum_m c_{j,m}\hat{\phi}'_m\left(\frac{n}{N\Delta}\right)\exp\left(-i2\pi nj/N\right)$$

$$= \sum_m\left[\frac{1}{N}\sum_{j=0}^{N-1}c_{j,m}\exp\left(-i2\pi nj/N\right)\right]\hat{\phi}'_m\left(\frac{n}{N\Delta}\right)$$

$$= \sum_m B_m(n)\hat{\phi}'_m\left(\frac{n}{N\Delta}\right), \tag{7.60}$$

which is our final general expression for the complex Fourier series coefficients $\alpha_n^{(\cdot)}$, Equation (7.33), of the transmission-error contributions from the tooth-working-surface-deviations of a single gear $(\cdot)$ described by Equation (7.22).

Equation (7.60) expresses the complex Fourier series coefficients, Equation (7.33), of the transmission-error contributions, Equation (7.22), of the working-surface-deviations of gear $(\cdot)$ as a superposition of contributions arising from the individual expansion function $\psi_{KCm}(y,z)$ used to represent the working-surface-deviations on the individual teeth $j=0, 1, \ldots, N-1$. The first term $B_m(n)$ in Equation (7.60) is the discrete Fourier transform, Equation (7.59), of the expansion coefficients $c_{j,m}$ taken with respect to tooth number j, and the second term $\hat{\phi}'_m\left(\frac{n}{N\Delta}\right)$ is the "mesh-attenuation function", Equation (7.57), determined by the expansion function $\psi_{KCm}(y,z)$ in Equation (7.50) and the reciprocal mesh stiffness, but is independent of tooth number j. This formulation allows the rotational-harmonic spectrum contribution to be computed for any specific deviation form $\psi_{KCm}(y,z)$ that is common to all teeth, but with different tooth "strengths" $c_{j,m}$ as indicated by Equation (7.50).

More importantly, by choosing the expansion functions $\psi_{KCm}(y,z)$ in Equation (7.50) to be a "complete set", for example, the two-dimensional normalized Legendre polynomials of Equation (3.18), the complex Fourier series coefficients of the transmission-error contributions arising from working-surface-deviations of any form, that may differ in an arbitrary manner from one tooth to another, can be computed by Equation (7.60).

It is important to distinguish the method of derivation of Equation (7.60) from the method used to arrive at Equation (5.16). In arriving at Equation (5.16), a heuristic physical argument was used to motivate use of the DFT, Equations (4.3) and (4.5), leading to the DFT, Equation (4.21) of the Legendre expansion coefficients. In contrast to this heuristic physical argument, the derivation of Equation (7.60), from first principles, *required* the DFT of the expansion coefficients, Equation (7.59). It then became apparent that this required use of the DFT was caused by the equispaced nature of tooth-working-surfaces, which enabled us to use the DFT, Equations (4.3) and (4.5), as the starting point of the rotational-harmonic analysis in Chapter 4, leading to the DFT, Equation (4.21) of the Legendre expansion coefficients, and the resultant tooth-deviation spectrum, Equation (4.30). But the derivation of Equation (7.60) showed that the DFT, Equation (7.59), is *required*, it was not heuristically arrived at. Moreover, this derivation ultimately yields the functional forms of the mesh-attenuation functions.

Evaluation of Generic Form of 'Mesh-Attenuation Functions' $\hat{\phi}'_m(g)$

Evaluation of the mesh-attenuation function $\hat{\phi}'_m(g)$, Equation (7.57), requires the Fourier transform $\hat{\psi}_{Km}(g)$, Equation (7.54), of $\psi_{Km}(s)$, which is the line-integral of the generic expansion function $\psi_{KCm}(y,z)$ over the line of tooth contact described by Equation (7.52), $z = \beta s + \gamma y$ as in Equation (7.19). The Fourier transform of the line-integral, Equations (7.24b, c), of a generic function $f_{Cj}(y, \beta s + \gamma y)$ over the line of tooth contact is carried out in Appendix 7.D. Again using the general notion of Equation (7.42), when the Fourier transform $\hat{f}_j(g)$ of $f_{\sim j}(s)$, Equation (7.24c), is applied to the Fourier transform of $\psi_{\sim Km}(s)$ in Equation (7.57), we obtain from Equation (7.D.35),

$$\hat{\psi}_{\sim Km}(g) = (L/D) \sec \psi_b \hat{\psi}_{KCm}\left[(L/A)g, (L/D)g\right], \tag{7.61}$$

where

$$\hat{\psi}_{KCm}(g_1, g_2) \triangleq \int\limits_{-\infty}^{\infty} \int\limits_{-\infty}^{\infty} \psi_{KCm}(y, z) \exp\left[-i2\pi(g_1 y + g_2 z)\right] dy dz \tag{7.62}$$

is the two-dimensional Fourier transform of the generic expansion function $\psi_{KCm}(y, z)$ in Equation (7.50). The parameters in Equation (7.61) are related to basic gear parameters by

$$\frac{L}{D} = \csc \phi, \quad \frac{L}{A} = \tan \psi_b, \tag{7.63a, b}$$

as shown by Equations (7.B.2) and (7.B.3), where ϕ and ψ_b are non-negative. The general results of Equations (7.61–7.63) permit evaluations of the mesh-attenuation function $\hat{\phi}_m(g)$ for any expansion function $\psi_{KCm}(y, z)$ used in Equation (7.50), once the complex Fourier series coefficients $\alpha_{(1/K)p}$, Equation (7.44), of the reciprocal mesh stiffness $K_M^{-1}(s)$ are obtained. Approximate and accurate methods for computing the complex Fourier series coefficients of the reciprocal mesh stiffness $K_M^{-1}(s)$ are derived in Sections 7.6 and 7.7 respectively.

Application to Gear-Health Monitoring

Fixed vibration transducers often are used to monitor gear systems in order to accomplish early detection of failing gears. When one or a few teeth begin to fail, the working-surfaces of the failing teeth are changed, for example, Mark, Reagor, and McPherson (2007), which becomes a source of changes in the vibration excitation (Mark et al., 2010). The methodology described by Equations (7.50), (7.57), (7.59), (7.60), and (7.61) provides a method to compute changes in the complex Fourier-series coefficients $\alpha_n^{(\cdot)}$ caused by any changes $c_{j,m} \psi_{KCm}(y, z)$ in the working-surfaces of the damaged teeth, specified by the teeth with non-zero indices $c_{j,m}$ in Equation (7.50). This capability enables one to compute in the frequency domain, by Equation (7.60), and in the time-domain, by Equations (7.60) and (7.32), changes in the vibration excitation caused by changes $c_{j,m} \psi_{KCm}(y, z)$ in the working-surfaces caused by tooth damage, for example, Mark and Reagor (2007). Normally, in such applications, only one or a few teeth would contribute non-zero contributions $c_{j,m} \psi_{KCm}(y, z)$, with resultant sparse contributions in forming the DFT in Equation (7.59).

Results for "Complete Set" of Generic Expansion Functions on Rectangular Contact Regions

Although, as described in the earlier chapters, the normalized two-dimensional Legendre polynomials of Equations (3.13) and (3.14) were chosen as the likely best general method for representing the tooth-working-surface-deviations, as in Equation (3.18), the results derived in the present chapter, up to this juncture, are applicable to any representations, Equation (7.50), of the tooth-working-surface-deviations $\eta_{KCj}(y, z)$. The only fundamental requirement for a general representation method is that the functions $\psi_{KCm}(y, z)$ constitute a "complete set", for example, Lanczos (1956, pp. 358–362), thereby allowing accurate representation of the stiffness weighted working-surface-deviations $\eta_{KCj}(y, z)$ of any form if a sufficient number of terms is used. In particular, there has been no explicit assumption that the tooth contact region on the working-surfaces is rectangular in the coordinates y, z. To proceed further, we shall assume a rectangular contact region.

Because there is a general requirement for representation of the integrands of each of Equations (7.10) and (7.11), we shall continue to use the general notation conventions of Equations (7.23) and (7.24), with the transformation $z = \beta s + \gamma y$ as in Equations (7.19–7.21). The requirement is a representation of the generic function $f_{Cj}(y, z)$ over the rectangular region $(-F/2) \leq y \leq (F/2)$, $(-D/2) \leq z \leq (D/2)$ of the working-surface illustrated in Figure 3.4. Letting the single index m, as in Equation (7.50), denote a pair of individual indices k, l, an expansion of the generic function $f_{Cj}(y, z)$ can be represented as

$$f_{Cj}(y, z) = \sum_{k=0}^{\infty} \sum_{\ell=0}^{\infty} c_{j,k\ell} \psi_{yk}(y) \psi_{z\ell}(z) \tag{7.64}$$

where the sets of expansion functions,

$$\psi_{yk}(y), \quad (-F/2) \leq y \leq (F/2), \quad k = 0, 1, 2, \ldots, \infty \tag{7.65}$$

$$\psi_{z\ell}(z), \quad (-D/2) \leq z \leq (D/2), \quad \ell = 0, 1, 2, \ldots, \infty \tag{7.66}$$

each is complete (Lanczos, 1956, pp. 358–362), orthogonal, and normalized, thereby satisfying

$$\frac{1}{F} \int_{-F/2}^{F/2} \psi_{yk}(y) \psi_{yk'}(y) dy = \begin{cases} 1, & k' = k \\ 0, & k' \neq k \end{cases} \tag{7.67}$$

$$\frac{1}{D} \int_{-D/2}^{D/2} \psi_{z\ell}(z) \psi_{z\ell'}(z) dz = \begin{cases} 1, & \ell' = \ell \\ 0, & \ell' \neq \ell \end{cases} \tag{7.68}$$

which are the same normalizations satisfied by the normalized Legendre polynomials of Equations (3.13) and (3.14), as shown in Equations (3.16) and (3.17). This is *not* the usual normalization used to describe "orthonormal" functions; hence, we have avoided use of the term "orthonormal." Exactly as shown by Equations (3.18–3.20), the expansion coefficients are obtained, using the normalized orthogonal properties, Equations (7.67), (7.68), as

$$c_{j,k\ell} = \frac{1}{FD} \int\limits_{-D/2}^{D/2} \int\limits_{-F/2}^{F/2} f_{Cj}(y,z)\psi_{yk}(y)\psi_{z\ell}(z)dydz. \tag{7.69}$$

As shown in the case of the normalized Legendre polynomials by Equations (3.21–3.23), the particular normalization of Equations (7.67) and (7.68) yields for the mean-square value of $f_{Cj}(y,z)$, averaged over the rectangular region of area FD,

$$\frac{1}{FD} \int\limits_{-D/2}^{D/2} \int\limits_{-F/2}^{F/2} f_{Cj}^2(y,z)dydz = \sum_{k=0}^{\infty} \sum_{\ell=0}^{\infty} c_{j,k\ell}^2, \tag{7.70}$$

which allows a direct meaningful interpretation of the squares of the expansion coefficients $c_{j,k\ell}^2$ as the contributions of the individual terms $k\ell$ to the mean-square value of the function $f_{Cj}(y,z)$ being represented.

We shall require the two-dimensional Fourier transform of the generic functions $f_{Cj}(y,z)$,

$$\hat{f}_{Cj}(g_1,g_2) \triangleq \int\limits_{-D/2}^{D/2} \int\limits_{-F/2}^{F/2} f_{Cj}(y,z)\exp\left[-i2\pi\left(g_1y + g_2z\right)\right]dydz. \tag{7.71}$$

The Fourier transforms of the individual expansion functions in Equation (7.64) are

$$\hat{\psi}_{yk}(g_1) \triangleq \int\limits_{-F/2}^{F/2} \psi_{yk}(y)\exp(-i2\pi g_1 y)dy \tag{7.72}$$

and

$$\hat{\psi}_{z\ell}(g_2) \triangleq \int\limits_{-D/2}^{D/2} \psi_{z\ell}(z)\exp(-i2\pi g_2 z)dz. \tag{7.73}$$

Forming the Fourier transforms of the individual terms in Equation (7.64), using Equations (7.72) and (7.73), there follows from Equation (7.71),

$$\hat{f}_{Cj}(g_1,g_2) = \sum_{k=0}^{\infty} \sum_{\ell=0}^{\infty} c_{j,k\ell}\hat{\psi}_{yk}(g_1)\hat{\psi}_{z\ell}(g_2). \tag{7.74}$$

Results Using Two-Dimensional Normalized Legendre Polynomials

The two-dimensional normalized Legendre polynomials of Equations (3.13) and (3.14) are a complete orthogonal set satisfying Equations (7.64–7.70). An especially useful property of the normalized Legendre polynomials is the fact that their Fourier transforms, Equations (7.72) and (7.73), can be expressed, very simply, in terms of known functions, spherical Bessel functions of the first kind, $j_n(\xi)$ defined by Equation (6.10),

$$j_n(\xi) = [\pi/(2\xi)]^{1/2} J_{n+1/2}(\xi), \tag{7.75}$$

where $J_{n+1/2}(\xi)$ is the Bessel function of the first kind of order $n + \frac{1}{2}$.

Consider the Fourier transform of $\psi_{yk}(y)$ given by Equation (7.72). Substituting $\xi = 2y/F$ into Equation (3.13), and the result into Equation (7.72) with $y = (F/2)\xi$, and therefore, $dy = (F/2)d\xi$, there follows from Equation (7.72),

$$\hat{\psi}_{yk}(g_1) = (2k + 1)^{1/2}(F/2) \int_{-1}^{1} P_k(\xi) \exp(-i\pi F g_1 \xi) d\xi. \tag{7.76}$$

Bateman (1954, p. 122), defines the Fourier transform of $f(\xi)$ as

$$g(y) = \int_{-\infty}^{\infty} f(\xi) \exp(-i\xi y) d\xi, \tag{7.77}$$

which when applied to the Legendre polynomial $f(\xi) = P_n(\xi)$ of degree n, gives according to formula 3.3(1) on p. 122 of the above-cited reference,

$$g(y) = (-1)^n i^n (2\pi)^{1/2} y^{-1/2} J_{n+1/2}(y). \tag{7.78}$$

Therefore, with $n = k$ and $y = \pi F g_1$, applying Equations (7.77) and (7.78) to the integration in Equation (7.76), we have for $\hat{\psi}_{yk}(g_1)$,

$$\hat{\psi}_{yk}(g_1) = (2k + 1)^{1/2} \frac{F}{2} (-1)^k i^k (2\pi)^{1/2} (\pi F g_1)^{-1/2} J_{k+\frac{1}{2}}(\pi F g_1) \tag{7.79a}$$

$$= i^k (-1)^k (F/2)^{1/2} g_1^{-1/2} (2k + 1)^{1/2} J_{k+\frac{1}{2}}(\pi F g_1). \tag{7.79b}$$

But from Equation (7.75) with $n = k$ and $\xi = \pi F g_1$,

$$J_{k+\frac{1}{2}}(\pi F g_1) = (2F g_1)^{1/2} j_k(\pi F g_1). \tag{7.80}$$

Substituting Equation (7.80) into Equation (7.79b), and simplifying, gives

$$\hat{\psi}_{yk}(g_1) = (-i)^k F(2k + 1)^{1/2} j_k(\pi F g_1). \tag{7.81}$$

In exactly the same manner, we obtain for $\hat{\psi}_{z\ell}(g_2)$,

$$\hat{\psi}_{z\ell}(g_2) = (-i)^\ell D(2\ell + 1)^{1/2} j_\ell(\pi D g_2). \tag{7.82}$$

Equations (7.81) and (7.82), when combined with Equation (7.74), allow representation of the two-dimensional Fourier transform $\hat{f}_{Cj}(g_1, g_2)$ of the generic function $f_{Cj}(y, z)$, Equation (7.64), as

$$\hat{f}_{Cj}(g_1, g_2) = FD \sum_{k=0}^{\infty} \sum_{\ell=0}^{\infty} c_{j,k\ell}(-i)^{k+\ell} \left[(2k+1)(2\ell+1) \right]^{1/2}$$
$$\times j_k(\pi F g_1) j_\ell(\pi D g_2). \tag{7.83}$$

7.4 Fourier-Series Using Legendre Representation of Working-Surface-Deviations

The above-described results, beginning with Equation (7.50), now will be specialized to representation of the locally-stiffness-weighted working-surface-deviations, Equation (7.28), using two-dimensional normalized Legendre polynomials (Mark, 1979).

Legendre Representation of Locally-Stiffness-Weighted Working-Surface-Deviations

According to Equations (7.28) and (7.50), we require a method for representing the locally-stiffness-weighted tooth deviations $\eta_{KCj}(y, z)$. Define the average value of the local tooth-pair stiffness $K_{TC}(y, z)$ per unit length of line of contact, averaged over the rectangular contact region of area FD, by

$$\overline{K}_{TC} \triangleq \frac{1}{FD} \int_{-D/2}^{D/2} \int_{-F/2}^{F/2} K_{TC}(y, z) dy dz. \tag{7.84}$$

Let the single index m in Equation (7.50) represent the pair of indices $k\ell$ of the normalized Legendre polynomials of Equations (3.13) and (3.14), and let the expansion functions $\psi_{KCm}(y, z)$ in Equation (7.50) take on the form

$$\psi_{KCk\ell}(y, z) = \overline{K}_{TC} \psi_{yk}(y) \psi_{z\ell}(z). \tag{7.85}$$

If we identify $K_{TC}(y, z)\eta_{Cj}(y, z)/\overline{K}_{TC}$ with $f_{Cj}(y, z)$ in Equation (7.64), there follows from Equation (7.69),

$$c_{j,k\ell} = \frac{1}{FD} \int_{-D/2}^{D/2} \int_{-F/2}^{F/2} \frac{K_{TC}(y, z)\, \eta_{Cj}(y, z)}{\overline{K}_{TC}} \psi_{yk}(y)\, \psi_{z\ell}(z)\, dy dz, \tag{7.86}$$

which yields the expansion coefficients $c_{j,m} = c_{j,k\ell}$. Then, from Equation (7.64), the locally-stiffness-weighted working-surface-deviations can be expressed as

$$\overline{K}_{TC}(y,z)\,\eta_{Cj}(y,z) = \overline{K}_{TC}\sum_{k=0}^{\infty}\sum_{\ell=0}^{\infty}c_{j,k\ell}\psi_{yk}(y)\,\psi_{z\ell}(z) \qquad (7.87)$$

since $\overline{K}_{TC}$ is a constant. But in the important case where the local tooth-pair stiffness is assumed to be a constant value $\overline{K}_{TC}$, that is, $K_{TC}(y,z) = \overline{K}_{TC}$, the expansion described by Equations (7.87) and (7.86) reduces, automatically, to the expansion of the unweighted working-surface-deviations $\eta_{Cj}(y,z)$ described by Equations (3.18) and (3.20):

$$\eta_{Cj}(y,z) = \sum_{k=0}^{\infty}\sum_{\ell=0}^{\infty}c_{j,k\ell}\psi_{yk}(y)\,\psi_{z\ell}(z) \qquad (7.88)$$

with

$$c_{j,k\ell} = \frac{1}{FD}\int_{-D/2}^{D/2}\int_{-F/2}^{F/2}\eta_{Cj}(y,z)\,\psi_{yk}(y)\,\psi_{z\ell}(z)\,dydz. \qquad (7.89)$$

Mesh-Attenuation Functions for Legendre Representation of Working-Surface-Deviations

According to Equation (7.57), to evaluate the mesh-attenuation function $\hat{\phi}'_m(g)$ for the normalized Legendre-polynomial representation, we require the function $\hat{\underset{\sim}{\psi}}_{Kk\ell}(g)$. With $k\ell$ replacing the index m, $\hat{\underset{\sim}{\psi}}_{Kk\ell}(g)$ is the Fourier transform, Equation (7.54), of the line integral $\underset{\sim}{\psi}_{Kk\ell}(s)$ defined by Equation (7.52). The general form of this Fourier transform is given by Equations (7.61) and (7.62). With $k\ell$ replacing m in Equation (7.62), there follows directly from Equations (7.62) and (7.85), and then from Equations (7.72), (7.73), (7.81), and (7.82),

$$\hat{\psi}_{KCk\ell}(g_1,g_2) = \overline{K}_{TC}\int_{-F/2}^{F/2}\psi_{yk}(y)\exp(-i2\pi g_1 y)\,dy$$

$$\times\int_{-D/2}^{D/2}\psi_{z\ell}(z)\exp(-i2\pi g_2 z)\,dz \qquad (7.90a)$$

$$= \overline{K}_{TC}FD(-i)^{k+\ell}\left[(2k+1)(2\ell+1)\right]^{1/2}$$

$$\times j_k(\pi Fg_1)j_\ell(\pi Dg_2), \qquad (7.90b)$$

where $\psi_{yk}(y)$ and $\psi_{z\ell}(z)$ are zero outside of the limits indicated in Equations (7.72) and (7.73), respectively. Then, by combining Equation (7.90b) with Equation (7.61), replacing m by $k\ell$, we have

$$\hat{\underset{\sim}{\psi}}_{Kk\ell}(g) = \overline{K}_{TC}FL\,sec\psi_b\,(-i)^{k+\ell}\,[(2k+1)\,(2\ell+1)]^{1/2}$$

$$\times j_k\,[\pi(FL/A)g]\,j_\ell\,[\pi Lg]. \tag{7.91}$$

But, as shown later by Equation (7.107), the mean total mesh stiffness $\overline{K}_M$ is related to the mean local tooth-pair stiffness per unit length of line of contact $\overline{K}_{TC}$ by

$$\overline{K}_M = \overline{K}_{TC}(FL/\Delta)\,sec\,\psi_b. \tag{7.92}$$

Furthermore, again replacing m by $k\ell$ in Equation (7.57), and g in Equation (7.91) by $g-p/\Delta$, as required by Equation (7.57), there follows from Equations (7.91) and (7.92),

$$\hat{\underset{\sim}{\psi}}_{Kk\ell}\left(g - \frac{p}{\Delta}\right) = \overline{K}_M\Delta\,(-i)^{k+\ell}\,[(2k+1)\,(2\ell+1)]^{1/2}$$

$$\times j_k\left[\pi\left(\frac{FL}{A\Delta}\right)(g\Delta - p)\right]j_\ell\left[\pi\left(\frac{L}{\Delta}\right)(g\Delta - p)\right]. \tag{7.93}$$

As shown by Equations (2.14) and (2.9), the parameters $FL/(A\Delta)$ and L/Δ are, respectively, the axial Q_a and transverse Q_t contact ratios of the rectangular region of tooth contact illustrated in Figure 2.6. Incorporating these parameter definitions into Equation (7.93), and replacing m by $k\ell$, as before, we have for the mesh-attenuation function of Equation (7.57), with $g = n/(N\Delta)$ as required by Equation (7.60),

$$\hat{\phi}'_{k\ell}\left(\frac{n}{N\Delta}\right) = (-i)^{k+\ell}\,[(2k+1)\,(2\ell+1)]^{1/2}$$

$$\times \sum_{p=-\infty}^{\infty}\overline{K}_M\alpha_{(1/K)p}j_k\left[\pi Q_a\left(\frac{n}{N} - p\right)\right]j_\ell\left[\pi Q_t\left(\frac{n}{N} - p\right)\right] \tag{7.94a}$$

$$\equiv \hat{\phi}_{k\ell}(n/N), \tag{7.94b}$$

where, as indicated earlier, $\hat{\phi}_{k\ell}(n/N)$ is independent of direct dependence on base pitch Δ.

In arriving at Equation (7.94), the contact region of the working-surfaces in the Cartesian y, z coordinate system, defined by Equations (3.3–3.7) and Figure 3.4, has been assumed to be rectangular with axial dimension F and radical dimension D, which determine the axial Q_a and transverse Q_t contact ratios in Equation (7.94). For this assumed contact region, the mesh-attenuation function described by Equation (7.94) is "exact" when the Legendre expansion coefficients of the stiffness weighted tooth deviations are evaluated by Equation (7.86). However, in the practically important case when the Legendre expansion coefficients are evaluated only from the working-surface-deviations by Equation (7.89), an additional assumption is employed of constant local tooth-pair stiffness per unit length of line of contact. The value of this local tooth-pair stiffness will be shown not to be required for evaluating

Equation (7.94) in this case, as was symbolically indicated by Equation (5.13). The approximation, Equation (6.3), to the mesh-attenuation function to $\hat{\phi}_{k\ell}(n/N)$ will be derived in Section 7.6, by using an *approximation* to $\overline{K}_M \alpha_{(1/K)p}$. Two methods for *accurate* computation of $\overline{K}_M \alpha_{(1/K)p}$, for use in Equation (7.94), will be derived in Section 7.7. One of these methods then is used in Section 7.8 to provide accurate evaluation of the mesh-attenuation functions and the Fourier series representation of the transmission-error contributions caused by the tooth-working-surface-deviations. The resultant formulas utilize only real (not complex) quantities.

Fourier Series Coefficients for Legendre Representation of Working-Surface-Deviations

Replacing m by $k\ell$ in Equation (7.60) yields the complex Fourier series coefficients of the transmission-error contributions arising from the tooth-working-surface-deviations of gear $(\cdot)$,

$$\alpha_n^{(\cdot)} = \sum_{k=0}^{\infty}\sum_{\ell=0}^{\infty} B_{k\ell}(n)\hat{\phi}_{k\ell}\left(\frac{n}{N}\right), \quad n = 0, \pm 1, \pm 2, \ldots \tag{7.95}$$

where $\hat{\phi}_{k\ell}(n/N)$ is given by Equations (7.94a, b), and with $k\ell$ replacing m, $B_{k\ell}(n)$ is given by Equation (7.59),

$$B_{k\ell}(n) = \frac{1}{N}\sum_{j=0}^{N-1} c_{j,k\ell}\exp\left(-i2\pi nj/N\right), \tag{7.96}$$

which is the DFT of the Legendre expansion coefficients, Equation (7.86), of the stiffness-weighted working-surface-deviations, or Equation (7.89) of the nonstiffness-weighted-deviations, where $\psi_{yk}(y)$ and $\psi_{z\ell}(z)$ are the normalized Legendre polynomials given by Equations (3.13) and (3.14), respectively.

Equations (7.95), (7.96), and (7.94), with Legendre coefficients evaluated by either of Equation (7.86) or (7.89), as appropriate, are our final general formulation for computation of the complex Fourier series coefficients of the transmission-error contributions arising from the working-surface-deviations of a generic helical or spur gear $(\cdot)$. Evaluation of $\overline{K}_M \alpha_{(1/K)p}$, for use in Equation (7.94) is carried out in Sections 7.5–7.7. Reduction of this formulation for computations involving only real (not complex) numbers is carried out in Section 7.8.

Physical/Mathematical Interpretation of Final Formulation

At each location y, z of the Cartesian coordinate system illustrated in Figure 3.4, the deviations of the working-surfaces of the N teeth on a gear constitute a set of N discrete values equispaced in roll-distance x with spacing Δ, the base pitch. Early in Chapter 4, it was shown that the natural counterpart to the Fourier series representation of a continuous periodic function with period $N\Delta$, Equations (4.1) and (4.2), when applied to a set of N equispaced discrete values with spacing

Δ, is the discrete finite Fourier transform (DFT), Equations (4.3) and (4.4). As a consequence of this equispaced discrete periodic behavior, the DFT counterpart to the complex Fourier series coefficients is periodic, with period in rotational harmonic n of $n = N$, the location of the tooth-meshing fundamental harmonic, as shown by Equation (4.13). Thus, the equispaced discrete nature of gear teeth is the source of the so-called sidebands located about the tooth-meshing harmonics caused by tooth-to-tooth variations in the working-surface-deviations.

When the Legendre representation, Equation (3.18), of the working-surface-deviations was introduced into the DFT of these deviations, Equation (4.5), the former DFT of the deviations, Equation (4.5), became Equation (4.20), with the result that the DFT was applied to the Legendre expansion coefficients, Equation (4.21), and the DFT of the working-surface-deviations at each location y, z was expressed by Equation (4.22).

When the gear $(\cdot)$ under investigation meshes with a mating gear, an averaging of the working-surface-deviations along all instantaneous lines of tooth contact takes place, as illustrated in Figure 2.6. These working-surface-deviations generally differ among the N teeth on the gear. The tremendous utility of the rotational harmonic representation given by Equation (4.22) is the separation of the tooth-to-tooth variability characterized by $B_{k\ell}(n)$, Equation (4.21), from the deviation pattern $\psi_{yk}(y)\,\psi_{z\ell}(z)$ that is common to all of the teeth for each pair of Legendre indices $k\ell$. Because this deviation pattern $\psi_{yk}(y)\,\psi_{z\ell}(z)$ for each $k\ell$ pair is common to all of the teeth, its attenuation that arises from the meshing action with a mating gear is unique to that deviation pattern characterized by the index pair $k\ell$. This attenuation is characterized by the "mesh-attenuation function" $\hat{\phi}_{k\ell}(n/N)$, as can be readily understood by comparing Equations (4.22) and (7.95).

The overall form of the mesh-attenuation function given by Equation (7.94) can be understood by comparing Equation (7.94) with the working-surface-deviation contribution to the transmission error given by Equation (7.22). Equation (7.22) expresses the transmission-error contribution as the product of contributions from working-surface-deviations given by the numerator and the reciprocal mesh stiffness $1/K_M(x)$, which is periodic with period Δ, the tooth-meshing period. The product given by the right-hand side of Equation (7.22) becomes a discrete convolution in the frequency domain, as shown by the summation over tooth-meshing harmonic p in Equation (7.94).

Legendre Representation Yields ''Simple'' Final Form

Apart from the discrete convolution that arises from the form of Equation (7.22), the form of $\hat{\phi}_{k\ell}(n/N)$ given by Equation (7.94) is very simple, requiring only the product of two spherical Bessel functions j_k and j_ℓ, with parameters Q_a and Q_t, the axial and transverse contact ratios of the prescribed rectangular tooth-contact region. The relative simplicity of this result arises from the very simple forms of the Fourier transforms of the normalized Legendre polynomials given by Equations (7.81) and (7.82). Other orthogonal function representations of working-surface-deviations, such as Fourier series, do not yield comparably simple Fourier transforms to those given by

Equations (7.81) and (7.82). The term $[(2k+1)(2l+1)]^{1/2}$ in Equation (7.94) arose from the normalization of the Legendre polynomials that allowed the direct interpretation of the squares of the expansion coefficients provided by Equation (3.23), and the direct interpretation of the nonattenuated rotational harmonic spectra $\left|B_{kl}(n)\right|^2$ provided by Equations (4.26), (4.27), and (4.30).

Thus, use of Legendre polynomials to represent tooth-working-surface-deviations offers three advantages over other potential representation methods: (i) As described in Section 3.4, the low-order Legendre terms very efficiently represent working-surface-deviations that arise from lack of coincidence between base-cylinder axis used in tooth generation and gear measurement or gear operational axis. (ii) When working-surface line-scanning measurements are made at the zeros of normalized Legendre polynomials, the formulas given by Equations (3.37) and (3.43) for the two-dimensional Legendre expansion coefficients *interpolate exactly* (Mark, 1983) the one-dimensional Legendre expansion coefficients generated along the individual line-scanning measurements, in addition to satisfying Gaussian quadrature implementation requirements. (iii) Use of the Legendre polynomial representation of working-surface-deviations has yielded the simple form for the mesh-attenuation functions given by Equation (7.94).

7.5 Fourier-Series Representation of Normalized Mesh Stiffness $K_M(s)/\overline{K}_M$

"Exact" evaluation of the mesh-attenuation functions $\hat{\phi}_{kl}(n/N)$, Equation (7.94), requires evaluation of the Fourier series coefficients of the reciprocal normalized mesh stiffness $\overline{K}_M \alpha_{(1/K)p}$, which are the Fourier series coefficients of $\overline{K}_M/K_M(s)$. In Section 7.6, an approximation to these coefficients is developed yielding the approximate mesh-attenuation function, Equation (6.3). In Section 7.7, two methods are developed for accurate evaluation of these coefficients from the Fourier series coefficients of the (non-reciprocal) normalized mesh stiffness $K_M(s)/\overline{K}_M$, which are determined here as a function of the local tooth-pair stiffness $K_{TC}(y,z)$. The result of this derivation then is used to establish validity of the important relationship for the mean mesh stiffness $\overline{K}_M$ given by Equation (7.92).

The total mesh stiffness $K_M(x)$, Equation (7.14), is the summation of the individual tooth-pair stiffnesses $\hat{K}_{Tj}(x)$, Equation (7.10), which are line integrations of the local tooth-pair stiffnesses per unit length of line of contact. The coordinate transformations of Equation (7.17–7.21) transform the line integrations, Equation (7.10), of the local tooth-pair stiffnesses to line-integrations of the local tooth-pair stiffnesses, Equation (7.26a), expressed in the Cartesian y, z coordinate system illustrated in Figure 3.4. When the resultant line-integrated local tooth-pair stiffness, Equation (7.26a), is expressed as a function of roll distance s, Equation (7.17), its roll-distance location is referenced to the center of the working-surface, as illustrated by the dashed line in Figure 3.2. This reference location is the same for every tooth-pair j, and therefore is independent of tooth-pair number j, as indicated by Equation (7.26b). As the individual tooth pairs move into and out of contact, as illustrated in Figure 2.6, the resultant transformation described by Equation (7.26b), when combined with Equation (7.14), allows the total mesh stiffness $K_M(x)$ to be represented as

$$K_M(x) = \sum_{j=-\infty}^{\infty} \underset{\sim}{K}_T(x - j\Delta) \equiv rep_\Delta \underset{\sim}{K}_T(x) \qquad (7.97)$$

where $rep_\Delta \underset{\sim}{K}_T(x)$ is the rep function, with repetition interval Δ, the base pitch, defined and developed in Appendix 7.C.

As shown by Equation (7.97), the mesh stiffness is periodic with repetition period Δ, the base-pitch tooth-meshing period. Using the *rep* function property of Equation (7.97), the complex Fourier series coefficients of the mesh stiffness $K_M(s)$ are shown by Equation (7.C.27) to be

$$\alpha_{Kp} \triangleq \frac{1}{\Delta} \int_{-\Delta/2}^{\Delta/2} K_M(s) \exp(-i2\pi ps/\Delta) ds \qquad (7.98a)$$

$$= (1/\Delta)\hat{\underset{\sim}{K}}_T(p/\Delta), \qquad (7.98b)$$

where $\hat{\underset{\sim}{K}}_T(p/\Delta)$, with $g = p/\Delta$, is the Fourier transform of $\underset{\sim}{K}_T(s)$, Equation (7.26),

$$\hat{\underset{\sim}{K}}_T(g) = \int_{-\infty}^{\infty} \underset{\sim}{K}_T(s) \exp(-i2\pi gs) ds, \qquad (7.99)$$

where the general notion of Equation (7.42) has been used.

Equation (7.26) is of the exact form of Equation (7.24b). Therefore, the Fourier transform $\hat{\underset{\sim}{K}}_T(g)$, Equation (7.99), of $\underset{\sim}{K}_T(s)$ can be evaluated by Equation (7.D.35) of Appendix 7.D from the two-dimensional Fourier transform of the local tooth-pair stiffness per unit length of line of contact $K_{TC}(y, z)$ expressed in the Cartesian y, z coordinate system of Figure 3.4:

$$\hat{K}_{TC}(g_1, g_2) = \int_{-D/2}^{D/2} \int_{-F/2}^{F/2} K_{TC}(y, z) \exp\left[-i2\pi(g_1 y + g_2 z)\right] dy dz. \qquad (7.100)$$

Applying Equation (7.D.35) to Equation (7.100) there follows directly for the Fourier transform, Equation (7.99),

$$\hat{\underset{\sim}{K}}_T(g) = (L/D) \sec \psi_b \hat{K}_{TC}\left[(L/A)g, (L/D)g\right], \qquad (7.101)$$

and therefore for the Fourier series coefficients, Equation (7.98),

$$\alpha_{Kp} = [L/(D\Delta)] \sec \psi_b \hat{K}_{TC}\left[pL/(A\Delta), pL/(D\Delta)\right], \qquad (7.102)$$

which follows directly from Equations (7.98) and (7.101). Using the definitions of axial Q_a and transverse Q_t contact ratios given by Equations (2.14) and (2.9), respectively, the Fourier series coefficients α_{Kp}, Equation (7.102), of the total mesh stiffness $K_M(s)$ can be expressed as

$$\alpha_{Kp} = (Q_t/D) \sec \psi_b \hat{K}_{TC}\left(pQ_a/F, pQ_t/D\right), \qquad (7.103)$$

where $\hat{K}_{TC}(g_1, g_2)$ is the two-dimensional Fourier transform, Equation (7.100), of the local tooth-pair stiffness per unit length of line of contact, and F and D determine the rectangular contact region illustrated in Figure 3.4.

As can be seen from Equation (7.33) with $n=0$, the average value of a periodic function is given by its $n=0$ complex Fourier series coefficient. Therefore, from Equation (7.98), the mean total mesh stiffness is

$$\overline{K}_M = (1/\Delta) \, \hat{\underset{\sim}{K}}_T(0). \qquad (7.104)$$

But, from Equation (7.101),

$$\hat{\underset{\sim}{K}}_T(0) = (L/D) \sec \psi_b \hat{K}_{TC}(0,0), \qquad (7.105)$$

and from Equation (7.100), the mean tooth-pair stiffness per unit length of line of contact is related to $\hat{K}_{TC}(0,0)$ by

$$\hat{K}_{TC}(0,0) = FD\overline{K}_{TC}, \qquad (7.106)$$

where $\overline{K}_{TC}$ is the average value of $K_{TC}(y,z)$ taken over the rectangular region of area FD. Inserting Equation (7.106) into Equation (7.105) relates the mean total mesh stiffness $\overline{K}_M$, Equation (7.104), to the mean tooth-pair stiffness per unit length of line of contact $\overline{K}_{TC}$ by

$$\overline{K}_M = \overline{K}_{TC}(FL/\Delta) \sec \psi_b, \qquad (7.107)$$

as pointed out earlier by Equation (7.92).

To obtain an expression for the complex Fourier series coefficients of the normalized total mesh stiffness $K_M(s)/\overline{K}_M$, we divide Equation (7.103) by Equation (7.104) after inserting Equation (7.105), yielding, with $Q_t = L/\Delta$, Equation (2.9),

$$\frac{\alpha_{Kp}}{\overline{K}_M} = \frac{\hat{K}_{TC}\left(pQ_a/F, pQ_t/D\right)}{\hat{K}_{TC}(0,0)}, \qquad (7.108)$$

where $\hat{K}_{TC}(g_1, g_2)$ is given by Equation (7.100). Equation (7.108) expresses the complex Fourier series coefficients of the normalized total mesh stiffness $K_M(s)/\overline{K}_M$ as a function of the two-dimensional Fourier transform, Equation (7.100), of the local tooth-pair stiffness $K_{TC}(y,z)$ per unit length of line of contact.

Legendre Representation of Local Tooth-Pair Stiffness $K_{TC}(y, z)$

A useful representation of the local tooth-pair stiffness per unit length of line of contact $K_{TC}(y, z)$ can be obtained by using the generic two-dimensional Legendre representation, Equation (7.64), where $\psi_{yk}(y)$ and $\psi_{z\ell}(z)$ are the normalized Legendre polynomials, Equations (3.13) and (3.14):

$$K_{TC}(y, z) = \sum_{k=0}^{\infty} \sum_{\ell=0}^{\infty} a_{k\ell} \psi_{yk}(y) \psi_{z\ell}(z). \qquad (7.109)$$

The expansion coefficients $a_{k\ell}$ are independent of tooth-pair number j, and according to Equation (7.69) are given by

$$a_{k\ell} = \frac{1}{FD} \int_{-D/2}^{D/2} \int_{-F/2}^{F/2} K_{TC}(y, z) \psi_{yk}(y) \psi_{z\ell}(z) dy dz. \qquad (7.110)$$

The two-dimensional Fourier transform of $K_{TC}(y, z)$, $\hat{K}_{TC}(g_1, g_2)$, is given by Equation (7.83) with $a_{k\ell}$ replacing $c_{j, k\ell}$,

$$\hat{K}_{TC}(g_1, g_2) = FD \sum_{k=0}^{\infty} \sum_{\ell=0}^{\infty} a_{k\ell}(-i)^{k+\ell} \left[(2k + 1)(2\ell + 1) \right]^{1/2}$$

$$\times j_k \left(\pi F g_1 \right) j_\ell \left(\pi D g_2 \right). \qquad (7.111)$$

But from Equations (6.19a, b), we obtain from Equation (7.111),

$$\hat{K}_{TC}(0, 0) = FDa_{00}. \qquad (7.112)$$

Combining Equations (7.111) and (7.112) with Equation (7.108) yields for the Fourier series coefficients of the normalized total mesh stiffness $K_M(s)/\overline{K}_M$,

$$\frac{\alpha_{Kp}}{\overline{K}_M} = \sum_{k=0}^{\infty} \sum_{\ell=0}^{\infty} \frac{a_{k\ell}}{a_{00}} (-i)^{k+\ell} \left[(2k + 1)(2\ell + 1) \right]^{1/2}$$

$$\times j_k \left(p\pi Q_a \right) j_\ell \left(p\pi Q_t \right), \qquad (7.113)$$

where the expansion coefficients $a_{k\ell}$ of the local tooth-pair stiffness per unit length of line of contact are given by Equation (7.110).

Local Stiffness Weighting of Working-Surface-Deviations

The principal utility of the methodology described in this book arises from the fact that the transmission-error contributions from working-surface-deviations are largely independent of local tooth-pair stiffness, as illustrated by Equation (5.13).

Nevertheless, it is possible to include local tooth-pair stiffness variability $K_{TC}(y,z)/\overline{K}_{TC}$ in computation of transmission error contributions, by including this variability in calculation of the expansion coefficients, Equation (7.86), which then affects the DFT spectra, Equation (7.96). By using Equation (7.109) to represent the local tooth-pair stiffness per unit length of line of contact, only a relatively small number of terms would provide a good representation of this local tooth-pair stiffness, which then would be used to weight the measured working-surface-deviations $\eta_{Cj}(y,z)$, as indicated by Equation (7.86). Because from Equations (3.8), (3.13), and (3.14), $\psi_{y0}(y)=1$ and $\psi_{z0}(z)=1$, and thus, from Equation (7.110),

$$a_{00} = \overline{K}_{TC}, \tag{7.114}$$

which is the mean local tooth-pair stiffness, the required local stiffness weighting for use in Equation (7.86) would be

$$\frac{K_{TC}(y,z)}{\overline{K}_{TC}} = \sum_{k=0}^{\infty}\sum_{\ell=0}^{\infty} \frac{a_{k\ell}}{a_{00}} \psi_{yk}(y)\,\psi_{z\ell}(z), \tag{7.115}$$

as can be seen from Equations (7.109) and (7.114).

7.6 Approximate Evaluation of Mesh-Attenuation Functions

Equation (7.95) expresses the complex Fourier series coefficients of the working-surface-deviation contributions to the transmission-error vibration excitation as a summation over the Legendre indices $k\ell$ of the product of the DFT, $B_{k\ell}(n)$, of the Legendre coefficients $c_{j,k\ell}$, Equation (7.96), and the mesh-attenuation functions, Equation (7.94). Both of these latter quantities are complex functions. Hence, to compute the Fourier coefficients, Equation (7.95), we are required to deal with the summation of products of complex functions. As can be seen from Equation (5.18), computation of the mean-square transmission-error spectrum of this vibration-excitation contribution also requires us to deal with these same complex quantities.

In this section, we provide a method for approximate evaluation of the mesh-attenuation function described by Equation (7.94) using complex quantities. The resulting approximation is Equation (6.3). This approximation to $\hat{\phi}_{k\ell}(n/N)$ is obtained from a first-order approximation to $\overline{K}_M \alpha_{(1/K)p}$, the Fourier series coefficients of the normalized reciprocal mesh stiffness – that is, the product of Equations (7.44) and (7.92). Two methods for accurate evaluation of $\overline{K}_M \alpha_{(1/K)p}$ then are derived in Section 7.7, thereby providing the means for accurate evaluations of the mesh-attenuation functions given by Equation (7.94). The resulting (complex) formulas for the mesh-attenuation functions, when combined with $B_{k\ell}(n)$ given by Equation (7.96), will allow accurate computation of the complex Fourier series coefficients, Equation (7.95), and the mean-square transmission error spectra, Equation (5.18). Formulas for the real and imaginary parts of the Fourier series coefficients, Equation (7.95), that involve only real (not complex) quantities, are derived in Section 7.8. These formulas also allow accurate computation of the mean-square transmission error spectrum, Equation (5.18), using computer programs capable of dealing only with real quantities.

Fourier Series Coefficients of Normalized Reciprocal Mesh Stiffness $\overline{K}_M/K_M(s)$

Because the expression of the transmission-error contribution described by Equation (7.22) involves division by the total mesh stiffness $K_M(x)$, the expression, Equation (7.94), for the mesh-attenuation functions involves the Fourier series coefficients $\alpha_{(1/K)p}$, Equation (7.44), of the reciprocal total mesh stiffness $K_M^{-1}(s)$ normalized by the mean total mesh stiffness, $\overline{K}_M$. According to Equation (7.44), the normalized Fourier series coefficients of this reciprocal mesh stiffness are

$$\overline{K}_M \alpha_{(1/K)p} = \frac{1}{\Delta} \int\limits_{-\Delta/2}^{\Delta/2} \left[\overline{K}_M/K_M(s) \right] \exp\left(-i2\pi ps/\Delta\right) ds, \tag{7.116}$$

since $\overline{K}_M$ is a constant. Yet, Equation (7.113) provides an expression for the Fourier series coefficients of the reciprocal $\left[K_M(s)/\overline{K}_M \right]$ to that in Equation (7.116). Hence, a method is required to compute the Fourier series coefficients, Equation (7.116), from the available expression, Equation (7.113), for the Fourier series coefficients of $K_M(s)/\overline{K}_M$. In particular, in the practically important case where the tooth-pair stiffness per unit length of line contact $K_{TC}(y,z)$ is assumed to be a constant value, $K_{TC}(y,z) = \overline{K}_{TC}$, there follows from the orthogonality of the normalized Legendre polynomials, Equations (3.16b) and (3.17b), that the only nonzero coefficient $\alpha_{k\ell}$ of Equation (7.110) is α_{00}, that is, $k=0, l=0$. Consequently, for this practically important case of constant local tooth-pair stiffness, one has from Equation (7.113),

$$\frac{\alpha_{Kp}}{\overline{K}_M} = j_0\left(p\pi Q_a\right) j_0\left(p\pi Q_t\right) \tag{7.117a}$$

$$= \frac{\sin\left(p\pi Q_a\right)}{p\pi Q_a} \frac{\sin\left(p\pi Q_t\right)}{p\pi Q_t}, \quad p = 0, \pm 1, \pm 2, \ldots \tag{7.117b}$$

according to Equation (6.11a). Because $\lim_{\xi \to 0} \frac{\sin \xi}{\xi} = 1$, this result yields $\alpha_{K0} = \overline{K}_M$, the mean total mesh stiffness. Also, for $p = \pm 1, \pm 2, \ldots$, for either integer contact ratio Q_a or Q_t, Equation (7.117b) yields $\alpha_{Kp} = 0$. This result is consistent with Figure 2.7, which shows that the total length of line of contact is a constant value if either Q_a or Q_t is an integer (or both are integers).

Approximation to Normalized Reciprocal Mesh Stiffness

To obtain the first-order approximation, Equation (6.3), to the mesh-attenuation functions, define the fluctuating contribution of the total mesh stiffness, about the mean mesh stiffness, by

$$\delta K_M(s) \triangleq K_M(s) - \overline{K}_M. \tag{7.118}$$

Then,

$$\frac{K_M(s)}{\overline{K}_M} = \frac{\overline{K}_M + \delta K_M(s)}{\overline{K}_M} \tag{7.119a}$$

$$= 1 + \frac{\delta K_M(s)}{\overline{K}_M}. \tag{7.119b}$$

We require the reciprocal of Equation (7.119) in Equation (7.116). Consider the Maclaurin series expansion of the reciprocal of Equation (7.119b),

$$\frac{\overline{K}_M}{K_M(s)} = \left[1 + \frac{\delta K_M(s)}{\overline{K}_M} \right]^{-1} \tag{7.120a}$$

$$= 1 - \frac{\delta K_M(s)}{\overline{K}_M} + \left[\frac{\delta K_M(s)}{\overline{K}_M} \right]^2 - \left[\frac{\delta K_M(s)}{\overline{K}_M} \right]^3 + \cdots, \quad \left| \frac{\delta K_M(s)}{\overline{K}_M} \right| < 1. \tag{7.120b}$$

As can be understood from Figure 2.6, for helical gears with several too many teeth simultaneously in contact, the fluctuation $\delta K_M(s)$ in total mesh stiffness is a small fraction of the mean mesh stiffness $\overline{K}_M$, that is, $|\delta K_M(s)/\overline{K}_M| \ll 1$. The largest fractional fluctuations of mesh stiffness $K_M(s)$ will occur in the case of spur gears. For example, for a spur gear with contact ratio $Q_t = 1.5$, part of the time two tooth pairs are in contact and part of the time only one tooth pair is in contact. Therefore, for this example, the maximum fractional mesh-stiffness fluctuation about the mean mesh stiffness will be about $\left[|\delta K_M(s)| / \overline{K}_M \right] \approx (1/2)/(3/2) = 1/3$. Utilizing only one correction term in Equation (7.120b) for this example gives $\left[\overline{K}_M / K_M(s) \right] \approx 1 - 1/3 = 2/3$. The next correction term in Equation (7.120b) is $1/9$. The exact value of Equation (7.120a) for this example is $\left(1 + \frac{1}{3} \right)^{-1} = 3/4$. Hence, the maximum fractional error by retaining only one correction term in Equation (7.120b), in this example, is $\{[(3/4)-(2/3)]/(3/4)\} = 1/9$, which is the value of the first neglected term in Equation (7.120b). In Appendix 7.E, it is shown that the fractional error in the geometric series, Equation (7.120b), always is given by the value of the first neglected term.

Therefore, for this spur gear example, a reasonable first-order approximation to $\overline{K}_M / K_M(s)$ is obtained by using only a single correction term in Equation (7.120b),

$$\frac{\overline{K}_M}{K_M(s)} \approx 1 - \frac{\delta K_M(s)}{\overline{K}_M}, \tag{7.121}$$

which will be seen to be the basis of the approximation, Equation (6.3), to the mesh-attenuation functions. The approximation given by Equation (7.121) for helical gears generally will be more accurate than the above-cited error for the spur-gear example.

First-Order Approximation to Fourier Series Coefficients

Incorporating Equation (7.118) into Equation (7.121), there follows

$$\frac{\overline{K}_M}{K_M(s)} \approx 1 - \frac{K_M(s) - \overline{K}_M}{\overline{K}_M} = 2 - \frac{K_M(s)}{\overline{K}_M}. \tag{7.122}$$

Incorporating Equation (7.122) into Equation (7.116) yields

$$\overline{K}_M \alpha_{(1/K)p} \approx 2\delta_{p,0} - \frac{1}{\Delta} \int_{-\Delta/2}^{\Delta/2} \left[K_M(s)/\overline{K}_M \right] \exp(-i2\pi ps/\Delta) ds \tag{7.123a}$$

$$= 2\delta_{p,0} - \left(\alpha_{Kp}/\overline{K}_M \right), \tag{7.123b}$$

where $\delta_{p,0}$ is the Kronecker delta symbol,

$$\delta_{p,0} = \begin{cases} 1, & p = 0 \\ 0, & p = \pm1, \pm2, \dots, \end{cases} \tag{7.124}$$

$\alpha_{Kp}/\overline{K}_M$ is given by Equation (7.113), and α_{Kp} by Equation (7.98a). The first term in Equation (7.123) involving $\delta_{p,0}$ is obtained from Equation (7.116) by using

$$\exp(-i2\pi ps/\Delta) = \cos(2\pi ps/\Delta) - i\sin(2\pi ps/\Delta). \tag{7.125}$$

The integration of the imaginary sine term is identically zero, and the integration of the cosine term yields $\sin(\pi p)/(\pi p)$, which is zero for $p = \pm1, \pm2, \dots$, and unity for $p = 0$, thereby yielding $\delta_{p,0}$.

We require, in Equation (7.123b), $\alpha_{Kp}/\overline{K}_M$, which is given by Equation (7.113). First consider evaluation of $\alpha_{Kp}/\overline{K}_M$ for $p = 0$. It follows directly from Equations (6.19a, b) that for $p = 0$, the only nonvanishing term in Equation (7.113) is the single term, $k = 0$, $l = 0$, yielding

$$\frac{\alpha_{K0}}{\overline{K}_M} = 1. \tag{7.126}$$

Hence, Equation (7.123b) can be re-expressed as

$$\overline{K}_M \alpha_{(1/K)p} \approx \begin{cases} 1, & p = 0 \\ -\alpha_{Kp}/\overline{K}_M, & p = \pm1, \pm2, \dots. \end{cases} \tag{7.127}$$

Therefore, by combining Equation (7.127) with Equation (7.94), and utilizing Equation (7.113) for $\alpha_{Kp}/\overline{K}_M$, $p = \pm1, \pm2, \cdots$, we obtain the following approximate expression for the mesh-attenuation functions:

$$\hat{\phi}_{k\ell}\,(n/N) \approx (-i)^{k+\ell}\,[(2k+1)\,(2\ell+1)]^{1/2}\,(j_k\,(\pi Q_a n/N)\,j_\ell\,(\pi Q_t n/N)$$

$$-\sum_{\substack{p=-\infty \\ \text{except} \\ p=0}}^{\infty}\left\{\left[\sum_{k'=0}^{\infty}\sum_{\ell'=0}^{\infty}\frac{a_{k'\ell'}}{a_{00}}\,(-i)^{k'+\ell'}\,[(2k'+1)\,(2\ell'+1)]^{1/2}\right.\right.$$

$$\left.\left.\times j_{k'}\,(\pi Q_a p)\,j_{\ell'}\,(\pi Q_t p)]\,j_k\left[\pi Q_a\left(\frac{n}{N}-p\right)\right]j_\ell\left[\pi Q_t\left(\frac{n}{N}-p\right)\right]\right\}\right). \quad (7.128)$$

The coefficients $a_{k'\ell'}$ are the coefficients of the tooth-pair stiffness per unit length of line of contact given by Equation (7.110) without primes. In the practically important case where this local stiffness is assumed to be a constant value, the only nonzero term $a_{k'\ell'}$ is a_{00}. Consequently, if the local tooth-pair stiffness per unit length of line of contact is assumed to be a constant value, the approximate expression, Equation (7.128), for the mesh-attenuation functions reduces to

$$\hat{\phi}_{k\ell}\,(n/N) \approx (-i)^{k+\ell}\,[(2k+1)\,(2\ell+1)]^{1/2}\,\{j_k\,(\pi Q_a n/N)\,j_\ell\,(\pi Q_t n/N)$$

$$-\sum_{\substack{p=-\infty \\ \text{except} \\ p=0}}^{\infty}j_o\,(\pi Q_a p)\,j_o\,(\pi Q_t p)\,j_k\left[\pi Q_a\left(\frac{n}{N}-p\right)\right]j_\ell\left[\pi Q_t\left(\frac{n}{N}-p\right)\right]\} \quad (7.129)$$

which is the approximate relationship for the mesh-attenuation functions given by Equation (6.3).

Apart from its dependence on the rotational-harmonic-number ratio n/N, for each Legendre term $k\ell$, the right-hand side of Equation (7.129) is dependent only on the axial Q_a and transverse Q_t contact ratios that characterize the assumed rectangular zone of tooth contact. Equation (7.129) is independent of tooth-pair stiffness, as illustrated by Equation (5.13).

The role played by the mesh-attenuation functions in the generation of transmission-error rotational harmonics was discussed in Chapter 6. In particular, if either Q_a or Q_t is an integer (other than $Q_a = 0$), the summation over p in Equation (7.129) vanishes, as can be seen from Equation (7.117b). This summation characterizes the amplitude modulating effect that arises from the division by the mesh stiffness $K_M(x)$ in Equation (5.2) and the modulating effect arising from division by the total length of line of contact $l(x)$ in Equation (5.13). When these quantities are constant, no resultant amplitude modulation will take place, but "sideband" contributions still will be potentially present due to the discrete nature of gear teeth and the existence of tooth-to-tooth variability in the tooth-working-surfaces.

7.7 Accurate Evaluation of Fourier-Series Coefficients of Normalized Reciprocal Mesh Stiffness $\overline{K}_M/K_M(s)$

Extension of Power-Series Method

It was seen above that a useful approximation to the normalized Fourier series coefficients $\overline{K}_M \alpha_{(1/K)p}$, Equation (7.116), for use in evaluating the mesh-attenuation functions, Equation (7.94), was obtained by including only the one correction term $\delta K_M(s)/\overline{K}_M$ in the series, Equation (7.120), with a fractional error equal to the first neglected term. Hence, an approximation to any desired accuracy of $\overline{K}_M/K_M(s)$ can be obtained by including additional terms in Equation (7.120), say

$$\frac{\overline{K}_M}{K_M(s)} \approx \sum_{m=0}^{L} \left(-\frac{\delta K_M(s)}{\overline{K}_M} \right)^m \tag{7.130}$$

which yields Equation (7.121) for $L=1$. Inserting Equation (7.118) into Equation (7.130) gives

$$\frac{\overline{K}_M}{K_M(s)} \approx \sum_{m=0}^{L} \left[1 - \frac{K_M(s)}{\overline{K}_M} \right]^m \tag{7.131a}$$

$$= \sum_{m=0}^{L} \sum_{\ell=0}^{m} \binom{m}{\ell} \left[-\frac{K_M(s)}{\overline{K}_M} \right]^{\ell} \tag{7.131b}$$

by the Binomial theorem (Korn and Korn, 1961, p. 1.4-1) where

$$\binom{m}{\ell} = \frac{m!}{(m-\ell)!\,\ell!} \tag{7.132}$$

are the binomial coefficients with m factorial defined for any positive integer m as

$$m! \triangleq 1 \times 2 \times 3 \ldots \times m \tag{7.133a}$$

and

$$0! \triangleq 1. \tag{7.133b}$$

For L larger than unity, the same power of $K_M(s)/\overline{K}_M$ will appear in the summation in Equation (7.131) more than once. However, once the truncation value L in Equation (7.130) has been chosen, this repetition of powers of $K_M(s)/\overline{K}_M$ can be eliminated, as shown below.

For any sum, we have

$$\sum_{m=0}^{L} \sum_{\ell=0}^{m} = \sum_{\ell=0}^{L} \sum_{m=\ell}^{L}. \tag{7.134}$$

Applying this relationship to Equation (7.131b) gives

$$\frac{\overline{K}_M}{K_M(s)} = \sum_{\ell=0}^{L}\sum_{m=\ell}^{L} \binom{m}{\ell}\left[-\frac{K_M(s)}{\overline{K}_M}\right]^{\ell}. \tag{7.135}$$

However, for the inner sum in Equation (7.135) one has (Korn and Korn, 1961, p. 21.5-2)

$$\sum_{m=\ell}^{L} \binom{m}{\ell} = \binom{L+1}{\ell+1}. \tag{7.136}$$

Therefore, from Equations (7.135) and (7.136),

$$\frac{\overline{K}_M}{K_M(s)} = \sum_{\ell=0}^{L} \binom{L+1}{\ell+1}\left[-\frac{K_M(s)}{\overline{K}_M}\right]^{\ell}, \tag{7.137}$$

which is the *approximation* to $\overline{K}_M/K_M(s)$ given by Equations (7.130) and (7.131), where $\binom{L+1}{\ell+1}$ is a binomial coefficient as defined by Equation (7.132). Each power of $K_M(s)/\overline{K}_M$ appears only once in Equation (7.137).

Let us now *define*

$$\left(\alpha_{Kp}/\overline{K}_M\right)^{[\ell]} \triangleq \frac{1}{\Delta}\int_{-\Delta/2}^{\Delta/2}\left[K_M(s)/\overline{K}_M\right]^{\ell}\exp\left(-i2\pi ps/\Delta\right)ds. \tag{7.138}$$

If we insert Equation (7.137) into Equation (7.116) and use the definition (Equation (7.138)) there follows directly

$$\overline{K}_M\alpha_{(1/K)p} \approx \sum_{\ell=0}^{L}\binom{L+1}{\ell+1}(-1)^{\ell}\left(\alpha_{Kp}/\overline{K}_M\right)^{[\ell]} \tag{7.139}$$

where the degree of the approximation is determined by the value of L chosen in Equations (7.130) and (7.131). For $l=1$, Equation (7.138) reduces to $\alpha_{Kp}/\overline{K}_M$ given by Equation (7.113).

Equation (7.138) describes the Fourier series coefficients of $\left[K_M(s)/\overline{K}_M\right]^{\ell}$, a multiple product of $K_M(s)/\overline{K}_M$. Therefore, if we denote a discrete convolution in harmonic number p by an asterisk, then from the convolution theorem for Fourier series, Equation (7.C.22), it follows that $\left(\alpha_{Kp}/\overline{K}_M\right)^{[\ell]}$ can be expressed as

$$\left(\frac{\alpha_{Kp}}{\overline{K}_M}\right)^{[\ell]} = \begin{cases} \overset{1}{\frac{\alpha_{Kp}}{\overline{K}_M}} * \overset{2}{\frac{\alpha_{Kp}}{\overline{K}_M}} * \cdots * \overset{\ell-1}{\frac{\alpha_{Kp}}{\overline{K}_M}}, & \ell > 1 \\[2ex] \dfrac{\alpha_{Kp}}{\overline{K}_M}, & \ell = 1 \\[2ex] \delta_{p,o}, & \ell = 0 \end{cases} \tag{7.140}$$

where $\delta_{p,o}$ is the Kronecker delta symbol, Equation (7.124). For $l > 1$, we have

$$\left(\frac{\alpha_{Kp}}{\overline{K}_M}\right)^{[\ell]} = \left(\frac{\alpha_{Kp}}{\overline{K}_M}\right)^{[\ell-1]} * \frac{\alpha_{Kp}}{\overline{K}_M}. \tag{7.141}$$

Hence, using Equation (7.141), for any chosen integer value of L, the $L-1$ convolutions required to evaluate Equation (7.139) can be generated successively by Equation (7.141). Formulas for evaluating such discrete convolutions, using only real quantities, are provided in Appendix 7.F.

As a check on Equation (7.139), let us evaluate it for the case $L=1$. Using Equations (7.132) and (7.133), we have for $L=1$, $l=0$, $\binom{2}{1}=2$, and for $L=1$, $l=1$, $\binom{2}{2}=1$. Hence, from Equation (7.139) using Equation (7.140), we have for $L=1$,

$$\overline{K}_M \alpha_{(1/K)p} \approx 2\delta_{p,0} - \left(\alpha_{Kp}/\overline{K}_M\right) \tag{7.142}$$

in agreement with the approximation, Equation (7.123b).

As an additional illustration of Equation (7.139), consider the case $L=2$. In this case we require, using Equations (7.132) and (7.133),

$$\binom{3}{1} = 3, \quad \binom{3}{2} = 3, \quad \binom{3}{3} = 1. \tag{7.143}$$

Therefore, by using Equations (7.143) and (7.140), Equation (7.139) for $L=2$ becomes

$$\overline{K}_M \alpha_{(1/K)p} \approx 3\delta_{p,o} - 3\frac{\alpha_{Kp}}{\overline{K}_M} + \frac{\alpha_{Kp}}{\overline{K}_M} * \frac{\alpha_{Kp}}{\overline{K}_M}$$

$$= 3\delta_{p,o} - 3\frac{\alpha_{Kp}}{\overline{K}_M} + \sum_{p'=-\infty}^{\infty} \frac{\alpha_{Kp'}}{\overline{K}_M} \frac{\alpha_{K(p-p')}}{\overline{K}_M}, \tag{7.144}$$

by using the discrete convolution, Equation (7.C.22). Equation (7.144) provides an improved approximation to $\overline{K}_M \alpha_{(1/K)p}$ for all $p=0, \pm1, \pm2, \ldots$

To summarize: For any positive integer choice of L, Equations (7.139–7.141) yield an approximation to $\overline{K}_M \alpha_{(1/K)p}$ which can be computed from $\alpha_{Kp}/\overline{K}_M$, given by Equation (7.113). The quantity $\alpha_{Kp}/\overline{K}_M$ can be computed for any local tooth-pair stiffness per unit length of line of contact, $K_{TC}(y,z)$, by Equations (7.110) and (7.113). In the practically important case where this local stiffness is assumed to be a constant value, $K_{TC}(y,z) = \overline{K}_{TC}$, $\alpha_{Kp}/\overline{K}_M$ is given by Equation (7.117), which is real and an even function of harmonic number p. The resultant value of $\overline{K}_M/\alpha_{(1/K)p}$, computed for each value of $p=0, \pm1, \pm2, \ldots$, then can be combined with Equation (7.94) to yield computations of the mesh-attenuation functions with improved accuracies. Larger values of L will produce more accurate computations of mesh-attenuation functions. Because in the practically important case where $K_{TC}(y,z)$ is assumed to be the constant value $\overline{K}_{TC}$, leading to Equation (7.117) for $\alpha_{Kp}/\overline{K}_M$, which is real and an even function of harmonic number $p = 0, \pm1, \pm2, \ldots$, the discrete convolutions required in computation of $\overline{K}_M \alpha_{(1/K)p}$ involve only real (not complex) numbers. Moreover, when $K_{TC}(y, z)$ is not assumed to be the constant value $\overline{K}_{TC}$, it can be shown that the resultant values of $\alpha_{Kp}/\overline{K}_M$ generally are quite well approximated by real even functions of harmonic number p.

Fractional Truncation Error

As pointed out in Appendix 7.E, the fractional truncation error in Equation (7.130) is the magnitude of the first neglected term $\left|\delta K_M(s)/\overline{K}_M\right|^{L+1}$. Because tooth-pair stiffness is approximately proportional to contact line length, as a practical matter, for spur gears, one will have, approximately,

$$\frac{\left|\delta K_M(s)\right|}{\overline{K}_M} \leq \frac{1}{Q_t}, \tag{7.145}$$

and for helical gears,

$$\frac{\left|\delta K_M(s)\right|}{\overline{K}_M} \leq \frac{1/2}{Q_a}, \tag{7.146}$$

as is readily apparent from Figure (2.7b), where Q_t and Q_a are, respectively, the transverse and axial contact ratios. Thus, using only contact ratios, it is possible to estimate the fractional truncation errors in Equations (7.130) and (7.139), and therefore, the required value of L in Equation (7.139) for any prescribed accuracy.

Linear Equation Method

There exists a large variety of computer programs capable of numerically solving systems of simultaneous linear algebraic equations with constant coefficients. A method is described below for accurately computing the real and imaginary components of the Fourier series coefficients, Equation (7.116), of the normalized reciprocal mesh stiffness $\overline{K}_M/K_M(s)$ from the real and imaginary components of the Fourier series coefficients of its reciprocal, $K_M(s)/\overline{K}_M$, which will be shown to provide the coefficients of the above-mentioned linear system of equations.

It is appropriate to introduce here some simplified notation. Define the Fourier series coefficients of the normalized reciprocal mesh stiffness, Equation (7.116), as

$$\alpha'_p \triangleq \frac{1}{\Delta} \int_{-\Delta/2}^{\Delta/2} \left[\overline{K}_M/K_M(s)\right] \exp\left(-i2\pi ps/\Delta\right) ds \tag{7.147a}$$

$$\equiv \overline{K}_M \alpha_{(1/K)p}. \tag{7.147b}$$

Then, by introducing Equation (7.125) into Equation (7.147a), α'_p can be expressed in terms of its real and imaginary components as

$$\alpha'_p = a_p - ib_p \tag{7.148}$$

where

$$a_p = \frac{1}{\Delta} \int_{-\Delta/2}^{\Delta/2} \left[\overline{K}_M/K_M(s)\right] \cos\left(2\pi ps/\Delta\right) ds \qquad (7.149a)$$

and

$$b_p = \frac{1}{\Delta} \int_{-\Delta/2}^{\Delta/2} \left[\overline{K}_M/K_M(s)\right] \sin\left(2\pi ps/\Delta\right) ds; \qquad (7.149b)$$

hence, a_p is an even function of p, and b_p is an odd function of p. In exactly the same manner, we require from Equation (7.98a) the Fourier series coefficients of the reciprocal to $\overline{K}_M/K_M(s)$, which are $\alpha_{Kp}/\overline{K}_M$,

$$\frac{\alpha_{Kp}}{\overline{K}_M} \equiv \alpha_p^\dagger \triangleq \frac{1}{\Delta} \int_{-\Delta/2}^{\Delta/2} \left[K_M(s)/\overline{K}_M\right] \exp\left(-i2\pi ps/\Delta\right) ds \qquad (7.150a)$$

$$= A_p - iB_p, \qquad (7.150b)$$

where

$$A_p = \frac{1}{\Delta} \int_{-\Delta/2}^{\Delta/2} \left[K_M(s)/\overline{K}_M\right] \cos\left(2\pi ps/\Delta\right) ds \qquad (7.151a)$$

and

$$B_p = \frac{1}{\Delta} \int_{-\Delta/2}^{\Delta/2} \left[K_M(s)/\overline{K}_M\right] \sin\left(2\pi ps/\Delta\right) ds; \qquad (7.151b)$$

hence, A_p is even in p and B_p is odd in p.

In the general case where the tooth-pair stiffness per unit length of line of contact $K_{TC}(y,z)$ is not a constant value, the coefficients $\left(\alpha_{Kp}/\overline{K}_M\right) \equiv \alpha_p^\dagger$ are given by Equation (7.113); and in the practically important case where $K_{TC}(y,z) = \overline{K}_{TC}$ is assumed to be constant, these coefficients are given by Equation (7.117). Therefore, the requirement is to obtain the coefficients $\alpha_p{'}$ from the coefficients $\alpha_p^\dagger$. It is shown below that the coefficients $\alpha_p{'}$ can be obtained from the coefficients $\alpha_p^\dagger$ by utilizing the known expression (Tolstov, 1962, pp. 123–125) for the Fourier series coefficients of the product of two periodic functions in terms of the Fourier series coefficients of the individual functions whose product is taken. The product of our two functions is

$$\left[\frac{K_M(s)}{\overline{K}_M}\right] \times \left[\frac{\overline{K}_M}{K_M(s)}\right] = 1. \qquad (7.152)$$

Additional notation is required to distinguish the Fourier series definitions of Tolstov (1962, pp. 12, 13) with period 2π from those above with period Δ, the base pitch. Define here the coefficients of Tolstov (1962, pp. 12, 13) as

$$a'_p = \frac{1}{\pi} \int_{-\pi}^{\pi} f(\xi) \cos(p\xi) d\xi, \quad p = 0, 1, 2, \cdots \tag{7.153a}$$

$$b'_p = \frac{1}{\pi} \int_{-\pi}^{\pi} f(\xi) \sin(p\xi) d\xi, \quad p = 0, 1, 2, \cdots. \tag{7.153b}$$

The corresponding Fourier series coefficients of this work are

$$A'_p = \frac{1}{\Delta} \int_{-\Delta/2}^{\Delta/2} f(s) \cos\left(2\pi ps/\Delta\right) ds, \quad p = 0, 1, 2, \cdots \tag{7.154a}$$

$$B'_p = \frac{1}{\Delta} \int_{-\Delta/2}^{\Delta/2} f(s) \sin\left(2\pi ps/\Delta\right) ds, \quad p = 0, 1, 2, \cdots. \tag{7.154b}$$

Comparing Equations (7.153) and (7.154) there follows

$$\xi = 2\pi s/\Delta, \tag{7.155}$$

and therefore,

$$\frac{ds}{\Delta} = \frac{d\xi}{2\pi}. \tag{7.156}$$

Consequently, the coefficients of Equations (7.154) are one-half those of Equations (7.153), that is,

$$A'_p = \frac{1}{2} a'_p, \quad B'_p = \frac{1}{2} b'_p. \tag{7.157a, b}$$

Hence, the coefficients of Equations (7.149a, b) and (7.151a, b) are one-half of those used in Tolstov (1962). Therefore, the required relations for the present work are obtained from Equations (6.3) and (6.4) of Tolstov (1962, pp. 124, 125) by dividing Equations (6.3) and (6.4) of that reference by 2 and ignoring the numerical coefficients of the results, which are the same for each term.

Using the notation of Equations (7.154a, b), the Fourier cosine coefficients, Equation (7.154a) of the right-hand side of the product in Equation (7.152) are $A'_o = 1$, $A'_p = 0$ for $p = 1, 2, \ldots$, and the Fourier sine coefficients, Equation (7.154b), are $B'_p = 0$ for $p = 0, 1, 2, \ldots$ Consequently, using the Kronecker delta symbol

$$\delta_{pq} = \begin{cases} 1, & p = q \\ 0, & \text{otherwise,} \end{cases} \tag{7.158}$$

there follows from Equations (6.3) and (6.4) on pp. 125 and 126 of Tolstov (1962) the following relationship for the Fourier series representation of the product,

Equation (7.152) in terms of the Fourier series coefficients, Equations (7.151a, b) and (7.149a, b) of $[K_M(s)/\overline{K}_M]$ and $[\overline{K}_M/\overline{K}_M(s)]$, respectively, for $L' = \infty$,

$$A_m a_0 + \sum_{n=1}^{L'} \left[(A_{n+m} + A_{n-m}) a_n + (B_{n+m} + B_{n-m}) b_n \right] = \delta_{m0},$$

$$m = 0, 1, 2, \cdots, L' \qquad (7.159)$$

$$B_m a_0 + \sum_{n=1}^{L'} \left[(B_{n+m} - B_{n-m}) a_n - (A_{n+m} - A_{n-m}) b_n \right] = 0,$$

$$m = 1, 2, \cdots, L'. \qquad (7.160)$$

The coefficients A_p and B_p in Equations (7.159) and (7.160) are the Fourier series coefficients, Equations (7.151a) and (7.151b), respectively, of $[K_M(s)/\overline{K}_M]$. They are the real and imaginary components of the complex Fourier series coefficients, Equation (7.113) or (7.117), as appropriate, as mentioned above. The coefficients A_p and B_p are more fully developed below, and thus may be regarded as known quantities. The coefficients a_n and b_n in Equations (7.159) and (7.160) are the Fourier coefficients a_p and b_p, $p = n$, Equations (7.149a) and (7.149b), respectively, and are the coefficients to be solved for from the coefficients A_p and B_p of Equations (7.151a) and (7.151b). Therefore, Equations (7.159) and (7.160) are a set of $2L' + 1$ simultaneous (real) linear algebraic equations to be solved for $a_0, a_1, a_2, \ldots, a_{L'}$, and $b_1, b_2, \ldots, b_{L'}$ from the coefficients A_p and B_p of Equations (7.151a) and (7.151b). (Except for differences in notation, the result given by Equations (7.159) and (7.160) is identical to the result obtained in Appendix 7.F.)

From Equation (7.98a) there follows for $p = 0$,

$$\overline{K}_M = \frac{1}{\Delta} \int_{-\Delta/2}^{\Delta/2} K_M(s) ds, \qquad (7.161)$$

and therefore from Equation (7.151a), $A_0 = 1$. Furthermore, from $\sin 0 = 0$, there follows from Equations (7.151b) and (7.149b), $B_0 = 0$ and $b_0 = 0$. Moreover, because the cosine functions in Equations (7.151a) and (7.149a) are even functions of p, and the sine functions in Equations (7.151b) and (7.149b) are odd functions of p, there follows $A_{-p} = A_p$, $a_{-p} = a_p$, $B_{-p} = -B_p$, $b_{-p} = -b_p$. Consequently, with $n = p$, as in Equations (7.159) and (7.160), these observations are summarized as

$$A_0 = 1, \quad B_0 = 0, \quad b_0 = 0, \qquad (7.162a-c)$$

and

$$A_{-n} = A_n, \quad B_{-n} = -B_n, \quad a_{-n} = a_n, \quad b_{-n} = -b_n \qquad (7.163a-d)$$

for use in Equations (7.159) and (7.160) with indices chosen as shown there. In particular, Equations (7.163c) and (7.163d) yield the solutions a_{-n} and b_{-n} from a_n and b_n obtained from the solutions of Equations (7.159) and (7.160).

General Expressions for Linear Equation Coefficients

A general expression for the complex Fourier series coefficients $\alpha_p^\dagger$, Equations (7.150) and (7.151), of $\left[K_M(s)/\overline{K}_M\right]$ is provided by Equation (7.113) as a function of the Legendre expansion coefficients $a_{k\ell}$ of the local tooth pair stiffness, Equations (7.109) and (7.110). Combining Equation (7.113) with Equations (7.150) and (7.151), and replacing the index p by index n, there follows

$$\alpha_n^\dagger = A_n - iB_n = \sum_{k=0}^{\infty} \sum_{\ell=0}^{\infty} (-i)^{k+\ell} \frac{a_{k\ell}}{a_{00}} [(2k+1)(2\ell+1)]^{1/2}$$

$$\times j_k \left(n\pi Q_a \right) j_\ell \left(n\pi Q_t \right). \tag{7.164}$$

However, from Antosiewicz (1964, p. 437),

$$j_m(y) = \begin{cases} \text{even in } y, & m = \text{even} \\ \text{odd in } y, & m = \text{odd}. \end{cases} \tag{7.165}$$

Moreover, the product of two even functions is even, the product of two odd functions is even, and the product of an even function and an odd function is odd. Therefore,

$$j_k(n\pi Q_a)j_\ell \left(n\pi Q_t \right) = \begin{cases} \text{even in } n, & k+\ell = \text{even} \\ \text{odd in } n, & k+\ell = \text{odd}. \end{cases} \tag{7.166}$$

Furthermore, for $k+l=even$,

$$(-i)^{k+\ell} = (-1)^{k+\ell} (i)^{k+\ell} = \left[(-1)^{\frac{1}{2}}\right]^{k+\ell} = (-1)^{\frac{k+\ell}{2}}, k+\ell = \text{even} \tag{7.167a}$$

and, for $k+l=odd$,

$$(-i)^{k+\ell} = (-1)^{k+\ell} (i)^{k+\ell} = -i \times (i)^{k+\ell-1} = -i \times (-1)^{\frac{k+\ell-1}{2}}, k+\ell = \text{odd}. \tag{7.167b}$$

From Equations (7.163a, b), A_n is even in n and B_n is odd in n. Therefore, from Equations (7.164–7.167),

$$A_n = \sum_{\substack{k=0 \\ k+\ell=even}}^{\infty} \sum_{\ell=0}^{\infty} (-1)^{\frac{k+\ell}{2}} \frac{a_{k\ell}}{a_{00}} [(2k+1)(2\ell+1)]^{1/2} j_k \left(n\pi Q_a \right) j_\ell \left(n\pi Q_t \right) \tag{7.168a}$$

and

$$B_n = \sum_{\substack{k=0 \\ k+\ell=odd}}^{\infty} \sum_{\ell=0}^{\infty} (-1)^{\frac{k+\ell-1}{2}} \frac{a_{k\ell}}{a_{00}} [(2k+1)(2\ell+1)]^{1/2} j_k \left(n\pi Q_a \right) j_\ell \left(n\pi Q_t \right) \tag{7.168b}$$

which are the real and imaginary parts of the complex Fourier series coefficients $\alpha_n^\dagger$ of Equations (7.150) and (7.164). The coefficient $a_{k\ell}$ is the two-dimensional Legendre expansion coefficient of the local stiffness per unit length of line of contact of a pair of meshing teeth provided by Equation (7.110).

Case of Constant Local Tooth-Pair Stiffness

In the practically important case where the tooth-pair stiffness per unit length of line of contact is assumed to be a constant value, the only nonzero coefficient in Equations (7.109) and (7.110) is the term $k=0$, $l=0$, that is, a_{00}. In this case, there follows from Equation (7.168a),

$$A_n = j_o\left(n\pi Q_a\right)j_o\left(n\pi Q_t\right) \qquad (7.169a)$$

$$= \frac{\sin\left(n\pi Q_a\right)}{n\pi Q_a}\frac{\sin\left(n\pi Q_t\right)}{n\pi Q_t}, \quad n=0,1,2,\cdots, \qquad (7.169b)$$

and from Equation (7.168b)

$$B_n = 0, \quad n=0,1,2,\cdots, \qquad (7.170)$$

where Equation (7.169b) is a consequence of Equation (6.11a). Equations (7.169) and (7.170) are consistent with Equation (7.117). Then, from Equation (7.170), the coefficient B_m of a_0, and the coefficients $(B_{n+m}-B_{n-m})$ of a_n for all n in Equation (7.160) are zero, as are the coefficients $(B_{n+m}+B_{n-m})$ of b_n for all n in Equation (7.159). Therefore, from Equation (7.170) there follows from Equations (7.159) and (7.160),

$$b_n = 0, \quad n=0,1,2,\cdots. \qquad (7.171)$$

Consequently, whenever Equation (7.170) is satisfied, the pair of Equations (7.159) and (7.160) reduce to the set of $L'+1$ simultaneous equations,

$$A_m a_o + \sum_{n=1}^{L'}\left[\left(A_{n+m}+A_{n-m}\right)a_n\right] = \delta_{mo},$$
$$m=0,1,2,\cdots,L' \qquad (7.172a)$$

for the $L'+1$ values $a_0, a_1, \ldots, a_{L'}$.

It is instructive to write out the first few of the set (Equation (7.172a)) using $A_{n-m}=A_{|n-m|}$, according to Equation (7.163a):

$$A_0 a_0 + \left(2A_1\right)a_1 + \left(2A_2\right)a_2 + \left(2A_3\right)a_3 + \cdots + \left(2A_{L'}\right)a_{L'} = 1$$
$$A_1 a_0 + \left(A_2+A_0\right)a_1 + \left(A_3+A_1\right)a_2 + \left(A_4+A_2\right)a_3 + \cdots + \left(A_{L'+1}+A_{L'-1}\right)a_{L'} = 0$$
$$A_2 a_0 + \left(A_3+A_1\right)a_1 + \left(A_4+A_0\right)a_2 + \left(A_5+A_1\right)a_3 + \cdots + \left(A_{L'+2}+A_{L'-2}\right)a_{L'} = 0$$
$$A_3 a_0 + \left(A_4+A_2\right)a_1 + \left(A_5+A_1\right)a_2 + \left(A_6+A_0\right)a_3 + \cdots + \left(A_{L'+3}+A_{L'-3}\right)a_{L'} = 0$$
$$\cdots$$
$$A_{L'} a_0 + \left(A_{L'+1}+A_{L'-1}\right)a_1 + \left(A_{L'+2}+A_{L'-2}\right)a_2 + \left(A_{L'+3}+A_{L'-3}\right)a_3 + \cdots$$
$$+ \left(A_{2L'}+A_0\right)\alpha_{L'} = 0. \qquad (7.172b)$$

Behavior of Equation Coefficients

The set of simultaneous equations Equations (7.159) and (7.160) are "exact" only as $L' \to \infty$. Let us therefore examine the behavior of the somewhat simplified set (Equations (7.172a, b)) which were determined from the condition, Equation (7.170), applicable to the practically important case where the tooth-pair stiffness per unit length of line of contact is assumed to be constant.

Equation (6.12) gives the asymptotic behavior of the spherical Bessel functions as a function of their argument ξ. Therefore, according to Equations (7.168a) and (7.169), for given axial Q_a and transverse Q_t contact ratios of helical gears, A_n is asymptotically (large n) proportional to $1/n^2$, and for spur gears asymptotically proportional to $1/n$, which follows for spur gears from Equations (6.19a), (7.169), and the fact that $Q_a = 0$.

For each unknown value a_n in the set (Equations (7.172a, b)), consider the coefficient $(A_{n+m} + A_{|n-m|})$. The asymptotic behavior of this coefficient is $1/(n+m)^2$ for helical gears and $1/(n+m)$ for spur gears. Consider the behavior of this coefficient for a typical unknown a_n, that is, the *column* of coefficients $(A_{n+m} + A_{|n-m|})$ for fixed n, as equation number m is increased. For fixed n, this coefficient asymptotically decreases as $1/m^2$ for helical gears and $1/m$ for spur gears. Because for any solution $a_0, a_1, a_2, \ldots, a_{L'}$ to the equation set (Equations (7.172a, b)), a_n is constant in each column of the equation set, the above-described asymptotically decreasing values of the equation coefficients $(A_{n+m} + A_{|n-m|})$ implies that for sufficiently large equation number $m = L'$, the contribution of the remaining (neglected) equations $m > L'$ in the set (Equations (7.172a, b)) is asymptotically negligible.

Next consider the behavior of the coefficients $(A_{n+m} + A_{|n-m|})$ in the set (Equations (7.172a, b)) for fixed equation number m but increasing designation n of the unknowns a_n, that is, moving *horizontally* with increasing n on one of the lines m in the equation set (Equations (7.172a, b)). For each fixed value of m, the coefficients $(A_{n+m} + A_{|n-m|})$ asymptotically decay as $1/n^2$ for helical gears and $1/n$ for spur gears. But the asymptotic decay of the terms $(A_{n+m} + A_{|n-m|}) a_n$ along each horizontal line also depends on the behavior of the solution a_n, for increasing value of n, which, of course, is determined by the coefficients of the linear set (Equations (7.172a, b)).

It is readily apparent from Figure 2.6 that for any helical or spur gear, the total mesh stiffness has a nonzero mean value $\overline{K}_M$ with fluctuations superimposed on this mean value. For a spur gear, as contact of a tooth is initiated or terminated, the instantaneous mesh stiffness at these transition instants possesses amplitude discontinuities (i.e., discrete steps); hence, the reciprocal mesh stiffness also possesses amplitude discontinuities at these same instants. As a consequence of these steps, the amplitude behavior of the Fourier series coefficients a_n of the normalized reciprocal mesh stiffness therefore will asymptotically (large n) decay as $1/n$ for spur gears (Papoulis, 1977, p. 191). In the case of helical gears, as a new tooth enters or exits the mesh, a *slope* discontinuity will take place in the normalized mesh stiffness at these transition instants. Consequently, as can be seen from a comparison of Equations (7.119b) and (7.120b), a corresponding slope discontinuity of opposite sign will take place in the reciprocal normalized mesh stiffness at these transition instants. Hence, (Papoulis, 1977, p. 191) the asymptotic behavior of the Fourier series coefficients a_n of the normalized reciprocal mesh stiffness in the case of helical gears will decay as $1/n^2$.

As a consequence of the above-described asymptotic behavior of the coefficients $(A_{n+m} + A_{|n-m|})$ and of the solutions a_n, as we move *horizontally* along equation number m in the set (Equations (7.172a,b)), the product $(A_{n+m} + A_{|n-m|})\, a_n$ will asymptotically decay as $1/n^2$ for spur gears and $1/n^4$ for helical gears. Therefore, in either case, for sufficiently large L', the added (neglected) terms beyond $n = L'$ will contribute negligibly.

Consequently, for sufficiently large L', a good approximation to the lower-order Fourier series coefficients $a_0, a_1, a_2, \ldots$ for $n \ll L'$ can be expected by numerically solving the equation set (Equations (7.172a,b)), where from Equation (7.163c), $a_{-n} = a_n$. By numerically solving the set (Equations (7.172a,b)) several times for increasing values of L', the convergence of the lower-order solutions $a_0, a_1, a_2, \ldots$ for $n \ll L'$ can be examined.

Because of the contrasting asymptotic behavior between the spur gear coefficients versus the helical gear coefficients, significantly larger values of L' for spur gears will be required to achieve accuracies comparable to those achievable for helical gears.

The above-described approach was utilized in the helical gear computations leading to the results shown in Figures 5.1, 6.7, 6.8, and 6.14. It is important to recognize that harmonic numbers n of the solutions a_n of the equation set (Equations (7.172a,b)) are *tooth-meshing harmonics*, not rotational harmonics. Therefore, accurate values of a_n beyond about $n = 5$ will seldom be required. Notice that none of the above-mentioned spectra of Figures 5.1, 6.7, 6.8, and 6.14 extends significantly beyond the third tooth-meshing harmonic.

Because the asymptotic decay in n of the coefficients B_n of Equation (7.168b) is the same as that of the coefficients A_n of Equation (7.168a), the above comments pertaining to the coefficients and solutions of Equations (7.172a,b) also apply to the more general set (Equations (7.159–7.160)).

7.8 Fourier-Series Representation of Working-Surface-Deviation Transmission-Error Contributions Utilizing only Real (Not-Complex) Quantities

In the foregoing pages, methods have been developed for accurate computation of the working-surface-deviation transmission-error Fourier-series coefficients utilizing complex quantities. Computer implementation is most efficiently carried out using only real quantities. Reduction of the earlier results, for implementation, using only real quantities is developed below.

The Fourier-series coefficients are given by Equation (7.95) in terms of the discrete Fourier transforms $B_{k\ell}(n)$, Equation (7.96), of the Legendre expansion coefficients, Equation (7.86) or (7.89), of the stiffness-weighted or non-stiffness-weighted working-surface-deviations, respectively, the latter also given by Equation (3.20). Evaluation of these expansion coefficients from line-scanning metrology measurements is given by Equations (3.37) and (3.43), with Equation (3.37) applying to lead measurements and Equation (3.43) applying to profile measurements. The remaining terms in Equation (7.95) are the mesh-attenuation functions $\hat{\phi}_{k\ell}(n/N)$ given by Equation (7.94) in terms of the Fourier series coefficients $\overline{K}_M \alpha_{(1/K)p}$ obtainable either from Equations (7.138–7.141) or Equations (7.147–7.172a, b), and spherical Bessel functions, $j_m(\xi)$, defined by Equation (6.10). Evaluation of the Fourier series coefficients $\overline{K}_M \alpha_{(1/K)p}$ using Equations (7.147–7.172a, b) is, likely, the easier method. But their evaluation by Equations (7.138–7.141) enables a priori estimates of truncation errors in Equations (7.130) and (7.139), according to Equations (7.145) and (7.146).

Because for each application, we will be dealing with working-surface measurements made on only a single gear, the superscript (·) designating each single gear of a meshing pair will no longer be utilized in the notation below.

Fourier Series Using Only Real Quantities

As described earlier, for example, Equation (3.1), the independent variable x utilized in this work is "roll distance" associated with base cylinders. Using x as independent variable, the rotation period of the gear under consideration is $N\Delta$, the base-cylinder circumference, where N is the number of teeth and Δ is the base pitch. The fundamental period of harmonics caused by working-surface-deviations is $N\Delta$; hence, the complex form of the Fourier series coefficients generated by the (real) transmission-error contribution $\zeta(x)$ from working-surface-deviations is

$$\alpha_n = \frac{1}{N\Delta} \int_{-N\Delta/2}^{N\Delta/2} \zeta(x) \exp\left[-i2\pi nx/(N\Delta)\right] dx \qquad (7.173a)$$

$$= \alpha_{ne} - i\alpha_{no} \qquad (7.173b)$$

where

$$\alpha_{ne} = \frac{1}{N\Delta} \int_{-N\Delta/2}^{N\Delta/2} \zeta(x) \cos\left[2\pi nx/(N\Delta)\right] dx \qquad (7.174a)$$

$$\alpha_{no} = \frac{1}{N\Delta} \int_{-N\Delta/2}^{N\Delta/2} \zeta(x) \sin\left[2\pi nx/(N\Delta)\right] dx \qquad (7.174b)$$

where subscripts e and o denote even and odd functions of n; $\cos[2\pi nx/(N\Delta)]$ is an even function of n and $\sin[2\pi nx/(N\Delta)]$ is an odd function of n. The fundamental identity

$$\exp\left(\pm i\theta\right) = \cos\theta \pm i \sin\theta \qquad (7.175)$$

has been used in going from Equation (7.173a) to Equations (7.173b) and (7.174a, b). Equation (7.173a) is the same as Equations (5.14) and (7.33), but without the superscript (·). As in Equations (5.15) and (7.32), the (real) transmission error contribution $\zeta(x)$ can be generated from the Fourier series coefficients α_n by

$$\zeta(x) = \sum_{n=-\infty}^{\infty} \alpha_n \exp\left[i2\pi nx/(N\Delta)\right]$$

$$= \sum_{n=-\infty}^{\infty} (\alpha_{ne} - i\alpha_{no})\left[\cos(2\pi nx/N\Delta) + i\sin(2\pi nx/N\Delta)\right] \tag{7.176a}$$

$$= \sum_{n=-\infty}^{\infty} \left\{\left[\alpha_{ne}\cos(2\pi nx/N\Delta) + \alpha_{no}\sin(2\pi nx/N\Delta)\right]\right.$$

$$\left. + i\left[\alpha_{ne}\sin(2\pi nx/N\Delta) - \alpha_{no}\cos(2\pi nx/N\Delta)\right]\right\} \tag{7.176b}$$

where Equations (7.173b) and (7.175) were used in going from the first line to the second line, and $i^2 = -1$ was used in going to Equation (7.176b). The product of an even function and an odd function is odd. Thus, the coefficient of the imaginary term in Equation (7.176b) is odd in n. For every positive integer value of n in the imaginary term in Equation (7.176b), there exists a negative integer value of n with coefficient that exactly cancels the coefficient of the positive value of n in the summation, thus leaving only the coefficient of the $n = 0$ term in the overall imaginary term in Equation (7.176b). But $\sin 0 = 0$, and therefore from Equation (7.174b), $\alpha_{0o} = 0$; hence, the imaginary term in Equation (7.176b) is identically zero. (This conclusion also can be arrived at directly because $\zeta(x)$ is real.) Turning to the real term in Equation (7.176b), the product of two even functions is even, and the product of two odd functions also is even. Hence, the real term in Equation (7.176b) is even in n. Therefore, $\zeta(x)$ can be generated as a function of roll-distance x by

$$\zeta(x) = \alpha_{0e} + 2\sum_{n=1}^{\infty}\left[\alpha_{ne}\cos(2\pi nx/N\Delta) + \alpha_{no}\sin(2\pi nx/N\Delta)\right] \tag{7.177}$$

since $\sin 0 = 0$ and $\cos 0 = 1$. Only required are the real Fourier series coefficients of Equations (7.174a, b) for $n = 0, 1, 2, \ldots$

Real Fourier-Series Coefficients

The complex Fourier series coefficients of the working-surface-deviation contributions to the transmission error are given by Equation (7.95),

$$\alpha_n = \sum_{k=0}^{\infty}\sum_{\ell=0}^{\infty} B_{k\ell}(n)\hat{\phi}_{k\ell}(n/N), \quad n = 0, \pm 1, \pm 2, \cdots \tag{7.178}$$

where, from Equation (7.96)

$$B_{k\ell}(n) = \frac{1}{N}\sum_{j=0}^{N-1} c_{j,k\ell} \exp\left(-i2\pi nj/N\right) \tag{7.179a}$$

$$= B_{k\ell}(n)_e - iB_{k\ell}(n)_o \tag{7.179b}$$

where, from Equation (7.175) and the fact that the Legendre expansion coefficients $c_{j,k\ell}$ and real, $B_{k\ell}(n)_e$ and $B_{k\ell}(n)_o$ both are real and are given by

$$B_{k\ell}(n)_e = \frac{1}{N}\sum_{j=0}^{N-1} c_{j,k\ell}\cos\left(2\pi nj/N\right) \tag{7.180}$$

and

$$B_{k\ell}(n)_o = \frac{1}{N}\sum_{j=0}^{N-1} c_{j,k\ell}\sin\left(2\pi nj/N\right). \tag{7.181}$$

The generic form of the mesh-attenuation function $\hat{\phi}_{k\ell}(n/N)$ is given by Equation (7.57) with m denoting $k\ell$. Because the reciprocal mesh stiffness $K_M^{-1}(s)$ and the (generic) expansion function $\underset{\sim}{\psi}_{Km}(s)$, Equation (7.52), both are real, it follows from Equations (7.57a), (7.42), and (7.175) that $\hat{\phi}_{k\ell}(n/N)$ can be expressed as

$$\hat{\phi}_{k\ell}(n/N) = \hat{\phi}_{k\ell}(n/N)_e - i\hat{\phi}_{k\ell}(n/N)_o \tag{7.182}$$

where $\hat{\phi}_{k\ell}(n/N)_e$ is real and an even function of n, and $\hat{\phi}_{k\ell}(n/N)_o$ is real and an odd function of n.

According to Equation (7.178), we require the product of $B_{k\ell}(n)$ and $\hat{\phi}_{k\ell}(n/N)$. Forming the product of the right-hand sides of Equations (7.179b) and (7.182), there follows directly using $i^2 = -1$,

$$\begin{aligned} B_{k\ell}(n)\hat{\phi}_{k\ell}(n/N) &= B_{k\ell}(n)_e\hat{\phi}_{k\ell}(n/N)_e - B_{k\ell}(n)_o\hat{\phi}_{k\ell}(n/N)_o \\ &\quad - i\left[B_{k\ell}(n)_e\hat{\phi}_{k\ell}(n/N)_o + B_{k\ell}(n)_o\hat{\phi}_{k\ell}(n/N)_e\right] \\ &= \left[B_{k\ell}(n)\hat{\phi}_{k\ell}(n/N)\right]_e - i\left[B_{k\ell}(n)\hat{\phi}_{k\ell}(n/N)\right]_o, \end{aligned} \tag{7.183}$$

where in going to the last line, we have defined using the above-described properties of products of even and odd functions,

$$\left[B_{k\ell}(n)\hat{\phi}_{k\ell}(n/N)\right]_e \triangleq B_{k\ell}(n)_e\hat{\phi}_{k\ell}(n/N)_e - B_{k\ell}(n)_o\hat{\phi}_{k\ell}(n/N)_o \tag{7.184}$$

and

$$\left[B_{k\ell}(n)\hat{\phi}_{k\ell}(n/N)\right]_o \triangleq B_{k\ell}(n)_e\hat{\phi}_{k\ell}(n/N)_o + B_{k\ell}(n)_o\hat{\phi}_{k\ell}(n/N)_e, \tag{7.185}$$

where subscripts denote even and odd functions as before. Hence, from Equations (7.173b), (7.178), and (7.183), we have

$$\alpha_{ne} = \sum_{k=0}^{\infty}\sum_{\ell=0}^{\infty}\left[B_{k\ell}(n)\hat{\phi}_{k\ell}(n/N)\right]_e \tag{7.186}$$

and

$$\alpha_{no} = \sum_{k=0}^{\infty}\sum_{\ell=0}^{\infty}\left[B_{k\ell}(n)\hat{\phi}_{k\ell}(n/N)\right]_o \tag{7.187}$$

where the summands are defined by Equations (7.184) and (7.185). The even and odd contributions of $B_{k\ell}(n)$ are given by Equations (7.180) and (7.181), respectively.

There remains evaluation of the even and odd contributions of the mesh-attenuation functions $\hat{\phi}_{k\ell}(n/N)$, Equation (7.182).

Mesh-Attenuation-Function Evaluation

The complex form of the mesh-attenuation functions is given by Equation (7.94). Define from Equation (7.94) the quantity

$$\underset{\sim k\ell}{\hat{\psi}'}\left(\frac{n}{N}-p\right) \triangleq (-i)^{k+\ell} [(2k+1)(2\ell+1)]^{1/2} j_k\left[\pi Q_a\left(\frac{n}{N}-p\right)\right] j_\ell\left[\pi Q_t\left(\frac{n}{N}-p\right)\right]. \tag{7.188}$$

Then, using the definitions (Equation (7.147)) and (Equation (7.188)), the mesh-attenuation function, Equation (7.94), can be expressed as the discrete convolution,

$$\hat{\phi}_{k\ell}(n/N) = \sum_{p=-\infty}^{\infty} \alpha'_p \underset{\sim k\ell}{\hat{\psi}'}\left(\frac{n}{N}-p\right). \tag{7.189}$$

Notice that the definition (7.188) is Equation (7.93) divided by $\overline{K}_M\Delta$, with the right-hand side evaluated at $g=n/(N\Delta)$; hence, its notation.

From Equations (7.147) and (7.149), the complex Fourier series coefficient α'_p can be expressed as

$$\alpha'_p = \left(\alpha'_p\right)_e - i\left(\alpha'_p\right)_o, \tag{7.190}$$

where $\left(\alpha'_p\right)_e$ and $\left(\alpha'_p\right)_o$ are real even and odd functions of p given, respectively, by Equations (7.149a) and (7.149b). These coefficients, a_p and b_p with $p=n$, are the solutions of the general set of linear equations, Equations (7.159) and (7.160), or as appropriate, the set (Equations (7.172a, b)) applicable to the case where the tooth-pair stiffness per unit length of line of contact is assumed to be constant. They also can be evaluated using Equations (7.138–7.141), as noted earlier.

Now consider the other term in Equation (7.189) given by Equation (7.188). This term has the same general form as the summand in the right-hand side of Equation (7.164), with n there replaced by $(n/N) - p$. Replacing either of these independent variables by g, it follows directly from Equations (7.165–7.167b) that $\underset{\sim k\ell}{\hat{\psi}'}(g)$ can be expressed as

$$\underset{\sim k\ell}{\hat{\psi}'}(g) = \underset{\sim k\ell}{\hat{\psi}'}(g)_e - i\underset{\sim k\ell}{\hat{\psi}'}(g)_o, \tag{7.191}$$

where $\underset{\sim k\ell}{\hat{\psi}'}(g)_e$ and $\underset{\sim k\ell}{\hat{\psi}'}(g)_o$ are real even and odd functions of g, respectively:

$$\underset{\sim k\ell}{\hat{\psi}'}(g)_e = (-1)^{\frac{k+\ell}{2}} [(2k+1)(2\ell+1)]^{1/2} j_k\left(\pi Q_a g\right) j_\ell\left(\pi Q_t g\right), k+\ell = \text{even} \tag{7.192a}$$

$$\underset{\sim k\ell}{\hat{\psi}'}(g)_o = (-1)^{\frac{k+\ell-1}{2}} [(2k+1)(2\ell+1)]^{1/2} j_k\left(\pi Q_a g\right) j_\ell\left(\pi Q_t g\right), k+\ell = \text{odd}. \tag{7.192b}$$

Depending on whether $k+l$ is even or odd, only one or the other of Equations (7.192a, b) will provide a nonzero contribution to Equation (7.191).

At this juncture, to utilize the results of Appendix 7.F, we require an abbreviated notation for the three quantities involved in the convolution, Equation (7.189). Denote

$$\alpha_g \triangleq \hat{\phi}_{kl}(g), \qquad \alpha'_p = \alpha'_p, \qquad \alpha''_g \triangleq \hat{\tilde{\psi}}'_{\sim kl}(g), \qquad (7.193a\text{-}c)$$

where the notation for α'_p remains unchanged. With this notation, the discrete convolution, Equation (7.189) is expressed, with $g = n/N$, as

$$\alpha_g = \sum_{p=-\infty}^{\infty} \alpha'_p \alpha''_{g-p} \qquad (7.194)$$

where, from Equations (7.182), (7.190), and (7.191), each term in the convolution, Equation (7.194), has the same form, that is,

$$\alpha_g = \alpha_g^e - i\alpha_g^o \qquad (7.195a)$$

$$\alpha'_p = \alpha_p'^e - i\alpha_p'^o \qquad (7.195b)$$

$$\alpha''_g = \alpha_g''^e - i\alpha_g''^o, \qquad (7.195c)$$

where each α function is real and an even or odd function of its subscript as indicated by the superscripts e and o, respectively. Then, by substituting Equations (7.195a–c) into Equation (7.194) and carrying out an algebraic manipulation described in Appendix 7.F, we have found that the discrete convolution, Equation (7.194), can be expressed in terms of the real even and odd parts of the three complex quantities, Equations (7.195a–c), as

$$\alpha_g^e = \alpha_0'^e \alpha_g''^e + \sum_{p=1}^{\infty} \left[\alpha_p'^e \left(\alpha_{p+g}''^e + \alpha_{p-g}''^e \right) + \alpha_p'^o \left(\alpha_{p+g}''^o + \alpha_{p-g}''^o \right) \right] \qquad (7.196a)$$

$$\alpha_g^o = \alpha_0'^e \alpha_g''^o + \sum_{p=1}^{\infty} \left[\alpha_p'^e \left(\alpha_{p+g}''^o - \alpha_{p-g}''^o \right) - \alpha_p'^o \left(\alpha_{p+g}''^e - \alpha_{p-g}''^e \right) \right], \qquad (7.196b)$$

Although the above results are expressed in terms of a variable denoted by the symbol α, the above results are not restricted to relationships involving Fourier series coefficients. However, when applied to Fourier series coefficients, and recognizing the relationships, Equations (7.157a, b), one observes that Equations (7.196a) and (7.196b) have the same exact form as Equations (6.3) and (6.4), respectively, of Tolstov (1962, pp. 124, 125). This is as it must be because the complex Fourier series of the product of two functions is the discrete convolution of their individual complex Fourier series coefficients, Equations (7.C.22a, b).

Comparing the identification of each of the three terms, Equations (7.193a–c), respectively, with the notation of Equations (7.195a–c), and then with Equations (7.182), (7.190), and (7.191), respectively, we have the identifications of the terms in Equations (7.196a) and (7.196b), with $g = n/N$,

$$\alpha_g^e = \hat{\phi}_{k\ell}\left(\frac{n}{N}\right)_e, \quad \alpha_g^o = \hat{\phi}_{k\ell}\left(\frac{n}{N}\right)_o, \tag{7.197a}$$

$$\alpha_p^{'e} = \left(\alpha_p'\right)_e, \quad \alpha_p^{'o} = \left(\alpha_p'\right)_o \tag{7.197b}$$

$$\alpha_g^{''e} = \underset{\sim k\ell}{\hat{\psi}'}\left(\frac{n}{N}\right)_e, \quad \alpha_g^{''o} = \underset{\sim k\ell}{\hat{\psi}'}\left(\frac{n}{N}\right)_o. \tag{7.197c}$$

Consequently, utilizing the identifications, Equations (7.197a–c) in Equations (7.196a) and (7.196b), there follows directly with $g = n/N$,

$$\hat{\phi}_{k\ell}\left(\frac{n}{N}\right)_e = \left(\alpha_0'\right)_e \underset{\sim k\ell}{\hat{\psi}'}\left(\frac{n}{N}\right)_e + \sum_{p=1}^{\infty}\left\{\left(\alpha_p'\right)_e\left[\underset{\sim k\ell}{\hat{\psi}'}\left(p+\frac{n}{N}\right)_e + \underset{\sim k\ell}{\hat{\psi}'}\left(p-\frac{n}{N}\right)_e\right]\right.$$
$$\left. + \left(\alpha_p'\right)_o\left[\underset{\sim k\ell}{\hat{\psi}'}\left(p+\frac{n}{N}\right)_o + \underset{\sim k\ell}{\hat{\psi}'}\left(p-\frac{n}{N}\right)_o\right]\right\} \tag{7.198a}$$

and

$$\hat{\phi}_{k\ell}\left(\frac{n}{N}\right)_o = \left(\alpha_0'\right)_e \underset{\sim k\ell}{\hat{\psi}'}\left(\frac{n}{N}\right)_o + \sum_{p=1}^{\infty}\left\{\left(\alpha_p'\right)_e\left[\underset{\sim k\ell}{\hat{\psi}'}\left(p+\frac{n}{N}\right)_o - \underset{\sim k\ell}{\hat{\psi}'}\left(p-\frac{n}{N}\right)_o\right]\right.$$
$$\left. - \left(\alpha_p'\right)_o\left[\underset{\sim k\ell}{\hat{\psi}'}\left(p+\frac{n}{N}\right)_e - \underset{\sim k\ell}{\hat{\psi}'}\left(p-\frac{n}{N}\right)_e\right]\right\}. \tag{7.198b}$$

Equation (7.94) provided an expression for the complex form of the mesh attenuation functions, $\hat{\phi}_{k\ell}(n/N)$; and Equation (7.182) expressed $\hat{\phi}_{k\ell}(n/N)$ in terms of its real and imaginary parts, which are, respectively, even and odd functions of n/N. These even and odd contributions, $\hat{\phi}_{k\ell}(n/N)_e$ and $\hat{\phi}_{k\ell}(n/N)_o$, are given by Equations (7.198a) and (7.198b). The even and odd contributions, $\left(\alpha_p'\right)_e$ and $\left(\alpha_p'\right)_o$ of the normalized reciprocal mesh stiffness α_p', Equations (7.190) and (7.147), (7.148), are, respectively a_p and b_p, Equations (7.149a) and (7.149b). These are obtained as the solutions of the simultaneous linear equations, Equations (7.159) and (7.160), or Equations (7.171) and (7.172a, b), as appropriate. They also can be evaluated using Equations (7.138–7.141). The functions $\underset{\sim k\ell}{\hat{\psi}'}(g)_e$ and $\underset{\sim k\ell}{\hat{\psi}'}(g)_o$ with $g = n/N$, $g = p + (n/N)$, or $g = p - (n/N)$, are given by Equations (7.192a, b). If $k + l = even$, only the $\underset{\sim k\ell}{\hat{\psi}'}(g)_e$ terms contribute to $\hat{\phi}_{k\ell}(n/N)$, and if $k + l = odd$, only the $\underset{\sim k\ell}{\hat{\psi}'}(g)_o$ terms contribute to $\hat{\phi}_{k\ell}(n/N)$.

The even and odd Fourier series coefficients of the working-surface-deviation contributions to the transmission error, Equation (7.177), are given by Equations (7.186) and (7.187), respectively. The right-hand sides of these equations are given by Equations (7.184) and (7.185), respectively, in terms of the DFTs given by Equations (7.180) and (7.181), and the even and odd mesh-attenuation function contributions given by Equations (7.198a) and (7.198b), respectively.

Transmission-Error Rotational-Harmonic rms Spectrum from Working-Surface-Deviations

Finally, the transmission-error rotational-harmonic rms one-sided spectrum from the working-surface-deviations on the gear under consideration is computed from the even and odd Fourier series coefficients, Equations (7.186) and (7.187), according to Equation (5.18a), by

$$[G_\zeta(n)]^{1/2} = \sqrt{2}\,|\alpha_n| = (2\alpha_n\alpha_n^*)^{1/2}$$
$$= [2\,(\alpha_{ne}^2 + \alpha_{no}^2)]^{1/2}, \qquad n = 1, 2, 3, \ldots \qquad (7.199)$$

where the asterisk denotes the complex conjugate, and Equation (7.173) was used.

Appendix 7.A. Integral Equation for and Interpretation of Local Tooth-Pair Stiffness $K_{Tj}(x, y)$ Per Unit Length of Line of Contact

The validity of the expression for the transmission error given by Equation (7.15) is dependent on the validity of $K_{Tj}(x, y)$, the local tooth-pair stiffness per unit length of line of contact, utilized in Equation (7.8), which describes the total loading $W_j(x)$ carried by tooth-pair j as a function of the combined local elastic deformation $u_j(x, y)$, Equation (7.4), along the line of contact of the mating pair of teeth. The validity of $K_{Tj}(x, y)$, as used in Equation (7.8), a method to compute it, and its interpretation, are established below (Mark, 1978).

Let $k_j(v, v'; x)$ denote the combined elastic deformation of tooth-pair j at a point v on the line of contact of the tooth pair due to a unit force applied at a point v' on the line of contact, where force and deformation both are measured in the direction defined by the intersection of the plane of contact and transverse plane, Figure 2.6, and where x denotes the roll-distance position of a generic mating tooth pair, as illustrated in Figure 2.6. The location of a line of contact on the tooth pair, Figure 3.2, is related to x, Figure 2.6, by Equation (7.17), $s = x - j\Delta$. The quantity $k_j(v, v'; x)$ is known as an "influence function" (Heaslet, 1962, p. 17–3; Fung, 1965, p. 19); $k_j(v, v'; x)$ is assumed to include bending, shear, and Hertzian deformation components. The unit force used to *define* $k_j(v, v'; x)$ is assumed to be applied uniformly to a circular area of diameter equal to the width of the contact line between the two gear teeth. Since the Hertzian deformation component of real teeth is a (weakly) non-linear function of loading because of dependence of the width of the contact line on loading, specification of the area of applied force implies that our influence function is defined for a given nominal loading condition.

The deformation $u_j'(v\,;x)$ of the tooth-pair j at location v on the line of contact, caused by a lineal force density $p_j(v';x)$ applied on the line of contact at points of application v', can be expressed using the influence function $k_j(v,v';x)$ by

$$u_j'(v;x) = \int k_j(v,v';x)p_j(v';x)dv',\qquad(7.A.1)$$

where dv' denotes differential distance measured along the line of contact and where the integral is taken over the portion of the line of contact contained in the zone of contact. Deformations $u_j(x,y)$ in Equation (7.8) and $u_j'(v\,;x)$ in Equation (7.A.1) are related by

$$u_j(x,y) = u_j'(y\sec\psi_b;x),\qquad(7.A.2)$$

where we note from Figure 2.6 that locations v along the line of contact are related to axial locations y by

$$v = y\sec\psi_b.\qquad(7.A.3)$$

Equation (7.A.1) can be regarded as a Fredholm integral equation of the first kind (Heaslet, 1962, p.17-1) for the lineal force density $p_j(v';x)$. However, if we resolve the integration in Equation (7.A.1) into the sum of a large number of increments $\delta v' \approx dv'$, we obtain, instead, a set of a large number of simultaneous linear algebraic equations for the force density $p_j(v';x)$ to be determined from the deformations $u_j'(v;x)$ and the *matrix* $k_j(v,v';x)$ which replaces the continuous influence function, and where, in this matrix formulation of Equation (7.A.1), the deformation "vector" $u_j'(v;x)$ normally is written on the right-hand side (e.g., Press *et al.*, 1999, p. 779). The resultant system of linear algebraic equations is positive-definite (Fung, 1965, pp. 4–9, 161); the influence function is symmetric, that is,

$$k_j(v,v';x) = k_j(v',v;x),\qquad(7.A.4)$$

and moreover, a unique inverse kernel (matrix) $k_j^{-1}(v,v'\,;x)$ exists which also is symmetric,

$$k_j^{-1}(v,v';x) = k_j^{-1}(v',v;x).\qquad(7.A.5)$$

Using the inverse kernel, the lineal force distribution $p_j(v\,;x)$ can be obtained for any deformation distribution $u_j'(v'\,;x)$ by using the inverse to Equation (7.A.1); (Fung, 1965, p. 9),

$$p_j(v;x) = \int k_j^{-1}(v,v',x)u_j'(v';x)dv'.\qquad(7.A.6)$$

Integration of $p_j(v;x)$ over the full line of contact yields the total force $W_j(x)$ transmitted by tooth pair j:

$$W_j(x) = \int p_j(v\,;x)\,dv$$

$$= \int \left[\int k_j^{-1}(v, v'; x) dv \right] u'_j(v'; x) dv'$$

$$= \int K'_{Tj}(x, v') u'_j(v'; x) dv' \qquad (7.A.7a, b)$$

where in going to the second line we have used Equation (7.A.6), and in going to the last line we have defined

$$K'_{Tj}(x, v') \triangleq \int k_j^{-1}(v, v'; x) dv. \qquad (7.A.8)$$

To put Equation (7.A.7b) into the form of Equation (7.8), we transform v' to y using $v' = y \sec \psi_b$, as in Equation (7.A.3), thereby obtaining, with $dv' = \sec \psi_b \, dy$,

$$W_j(x) = \sec \psi_b \int_{y_A(x)}^{y_B(x)} K'_{Tj}(x, y \sec \psi_b) u'_j(y \sec \psi_b; x) dy. \qquad (7.A.9)$$

From Equations (7.8), (7.A.2), and (7.A.9), we have

$$K_{Tj}(x, y) = K'_{Tj}(x, y \sec \psi_b)$$

$$= \int k_j^{-1}(v, y \sec \psi_b; x) dv, \qquad (7.A.10a, b)$$

where Equation (7.A.8) was used in going to the second line.

Equation (7.A.10b) expresses the required local tooth-pair stiffness per unit length of line of contact $K_{Tj}(x, y)$ in terms of the inverse influence function $k_j^{-1}(v, v'; x)$ with $v' = y \sec \psi_b$, which is known to exist (Fung, 1965, p. 9). However, it is shown below that this inverse influence function $k_j^{-1}(v, v'; x)$ need not be evaluated in order to determine the required $K_{Tj}(x, y)$. Let $\bar{p}_j(v'; x)$ be the solution to Equation (7.A.1) for a deformation $u'_j(v; x) \equiv \bar{u}$, which is a *constant value* independent of location v on the line of contact, that is,

$$\bar{u} = \int k_j(v, v'; x) \bar{p}_j(v'; x) dv'. \qquad (7.A.11)$$

Then, from Equation (7.A.6) and the fact that $\bar{u}$ is a constant independent of location v' on the line of contact, we have

$$\bar{p}_j(v; x) = \bar{u} \int k_j^{-1}(v, v'; x) dv'. \qquad (7.A.12)$$

But, from the symmetric property, Equation (7.A.5), of the inverse influence function, there follows from Equations (7.A.5) and (7.A.12),

$$\bar{p}_j(v; x)/\bar{u} = \int k_j^{-1}(v', v; x) dv', \qquad (7.A.13)$$

which for $v = y \sec \psi_b$, Equation (7.A.3) becomes

$$\bar{p}_j(y \sec \psi_b; x)/\bar{u} = \int k_j^{-1}(v', y \sec \psi_b; x)dv'. \tag{7.A.14}$$

Comparing Equations (7.A.10b) and (7.A.14) then yields

$$K_{Tj}(x, y) = \bar{p}_j(y \sec \psi_b; x)/\bar{u}. \tag{7.A.15}$$

Because $\bar{p}_j(v'; x)$ is the lineal force density (force per unit length of line of contact), the right-hand side of Equation (7.A.15) has the dimension of force divided by length squared, which is the same dimension as $K_{Tj}(x, y)$, most easily seen from Equation (7.8).

It follows from Equations (7.A.11) and (7.A.15) that the lineal force density $\bar{p}_j(v'; x)$ that yields a constant deformation $\bar{u}$ along the line of contact provides the local tooth-pair stiffness $K_{Tj}(x, y)$ per unit length of line of contact. Equation (7.A.11) is the integral equation whose solution yields the local tooth-pair stiffness with the aid of Equation (7.A.15).

Notice that the above-described solution for $K_{Tj}(x, y)$ by Equations (7.A.15) and (7.A.11) requires only the original influence function $k_j(v, v'; x)$ of Equation (7.A.1) and *not* its inverse $k_j^{-1}(v', v; x)$. The solution given by Equation (7.A.15) makes intuitive sense – for example, at the ends of a line of tooth-pair contact on a helical gear-pair where the local tooth stiffness is less than near the center of the contact region, less local loading $\bar{p}_j$ will be required to provide a constant deformation $\bar{u}$ than near the center of the contact region.

Welker (1996) and Jankowich (1997) have developed a method for tailoring finite-element dimensions along lines of contact to yield accurate Hertzian deformation contributions, and Alulis (1999) has utilized these results in developing a method for accurately computing the influence function described by Equations (7.A.11) and (7.A.15).

Appendix 7.B. Transformation of Tooth-Contact-Line Coordinates to Cartesian Working-Surface Coordinates

Figure 2.6 shows the plane of contact of a pair of meshing helical gears. The base helix angle ψ_b shown there is the helix angle of the lower gear. It is a "right-hand" helix (Drago, 1988, pp. 483, 484; Colbourne, 1987, p. 319). Two external helical gears operating on parallel axes must be of opposite hand. The figure on page 309 of Colbourne (1987) nicely illustrates this fact. Hence, when properly interpreted, Figure 2.6 illustrates a generic pair of meshing external helical gears.

Figure 7.1 shows a sketch of the lower gear in Figure 2.6 meshing with the basic rack (Buckingham, 1949, p. 153; Colbourne, 1987, p. 309) of the upper gear of Figure 2.6. The basic rack of an involute gear is the infinite-radius counterpart to that gear with the same transverse pressure angle ϕ, base pitch Δ, base helix angle ψ_b, lines of contact, and zone of contact as its finite-radius counterpart. Hence, the contact geometry of the finite-radius gear and its infinite-radius basic-rack counterpart are identical. Dealing with the basic rack of the upper gear of Figure 2.6 simplifies the required derivation involving the contact geometry.

Recall from Figure 3.3 the $R_b \epsilon$ is roll distance along the fictitious belt drive riding on the two base cylinders. The height D on the tooth-working-surfaces of the active contact region defined by Equation (3.7) is, simply, the projection of that contact region onto the line of shaft centers illustrated in Figure 3.3. This height D is illustrated in Figure 3.2, and is shown again in the upper portion of Figure 7.1.

The "width" of the rectangular contact zone shown in the upper portion of Figure 2.6 is L. This "width" L is the roll distance within the zone of contact, as can be seen from the projection of L onto the plane of contact in the lower portion of Figure 2.6. Hence, recognizing the roll-angle upper and lower limits, ϵ_t and ϵ_r respectively, of the contact region in Equation (3.7), we have

$$L = R_b \left(\epsilon_t - \epsilon_r \right). \tag{7.B.1}$$

Consequently, from Equations (3.7) and (7.B.1), we have

$$\frac{D}{L} = \sin \phi. \tag{7.B.2}$$

From Figure 2.6, distance A is the axial projection of a full length of line of contact. Consequently, from Figure 2.6, we have

$$\frac{L}{A} = \tan \psi_b, \tag{7.B.3}$$

as indicated earlier in Equation (2.12). Distance A also is shown in Figure 7.1.

As in Figure 3.2, the upper-right view in Figure 7.1 of the tooth is the projection of the tooth-working-surface onto the plane of the two parallel gear shafts, Figure 3.3b. The angle θ in Figure 7.1 is the angle between the horizontal y axis and the line of contact projection shown there. From D and A, both shown in Figure 7.1, there follows

$$\frac{D}{A} = \tan \theta. \tag{7.B.4}$$

Combining Equations (7.B.2) and (7.B.3) with Equation (7.B.4) there follows

$$\tan \theta = \frac{D}{A} = \sin \phi \tan \psi_b. \tag{7.B.5}$$

As shown earlier in Figure 3.2, as the gears rotate, the line of contact between two mating teeth moves parallel to itself along the tooth-working-surfaces. Two such lines of contact are shown in the upper-right projection in Figure 7.1. The dashed line of contact passes through the origin $y = 0$, $z = 0$ of the Cartesian y, z coordinate system; the solid line of contact represents a generic line-of-contact position. The radial z coordinate difference of the location of the solid generic line-of-contact location from the location passing through the origin is δz, as shown in the upper-right projection.

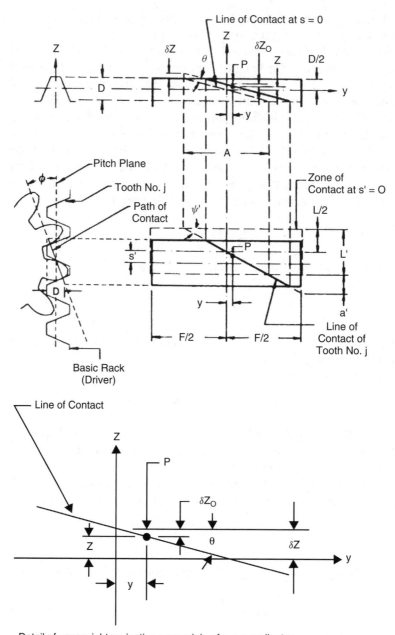

Detail of upper-right projection near origin of y, z coordinates

Figure 7.1 Four orthogonal views of the lower gear in Figure 2.6 meshing with the basic rack of the upper gear in Figure 2.6 (Adapted from Mark (1978))

As discussed earlier in the explanation of Equation (7.17) and its relation to Figure 3.2, at roll-distance location $s = 0$ of a generic mating tooth pair, the line of contact passes through the Cartesian tooth-working-surface coordinate origin $y = 0$, $z = 0$. Since s describes roll-distance motion along the path of contact in the lower-left projection in Figure 7.1, and δz is a radial coordinate normal to the pitch plane, it follows from the lower-left projection in Figure 7.1 that

$$\frac{\delta z}{s} = \sin \phi. \tag{7.B.6}$$

Referring again to the upper-right projection in Figure 7.1, δz_0 describes the radial distance between the point p on the solid line of contact and the radial location where the solid line of contact intersects the z-axis at $y = 0$. Therefore,

$$\frac{\delta z_0}{y} = \tan \theta. \tag{7.B.7}$$

Then, the radial z coordinate of a generic point p on the solid line of contact is

$$z = \delta z - \delta z_0. \tag{7.B.8}$$

Hence, from Equations (7.B.6), (7.B.7), and (7.B.5), there follows from Equation (7.B.8),

$$z = s \sin \phi - y \tan \theta$$

$$= s \sin \phi - y \sin \phi \tan \psi_b. \tag{7.B.9a, b}$$

Equation (7.B.9b) relates the Cartesian y, z coordinates of a generic point p on the line of contact on the tooth shown in Figure 7.1 to roll-distance-location s relative to the origin $y = 0$, $z = 0$, as a function of basic gear parameters, transverse pressure angle ϕ and base helix angle ψ_b of the lower follower gear illustrated in Figures 2.6 and 7.1.

It is now necessary to relate the above result to the four possible applications of a single helical gear – that is, right-hand versus left-hand helix, and driver versus follower for either right-hand or left-hand helix. Because the helix angle ψ_b in Figures 2.6 and 7.1, and Equation (7.B.9b), is that of the lower right-hand follower (RHF) gear in Figure 2.6, it will be convenient to do right-hand helix follower first.

Right-Hand Helix Follower

The tooth shown in the upper-right projection in Figure 7.1 is a tooth of the upper gear in Figure 2.6. The lower gear in Figure 2.6 is a right-hand helix follower. But meshing gears have tip to root contact. Thus, to apply Equation (7.B.9b) to the RHF case, a change in the sign of z is required, that is, for RHF,

$$z = -s \sin \phi + y \sin \phi \tan \psi_b. \tag{7.B.10}$$

It is important to notice that the direction of increasing y shown in Figures 2.6 and 7.1 is the same.

Right-Hand Helix Driver

This case remains the lower gear in Figure 2.6, but acting as a driver rather than a follower. This lower gear becomes a driver by changing the location of the plane of contact shown there which is accomplished, exactly, by changing the sign of the pressure angle ϕ. But $\sin \phi$ is an odd function, that is, $\sin (-\phi) = -\sin \phi$. Making this change in Equation (7.B.10) yields for the case of a right-hand helical driver (RHD),

$$z = s \sin \phi - y \sin \phi \tan \psi_b. \tag{7.B.11}$$

Left-Hand Helix Driver

Because it is necessary for an external left-hand helical gear to mesh with a right-hand helical gear, the upper gear in Figure 2.6 has a left-hand helix, which is of opposite sign to a right-hand helix. But $\tan (-\psi_b) = -\tan \psi_b$. Hence, reversing the sign of ψ_b in Equation (7.B.9b) yields for the left-hand driver (LHD) case,

$$z = s \sin \phi + y \sin \phi \tan \psi_b. \tag{7.B.12}$$

Left-Hand Helix Follower

As mentioned earlier, a change from driver to follower requires a change in the sign of the pressure angle ϕ shown in Figure 2.6. This, in turn, changes the sign of $\sin \phi$ in Equation (7.B.12). Therefore, for the case of a left-hand helix follower (LHF) it follows from Equation (7.B.12) that

$$z = -s \sin \phi - y \sin \phi \tan \psi_b. \tag{7.B.13}$$

Recall that the sine and tangent functions both are odd functions – that is, $\sin (-\phi) = -\sin \phi$ and $\tan (-\psi_b) = -\tan \psi_b$. Hence, both driver results above are identical except for the sign of helix angle ψ_b; and both follower results above also are identical except for sign of helix angle, as one would expect. Moreover, both right-hand results are identical except for the sign of the pressure angle ϕ, and both left-hand results are identical except for the sign of the pressure angle, also as one would expect. *In the above tabulations, and those to follow, ϕ and ψ_b are taken as positive.*

Generic Representation

The generic representation used in Equations (7.18–7.21) is

$$y = y \tag{7.B.14}$$

$$z = \beta s + \gamma y \tag{7.B.15}$$

Table 7.B.1 Rearranged summary of Equations (7.B.10) – (7.B.13)

RHD	$z = s \sin \phi - y \sin \phi \tan \psi_b$
LHD	$z = s \sin \phi + y \sin \phi \tan \psi_b$
RHF	$z = -s \sin \phi + y \sin \phi \tan \psi_b$
LHF	$z = -s \sin \phi - y \sin \phi \tan \psi_b$

Table 7.B.2 Signs for β and γ

	Sign for β	Sign for γ
RHD	+	−
LHD	+	+
RHF	−	+
LHF	−	−

with

$$\beta = \pm \sin \phi \tag{7.B.16}$$

$$\gamma = \pm \sin \phi \tan \psi_b \tag{7.B.17a}$$

$$= \pm \sin \phi \cos \phi \tan \psi \tag{7.B.17b}$$

where the signs for β and γ are determined from Table 7.B.1 as displayed in Table 7.B.2.

Appendix 7.C. Fourier Transform and Fourier Series

Fourier Transform

Let $\hat{G}(g)$ be the Fourier transform of a function $G(x)$. Then $\hat{G}(g)$ and $G(x)$ are uniquely related by

$$\hat{G}(g) \triangleq \int_{-\infty}^{\infty} G(x) \exp(-i2\pi gx) dx \tag{7.C.1}$$

and

$$G(x) = \int_{-\infty}^{\infty} \hat{G}(g) \exp(i2\pi gx) dg. \tag{7.C.2}$$

Fourier Series

Now let $G(x)$ be a periodic function with period Δ, that is,

$$G(x + j\Delta) = G(x), \qquad j = \dots -2, -1, 0, 1, 2, \dots . \tag{7.C.3}$$

Let α_p be the pth Fourier series coefficient of $G(x)$,

$$\alpha_p \triangleq \frac{1}{\Delta} \int_{-\Delta/2}^{\Delta/2} G(x) \exp(-i2\pi px/\Delta) dx , \quad p = 0, \pm 1, \pm 2, \dots . \tag{7.C.4}$$

Then $G(x)$ is related to the α_p by

$$G(x) = \sum_{p=-\infty}^{\infty} \alpha_p \exp(i2\pi px/\Delta). \tag{7.C.5}$$

Dirac Delta ''Function''

Define the Dirac delta "function" $\delta(g)$ by

$$\delta(g) = \begin{cases} \infty, & g = 0 \\ 0, & g \neq 0 \end{cases} \tag{7.C.6a}$$

and

$$\int_{-\infty}^{\infty} \delta(g)\, dg = \int_{-\epsilon/2}^{\epsilon/2} \delta(g) dg = 1, \tag{7.C.6b}$$

for any small positive ϵ. Then for any continuous function $f(g)$,

$$\int_{-\infty}^{\infty} f(g)\delta(g - v) dg = f(v). \tag{7.C.6c}$$

Fourier Transform of Fourier Series

Suppose $G(x)$ is periodic with period Δ. If

$$\hat{G}(g) = \sum_{p=-\infty}^{\infty} \alpha_p \, \delta(g - p/\Delta) \tag{7.C.7}$$

where $\delta(\cdot)$ is the Dirac delta "function", then by substituting Equation (7.C.7) into Equation (7.C.2), there follows

$$G(x) = \sum_{p=-\infty}^{\infty} \alpha_p \int_{-\infty}^{\infty} \delta(g - p/\Delta) \exp(i2\pi gx) dg$$

$$= \sum_{p=-\infty}^{\infty} \alpha_p \exp\left(\frac{i2\pi px}{\Delta}\right), \tag{7.C.8a, b}$$

which, according to Equation (7.C.5), is the Fourier series representation of the periodic function $G(x)$. Hence, Equation (7.C.7) describes the Fourier transform of the periodic function $G(x)$ in terms of the Fourier series coefficients α_p of $G(x)$.

Fourier Transform of the Product of Two Functions

Let $G(x)$ be the product of two functions $G'(x)$ and $G''(x)$, that is,

$$G(x) = G'(x)G''(x), \tag{7.C.9}$$

and let $\hat{G}'(g)$ and $\hat{G}''(g_1)$ be the Fourier transforms of $G'(x)$ and $G''(x)$, respectively, that is,

$$\hat{G}'(g) \triangleq \int_{-\infty}^{\infty} G'(x)\exp(-i2\pi gx)dx \qquad (7.\text{C}.10)$$

$$\hat{G}''(g_1) \triangleq \int_{-\infty}^{\infty} G''(x)\exp(-i2\pi g_1 x)dx. \qquad (7.\text{C}.11)$$

Then, from Equation (7.C.2), the inverse Fourier transform of $\hat{G}''(g_1)$, Equation (7.C.11), is

$$G''(x) = \int_{-\infty}^{\infty} \hat{G}''(g_1)\exp(i2\pi g_1 x)dg_1. \qquad (7.\text{C}.12)$$

The Fourier transform, Equation (7.C.1) of the product, Equation (7.C.9), is

$$\hat{G}(g) = \int_{-\infty}^{\infty} G'(x)G''(x)\exp(-i2\pi gx)dx. \qquad (7.\text{C}.13)$$

Substitute Equation (7.C.12) into Equation (7.C.13), then reverse the order of integration,

$$\hat{G}(g) = \int_{-\infty}^{\infty} G'(x) \int_{-\infty}^{\infty} \hat{G}''(g_1)\exp(i2\pi g_1 x)dg_1 \exp(-i2\pi gx)dx \qquad (7.\text{C}.14\text{a})$$

$$= \int_{-\infty}^{\infty} \hat{G}''(g_1) \int_{-\infty}^{\infty} G'(x) \exp\left[-i2\pi (g - g_1) x\right] dx dg_1 \qquad (7.\text{C}.14\text{b})$$

$$= \int_{-\infty}^{\infty} \hat{G}''(g_1)\hat{G}'(g - g_1)dg_1, \qquad (7.\text{C}.14\text{c})$$

where the definition, Equation (7.C.1) applied to $G'(x)$ was used in going to the last line. Equation (7.C.14c) describes the convolution of the Fourier transforms of the two functions $G'(x)$ and $G''(x)$ whose product was formed in Equation (7.C.13). Equation (7.C.14c) is the Frequency Convolution Theorem for Fourier transforms.

Fourier Series of the Product of Two Periodic Functions with the Same Period

Now let $G'(x)$ and $G''(x)$ in Equation (7.C.9) each be periodic with the same period Δ as in Equations (7.C.3–7.C.5). With

$$w \triangleq \exp(i2\pi x/\Delta) \qquad (7.\text{C}.15)$$

it follows from Equation (7.C.5) that $G'(x)$ and $G''(x)$ each can be expressed as

$$G'(x) = \sum_{m=-\infty}^{\infty} \alpha'_m w^m \tag{7.C.16}$$

and

$$G''(x) = \sum_{n=-\infty}^{\infty} \alpha''_n w^n \tag{7.C.17}$$

where α'_m and α''_n are the Fourier series coefficients Equation (7.C.4) of $G'(x)$ and $G''(x)$, respectively. Let the Fourier series coefficients of the product,

$$G(x) = G'(x)G''(x), \tag{7.C.18}$$

be denoted by α_p. Then, using Equation (7.C.5) and Equations (7.C.15–7.C.17), we have for the Fourier series representation of Equation (7.C.18),

$$\sum_{p=-\infty}^{\infty} \alpha_p w^p = \sum_{m=-\infty}^{\infty} \alpha'_m w^m \sum_{n=-\infty}^{\infty} \alpha''_n w^n \tag{7.C.19a}$$

$$= \sum_{m=-\infty}^{\infty} \alpha'_m \sum_{n=-\infty}^{\infty} \alpha''_n w^{m+n}. \tag{7.C.19b}$$

Let $p = m + n$, then $n = p - m$. But for any value of m the infinite summation over n in Equation (7.C.19b) becomes an infinite summation over p, that is,

$$\sum_{p=-\infty}^{\infty} \alpha_p w^p = \sum_{m=-\infty}^{\infty} \alpha'_m \sum_{p=-\infty}^{\infty} \alpha''_{p-m} w^p$$

$$= \sum_{p=-\infty}^{\infty} \left(\sum_{m=-\infty}^{\infty} \alpha'_m \alpha''_{p-m} \right) w^p. \tag{7.C.20}$$

Hence, re-introducing Equations (7.C.5) and (7.C.15) into Equation (7.C.20), there follows

$$G(x) = \sum_{p=-\infty}^{\infty} \left(\sum_{m=-\infty}^{\infty} \alpha'_m \alpha''_{p-m} \right) \exp\left(\frac{i2\pi px}{\Delta} \right)$$

$$= \sum_{p=-\infty}^{\infty} \alpha_p \exp(i2\pi px/\Delta), \tag{7.C.21}$$

where

$$\alpha_p = \sum_{m=-\infty}^{\infty} \alpha'_m \alpha''_{p-m} \tag{7.C.22a}$$

$$= \alpha'_p * \alpha''_p, \tag{7.C.22b}$$

where the asterisk denotes the convolution operation, Equation (7.C.22a). Equation (7.C.22) describes the discrete convolution of the Fourier series coefficients

α'_m and α''_n of the periodic functions $G'(x)$ and $G''(x)$ of Equations (7.C.16) and (7.C.17) whose product was taken in Equation (7.C.18). Equation (7.C.22) is the Frequency Convolution Theorem for Fourier series. Formulas for evaluation of discrete convolutions, utilizing only real quantities, are provided in Appendix 7.F.

Fourier Series Coefficients of the rep Function

The "rep function" introduced by Woodward (1964, p. 28) has been useful due to the periodic character of meshing gear teeth. For a generic periodic function $v(x)$, which repeats at intervals Δ, define

$$rep_\Delta v(x) \triangleq \sum_{j=-\infty}^{\infty} v(x - j\Delta). \tag{7.C.23}$$

Because $rep_\Delta\ v(x)$ is periodic with period Δ, it can be represented by Fourier series, that is, by Equation (7.C.5),

$$\sum_{j=-\infty}^{\infty} v(x - j\Delta) = \sum_{p=-\infty}^{\infty} \alpha_p \exp(i2\pi px/\Delta). \tag{7.C.24}$$

Defining the Fourier transform, Equation (7.C.1), of $v(x)$ as

$$\hat{v}(g) \triangleq \int_{-\infty}^{\infty} v(x)\exp(-i2\pi gx)dx, \tag{7.C.25}$$

there follows from the Poisson sum formula (Kammler, 2000, p. 50),

$$\sum_{j=-\infty}^{\infty} v(x - j\Delta) = \frac{1}{\Delta} \sum_{p=-\infty}^{\infty} \hat{v}(p/\Delta)\exp(i2\pi px/\Delta). \tag{7.C.26}$$

By comparing Equations (7.C.24) and (7.C.26) with Equation (7.C.23) it follows that the Fourier series coefficients α_p of $rep_\Delta v(x)$ can be expressed in terms of the Fourier transform $\hat{v}(g)$ of $v(x)$ by

$$\alpha_p = \frac{1}{\Delta}\hat{v}(p/\Delta). \tag{7.C.27}$$

Appendix 7.D. Fourier Transform of Scanning Content of the Line Integral (Equation (7.24b, c)) $\underline{f}_j(s) \triangleq \sec \psi_b \int_{y_A(s)}^{y_B(s)} f_{Cj}(y, \beta s + \gamma y)\, dy$

Consider a generic function $f_{Cj}(y, z)$ defined on the Cartesian tooth-coordinate system illustrated in Figure 3.2. The function defined by Equations (7.24b, c),

$$\underline{f}_j(s) \triangleq \sec\psi_b \int_{y_A(s)}^{y_B(s)} f_{Cj}(y, \beta s + \gamma y)dy, \tag{7.D.1}$$

describes the integral of $f_{Cj}(y, z)$ over the line of contact shown in Figure 3.2 as a function of line-of-contact roll-distance location s, where $s = 0$ designates the location of the line of contact when it passes through the origin $y = 0$, $z = 0$, shown dashed in Figure 3.2. The Fourier transform $\hat{f}_{\sim j}(g)$ of the scanning function $f_{\sim j}(s)$ defined by Equation (7.D.1) is required, where

$$\hat{f}_{\sim j}(g) \triangleq \int_{-\infty}^{\infty} f_{\sim j}(s) \exp(-i2\pi g s) ds. \qquad (7.D.2)$$

It is required to determine this Fourier transform as a function of $f_{Cj}(y, z)$ found in Equation (7.D.1) with

$$z = \beta s + \gamma y. \qquad (7.D.3)$$

It will be seen that this can be accomplished by obtaining $\hat{f}_{\sim j}(g)$, Equation (7.D.2), as a function of the two-dimensional Fourier transform of the function $f_{Cj}(y, z)$, that is,

$$\hat{f}_{Cj}(g_1, g_2) \triangleq \int_{-\infty}^{\infty} \int_{-\infty}^{\infty} f_{Cj}(y, z) \exp\left[-i2\pi (g_1 y + g_2 z)\right] dy dz. \qquad (7.D.4)$$

The inverse two-dimensional Fourier transform of Equation (7.D.4) is

$$f_{Cj}(y, z) = \int_{-\infty}^{\infty} \int_{-\infty}^{\infty} \hat{f}_{Cj}(g_1, g_2) \exp\left[i2\pi (g_1 y + g_2 z)\right] dg_1 dg_2, \qquad (7.D.5)$$

and with

$$z = \beta s + \gamma y \qquad (7.D.6)$$

in Equation (7.D.1),

$$f_{Cj}(y, \beta s + \gamma y) = \int_{-\infty}^{\infty} \int_{-\infty}^{\infty} \hat{f}_{Cj}(g_1, g_2) \exp\left\{i2\pi \left[g_1 y + g_2 (\beta s + \gamma y)\right]\right\} dg_1 dg_2. \qquad (7.D.7)$$

Substituting Equation (7.D.7) into Equation (7.D.1) gives

$$f_{\sim j}(s) = \sec\psi_b \int_{y_A(s)}^{y_B(s)} \int_{-\infty}^{\infty} \int_{-\infty}^{\infty} \hat{f}_{Cj}(g_1, g_2)$$
$$\times \exp\left\{i2\pi \left[g_1 y + g_2 (\beta s + \gamma y)\right]\right\} dg_1 dg_2 dy, \qquad (7.D.8)$$

and substituting this result into Equation (7.D.2) gives for the Fourier transform $\hat{f}_{\sim j}(s)$ of the scanning content function $f_{\sim j}(s)$,

$$\hat{f}_{\sim j}(s) = \sec\psi_b \int_{-\infty}^{\infty} \int_{y_A(s)}^{y_B(s)} \int_{-\infty}^{\infty} \int_{-\infty}^{\infty} \hat{f}_{Cj}(g_1, g_2)$$
$$\times \exp\left\{i2\pi \left[g_1 y + g_2 (\beta s + \gamma y) - gs\right]\right\} dg_1 dg_2 dy ds. \qquad (7.D.9)$$

At this juncture, it is important to understand the roles of the limits of integration of the fourfold integration in Equation (7.D.9). The inner two integrations over g_1 and g_2 pertain to the Fourier-transform pair, Equations (7.D.4) and (7.D.5). Hence, this two-dimensional Fourier transform automatically correctly represents the function $f_{Cj}(y, z)$ over the rectangular region $(-F/2) < y < (F/2)$, $(-D/2) < z < (D/2)$ illustrated in Figure 3.2. The outer two integrations over the spatial variables y and s could be used to control or limit the region within the working-surface, Figure 3.2, over which the line-integral, Equation (7.D.1), is taken. The role of these outer two integrations will become more obvious below after they are transformed to the rectangular y, z coordinate system of Figure 3.2.

Let us now interchange the orders of integration in Equation (7.D.9) and collect coefficients of s and y in the exponential function,

$$\hat{f}_{\sim j}(g) = sec\psi_b \int\limits_{-\infty}^{\infty} \int\limits_{-\infty}^{\infty} \hat{f}_{Cj}(g_1, g_2)$$

$$\times \int\limits_{-\infty}^{\infty} \int\limits_{y_A(s)}^{y_B(s)} \exp\left\{-i2\pi\left[(g - \beta g_2)s - (g_1 + \gamma g_2)y\right]\right\} dy ds dg_1\, dg_2. \qquad (7.D.10)$$

We now transform the inner integral over y and s to the Cartesian tooth coordinates y, z by using the transformation, Equations (7.18) and (7.19), that is,

$$y = y \qquad s = z/\beta - (\gamma/\beta)y. \qquad (7.D.11a, b)$$

This transformation requires the Jacobian (Hildebrand, 1976, p. 353) of the transformation,

$$dy ds = \left|\frac{\partial(y, s)}{\partial(y, z)}\right| dy dz, \qquad (7.D.12)$$

where

$$\left|\frac{\partial(y, s)}{\partial(y, z)}\right| = \begin{vmatrix} \dfrac{\partial y}{\partial y} & \dfrac{\partial y}{\partial z} \\[2mm] \dfrac{\partial s}{\partial y} & \dfrac{\partial s}{\partial z} \end{vmatrix} = \begin{vmatrix} 1 & 0 \\[1mm] -\dfrac{\gamma}{\beta} & \dfrac{1}{\beta} \end{vmatrix} = \frac{1}{\beta} \qquad (7.D.13)$$

by evaluating the above determinants. But, as mentioned above, the outer two integrations in Equation (7.D.9), which now are the inner two integrations in Equation (7.D.10), could be used to control or limit the region within the tooth surface, Figure 3.2, over which the line integral is taken. With the transformation from the y, s coordinates of Equation (7.D.10) to the Cartesian y, z coordinates illustrated in Figure 3.2, this region is described in the Cartesian y, z coordinates. Denote this region by the symbol Ω_C. Then, when the inner double integral in Equation (7.D.10) is expressed in the y, z Cartesian coordinates by substituting for s, as indicated by Equation (7.D.11b), also replacing $dy ds$ by $dy ds = \beta^{-1} dy dz$, as indicated by Equations (7.D.12) and (7.D.13), and denoting the controlling region of integration in the Cartesian y, z coordinates by Ω_C, the resultant integral expression for $\hat{\underset{\sim}{f}}_j(g)$ becomes

$$
\hat{\underset{\sim}{f}}_j(g) = \beta^{-1}sec\psi_b \int_{-\infty}^{\infty}\int_{-\infty}^{\infty} \hat{f}_{Cj}(g_1, g_2)
$$

$$
\times \int\int_{\Omega_C} \exp\left\{-i2\pi\left[(g - \beta g_2)\left(\frac{z}{\beta} - \frac{\gamma y}{\beta}\right) - (g_1 + \gamma g_2)\,y\right]\right\} dydzdg_1\,dg_2.
$$

$$
= \beta^{-1}sec\psi_b \int_{-\infty}^{\infty}\int_{-\infty}^{\infty} \hat{f}_{Cj}(g_1, g_2)
$$

$$
\times \int\int_{\Omega_C} \exp\left\{-i2\pi\left[-\left(\frac{\gamma g}{\beta} + g_1\right)y + \left(\frac{g}{\beta} - g_2\right)z\right]\right\} dydzdg_1\,dg_2. \qquad (7.D.14)
$$

The geometry of points on the line of contact shown in Figure 7.1 is described by Equation (7.B.9b). Consequently, from Equations (7.D.6), (7.B.9b), and (7.B.2), we have

$$
\beta = \sin\phi = D/L, \qquad (7.D.15)
$$

and from Equations (7.D.6), (7.B.9b), (7.B.2), and (7.B.3), we also have

$$
\gamma = -\sin\phi\tan\psi_b
$$

$$
= -(D/L)(L/A) = -D/A. \qquad (7.D.16a)
$$

Hence from Equations (7.D.15) and (7.D.16a),

$$
\frac{\gamma}{\beta} = -\frac{L}{A}. \qquad (7.D.16b)
$$

Substituting Equations (7.D.15) and (7.D.16b) into Equation (7.D.14) yields

$$
\hat{\underset{\sim}{f}}_j(g) = \frac{L}{D}sec\psi_b \int_{-\infty}^{\infty}\int_{-\infty}^{\infty} \hat{f}_{Cj}(g_1, g_2)
$$

$$
\times \int\int_{\Omega_C} \exp\left\{-i2\pi\left[\left(\frac{Lg}{A} - g_1\right)y + \left(\frac{Lg}{D} - g_2\right)z\right]\right\} dydzdg_1\,dg_2. \qquad (7.D.17)
$$

Let us now interpret the result, Equation (7.D.17). Define (Papoulis, 1968, p. 93) a zero-one function $\Omega_C(y, z)$ as

$$
\Omega_C(y, z) \triangleq \begin{cases} 1, & \text{inside contact region} \\ 0, & \text{outside contact region,} \end{cases} \qquad (7.D.18)
$$

which is defined for the Cartesian tooth coordinates y, z of Figure 3.2. Thus, multiplying a generic function $f_{Cj}(y, z)$ defined on the Cartesian tooth coordinates of Figure 3.2 by $\Omega_C(y, z)$, Equation (7.D.18), could be used to limit the region of integration originally described by Equation (7.D.1), that is, instead of Equation (7.D.1),

$$\underset{\sim}{f}_j(s) = sec\psi_b \int_{-\infty}^{\infty} \Omega_C(y, \beta s + \gamma y) f_{Cj}(y, \beta s + \gamma y) dy. \tag{7.D.19}$$

Let us now form the two-dimensional Fourier transform of the function $\Omega_C(y, z)$, Equation (7.D.18), as was carried out earlier for $f_{Cj}(y, z)$ by Equation (7.D.4),

$$\widehat{\Omega}_C(g', g'') \triangleq \int_{-\infty}^{\infty}\int_{-\infty}^{\infty} \Omega_C(y, z) exp[-i2\pi(g'y + g''z)] dy dz \tag{7.D.20a}$$

$$\equiv \int\int_{\Omega_C} exp[-i2\pi(g'y + g''z)] dy dz, \tag{7.D.20b}$$

where Ω_C denotes the region in the y, z coordinates where the zero-one function, Equation (7.D.18), is unity. We now observe that the double integral over y, z in Equation (7.D.17) has the same exact form as Equation (7.D.20b) with

$$g' = \frac{L}{A}g - g_1 \qquad g'' = \frac{L}{D}g - g_2. \tag{7.D.21a, b}$$

Therefore, using the definition of $\widehat{\Omega}_C(g', g'')$ given by Equation (7.D.20), the Fourier transform $\underset{\sim}{\hat{f}}_j(g)$, Equation (7.D.17), can be expressed as

$$\underset{\sim}{\hat{f}}_j(g) = \frac{L}{D} sec\psi_b \int_{-\infty}^{\infty}\int_{-\infty}^{\infty} \hat{f}_{Cj}(g_1, g_2) \widehat{\Omega}_C\left(\frac{L}{A}g - g_1, \frac{L}{D}g - g_2\right) dg_1 dg_2, \tag{7.D.22}$$

which we observe to be of a modified form of a two-dimensional frequency convolution.

Our principal interest is the case where the full rectangular region of the working-surfaces is utilized. In this case, $\Omega_C(y, z)$ can be taken to be infinite in both dimensions y and z. Consider the case where $\Omega_C(y, z)$ is a rectangular region of width F' in the axial y direction and height D' in the radial z direction. Then, $\Omega_C(y, z)$ can be expressed in this case as

$$\Omega_C(y, z) = rect(y/F')rect(z/D') \tag{7.D.23}$$

where the rectangular function is defined (Woodward, 1964, p. 29) as

$$rect(x) \triangleq \begin{cases} 1, & |x| < 1/2 \\ 0, & |x| > 1/2. \end{cases} \tag{7.D.24}$$

The case of principal interest is, then, the limiting case, $F' \to \infty$, $D' \to \infty$. For the case of Equation (7.D.23), we have

$$\Omega_C(y,z) = \begin{cases} 1, & (-F'/2) < y < (F'/2), \quad (-D'/2) < z < (D'/2) \\ 0, & \text{outside this rectangular region .} \end{cases} \tag{7.D.25}$$

The two-dimensional Fourier transform of $\Omega_C(y,z)$, Equation (7.D.23), is, according to Equations (7.D.20), (7.D.24), and (7.D.25),

$$\widehat{\Omega}_C(g',g'') = \int_{-\infty}^{\infty} \int_{-\infty}^{\infty} rect(y/F')rect(z/D')\exp[-i2\pi(g'y + g''z)]dydz \tag{7.D.26a}$$

$$= \int_{-F'/2}^{F'/2} \exp(-i2\pi g'y)dy \int_{-D'/2}^{D'/2} \exp(-i2\pi g''z)dz \tag{7.D.26b}$$

$$= \int_{-F'/2}^{F'/2} \cos(2\pi g'y)dy \int_{-D'/2}^{D'/2} \cos(2\pi g''z)dz , \tag{7.D.26c}$$

obtained by writing

$$\exp(-i2\pi g'y) = \cos(2\pi g'y) - i\sin(2\pi g'y) \tag{7.D.27a}$$

$$\exp(-i2\pi g''z) = \cos(2\pi g''z) - i\sin(2\pi g''z), \tag{7.D.27b}$$

and observing that the sine functions are odd functions of their arguments; hence, their integrations vanish over the even regions of the integrations in Equation (7.D.26b). The elementary integrations in Equation (7.D.26c) yield

$$\widehat{\Omega}_C(g',g'') = \frac{\sin(\pi F'g')}{\pi g'} \frac{\sin(\pi D'g'')}{\pi g''} \tag{7.D.28a}$$

$$= F'sinc(F'g')D'sinc(D'g'') \tag{7.D.28b}$$

where (Woodward, 1964, p. 29; Bracewell, 1965, p. 62) the function $sinc\ \xi$ is defined as

$$sinc\ \xi \triangleq \frac{\sin(\pi\xi)}{\pi\xi}, \tag{7.D.29}$$

which satisfies (Woodward, 1964, p. 29; Bracewell, 1965, p. 62)

$$\int_{-\infty}^{\infty} sinc\ \xi\ d\xi = 1 \tag{7.D.30}$$

as is readily shown using the known integral (Korn and Korn, 1961, Appendix E, integral 371)

$$\int_0^\infty \frac{\sin ax}{x}\,dx = \frac{\pi}{2}\,,\, a > 0 \qquad (7.D.31)$$

since $\sin (ax)/x$ is an even function of x.

Let $\xi = F'g'$ and $d\xi = F'\,dg'$. Then, from Equation (7.D.30),

$$\int_{-\infty}^\infty F'\,sinc(F'g')dg' = \int_{-\infty}^\infty sinc\,\xi\,d\xi = 1. \qquad (7.D.32)$$

Therefore, because $D'sinc\,(D'\,g'')$ has the same exact form as the integrand in Equation (7.D.32), it follows from Equations (7.D.28b) and (7.D.32) that for any values of F' and D', we have by integrating Equations (7.D.28a, b),

$$\int_{-\infty}^\infty \int_{-\infty}^\infty \widehat{\Omega}_C(g', g'')dg'dg'' = 1. \qquad (7.D.33)$$

But, the *nominal* width of $sinc\,\xi$ is proportional to the value of the spread between the first zeros of $sinc\,\xi$, which occur at $|\xi| = 1$, from Equation (7.D.29). Therefore, in the limit $F' \to \infty$, the integrand, $F'sinc\,(F'\,g')$ in Equation (7.D.32) approaches the Dirac delta function, $\delta(g')$, Equations (7.C.6a, b), and therefore, $F'\,sinc\,(F'\,g')D'\,sinc\,(D'\,g'')$, approaches with $F' \to \infty$, $D' \to \infty$, a two-dimensional Dirac delta function, that is, from Equations (7.D.28a, b),

$$\lim_{\substack{F'\to\infty \\ D'\to\infty}} \widehat{\Omega}_C(g', g'') = \delta(g')\delta(g''). \qquad (7.D.34)$$

Consequently, from Equations (7.D.22) and (7.D.34) there follows with $F' \to \infty$, and $D' \to \infty$,

$$\hat{f}_{\sim j}(g) = \frac{L}{D}\sec\psi_b \int_{-\infty}^\infty\int_{-\infty}^\infty \hat{f}_{Cj}(g_1, g_2)\,\delta\left(\frac{L}{A}g - g_1\right)\delta\left(\frac{L}{D}g - g_2\right)dg_1 dg_2$$

$$= \frac{L}{D}\sec\psi_b \hat{f}_{Cj}\left(\frac{L}{A}g, \frac{L}{D}g\right), \qquad (7.D.35)$$

according to the two-dimensional version of Equation (7.C.6c). Equation (7.D.35) provides the Fourier transform of the line integral

$$f_{\sim j}(s) = \sec\psi_b \int_{y_A(s)}^{y_B(s)} f_{Cj}(y, \beta s + \gamma y)dy, \qquad (7.D.36)$$

Equations (7.24b, c), where parameters L/D and L/A are given by Equations (7.63a, b). Evaluation of $\widehat{\Omega}_C(g', g'')$ for an elliptical contact region is given in Mark (1978).

Appendix 7.E. Fractional Error in Truncated Infinite Geometric Series

A truncated infinite geometric series with $n+1$ terms can be expressed as (Korn and Korn, 1961, p. 1.3-1)

$$S_n \triangleq \sum_{j=0}^{n} r^j = \frac{1 - r^{n+1}}{1 - r}. \tag{7.E.1}$$

Consequently, provided $|r| < 1$, with an infinite number of terms, $n = \infty$, Equation (7.E.1) becomes

$$S_\infty = \frac{1}{1 - r}, \qquad |r| < 1. \tag{7.E.2}$$

The error in truncating the infinite geometric series by using only $n+1$ terms therefore is

$$S_\infty - S_n = \frac{1 - \left(1 - r^{n+1}\right)}{1 - r} = \frac{r^{n+1}}{1 - r}. \tag{7.E.3}$$

The fractional error in the truncated series therefore is

$$\frac{S_\infty - S_n}{S_\infty} = r^{n+1} \tag{7.E.4}$$

which is the first neglected terms in S_n, Equation (7.E.1). If r is negative, this first neglected term will alternate in sign as n is increased.

Appendix 7.F. Evaluation of Discrete Convolution of Complex Quantities Using Real Quantities

Using the identification of terms given by Equations (7.193a–c), the mesh-attenuation function given by Equation (7.189), with $g = n/N$, is given by the discrete convolution, Equation (7.194):

$$\alpha_g = \sum_{p=-\infty}^{\infty} \alpha'_p \, \alpha''_{g-p} \tag{7.F.1}$$

where, according to Equations (7.195a–c), each term in Equation (7.F.1) is of the same form,

$$\alpha_g = \alpha_g^e - i\alpha_g^o \tag{7.F.2a}$$

$$\alpha'_p = \alpha_p'^{e} - i\alpha_p'^{o} \tag{7.F.2b}$$

$$\alpha''_g = \alpha_g''^{e} - i\alpha_g''^{o}, \tag{7.F.2c}$$

where each α function is real and an even or odd function of its subscript, as indicated by the superscripts e or o, respectively.

We first note that $\alpha_g'' \triangleq \hat{\underset{\sim}{\psi}}_{k\ell}(g)$, Equation (7.193c), is, ultimately, the result of a Fourier transform, Equation (7.54), with $m = k\ell$,

$$\hat{\underset{\sim}{\psi}}_{Km}(g) \triangleq \int_{-\infty}^{\infty} \underset{\sim}{\psi}_{Km}(s) \left[\cos(2\pi gs) - i \sin(2\pi gs)\right] ds$$

$$= \hat{\underset{\sim}{\psi}}_{Km}(g)_e - i\hat{\underset{\sim}{\psi}}_{Km}(g)_o \qquad (7.F.3)$$

from which it follows that $\hat{\underset{\sim}{\psi}}_{Km}(0)_o \equiv 0$, where subscripts e and o denote even and odd functions of g, as before. This fact also follows directly from Equation (7.192b) and Figure 6.2. Moreover, from Equations (7.193b), (7.148), and (7.149b), $b_o = 0$. Hence, using the notation of Equations (7.195b, c), these results are

$$\alpha_0^{'o} = 0, \quad \alpha_0^{''o} = 0. \qquad (7.F.4a,b)$$

Furthermore, from the definitions of even and odd functions of the subscripts, there follows

$$\alpha_{-p}^{'e} = \alpha_p^{'e}, \quad \alpha_{-p}^{'o} = -\alpha_p^{'o} \qquad (7.F.5a,b)$$

$$\alpha_{-g}^{''e} = \alpha_g^{''e}, \quad \alpha_{-g}^{''o} = -\alpha_g^{''o}. \qquad (7.F.6a,b)$$

Substituting Equations (7.F.2a–c) into Equation (7.F.1) we have

$$\alpha_g^e - i\alpha_g^o = \sum_{p=-\infty}^{\infty} \left(\alpha_p^{'e} - i\alpha_p^{'o}\right)\left(\alpha_{g-p}^{''e} - i\alpha_{g-p}^{''o}\right). \qquad (7.F.7)$$

Taking the product of the terms on the right hand side, and using $i^2 = -1$, the real and imaginary parts of both sides of Equation (7.F.7) can be separated, yielding

$$\alpha_g^e = \sum_{p=-\infty}^{\infty} \left(\alpha_p^{'e}\alpha_{g-p}^{''e} - \alpha_p^{'o}\alpha_{g-p}^{''o}\right) \qquad (7.F.8)$$

and

$$\alpha_g^o = \sum_{p=-\infty}^{\infty} \left(\alpha_p^{'e}\alpha_{g-p}^{''o} + \alpha_p^{'o}\alpha_{g-p}^{''e}\right). \qquad (7.F.9)$$

Treating first α_g^e, given by Equation (7.F.8), we have by using Equations (7.F.6a, b), that is, $\alpha_{g-p}^{''e} = \alpha_{p-g}^{''e}$ and $\alpha_{g-p}^{''o} = -\alpha_{p-g}^{''o}$,

$$\alpha_g^e = \sum_{p=-\infty}^{\infty} \left(\alpha_p^{'e}\alpha_{p-g}^{''e} + \alpha_p^{'o}\alpha_{p-g}^{''o}\right). \qquad (7.F.10)$$

Furthermore, using Equation (7.F.4a), Equations (7.F.5a, b), and Equations (7.F.6a, b), that is, $\alpha''^e_{-p-g} = \alpha''^e_{p+g}$ and $\alpha''^o_{-p-g} = -\alpha''^o_{p+g}$, the portion of the sum in Equation (7.F.10) over negative p can be combined with the portion over positive p to yield

$$\alpha^e_g = \alpha'^e_0 \alpha''^e_g + \sum_{p=1}^{\infty} \left[\alpha'^e_p \left(\alpha''^e_{p-g} + \alpha''^e_{p+g} \right) + \alpha'^o_p \left(\alpha''^o_{p-g} + \alpha''^o_{p+g} \right) \right]$$

$$= \alpha'^e_0 \alpha''^e_g + \sum_{p=1}^{\infty} \left[\alpha'^e_p \left(\alpha''^e_{p+g} + \alpha''^e_{p-g} \right) + \alpha'^o_p \left(\alpha''^o_{p+g} + \alpha''^o_{p-g} \right) \right] \qquad (7.F.11)$$

which is Equation (7.196a).

Treating α^o_p, Equation (7.F.9) in similar fashion, we have by using Equations (7.F.4a) and (7.F.5b),

$$\alpha^o_g = \alpha'^e_0 \alpha''^o_g + \sum_{p=1}^{\infty} \left[\alpha'^e_p \left(\alpha''^o_{g-p} + \alpha''^o_{g+p} \right) + \alpha'^o_p \left(\alpha''^e_{g-p} - \alpha''^e_{g+p} \right) \right] \qquad (7.F.12)$$

and further, by using Equations (7.F.6a, b), that is, $\alpha''^o_{g-p} = -\alpha''^o_{p-g}$ and $\alpha''^e_{g-p} = \alpha''^e_{p-g}$, there follows from Equation (7.F.12),

$$\alpha^o_g = \alpha'^e_0 \alpha''^o_g + \sum_{p=1}^{\infty} \left[\alpha'^e_p \left(\alpha''^o_{p+g} - \alpha''^o_{p-g} \right) - \alpha'^o_p \left(\alpha''^e_{p+g} - \alpha''^e_{p-g} \right) \right], \qquad (7.F.13)$$

which is Equation (7.196b).

References

Alulis, M.F. (1999) *A method for finite element computation of compliance influence functions of helical gear teeth.* MS thesis in Mechanical Engineering, The Pennsylvania State University, University Park, Pennsylvania.

Antosiewicz, H.A. (1964) Bessel functions of fractional order, in *Handbook of Mathematical Functions With Formulas, Graphs, and Mathematical Tables*, Chapter 10 (eds M. Abramowitz and I.A. Stegun), U.S. Government Printing Office, Washington, DC. Republished by Dover, Mineola, NY.

Bateman, H. (1954) in *Tables of Integral Transforms*, vol. **1** (ed. A. Erdelyi), McGraw-Hill, New York.

Bracewell, R. (1965) *The Fourier Transform and Its Applications*, McGraw-Hill, New York.

Buckingham, E. (1949) *Analytical Mechanics of Gears*, McGraw-Hill, New York. Republished by Dover, New York.

Colbourne, J.R. (1987) *The Geometry of Involute Gears*, Springer-Verlag, New York.

Drago, R.J. (1988) *Fundamentals of Gear Design*, Butterworths, Boston.

Fung, Y.C. (1965) *Foundations of Solid Mechanics*, Prentice-Hall, Englewood Cliffs, NJ.

Hildebrand, F.B. (1976) *Advanced Calculus for Applications*, 2nd edn, Prentice-Hall, Englewood Cliffs, NJ.

Jankowich, E.M. (1997) *A method for accurate finite-element computation of the contact deformation of helical gear teeth*. MS thesis in Mechanical Engineering, The Pennsylvania State University, University Park, Pennsylvania.

Kammler, D.W. (2000) *A First Course in Fourier Analysis*, Prentice-Hall, Upper Saddle River, NJ.

Korn, G.A. and Korn, T.M. (1961) *Mathematical Handbook for Scientists and Engineers*, McGraw-Hill, New York. Republished by Dover, Mineola, NY.

Heaslet, M.A. (1962) Integral equations, in *Handbook of Engineering Mechanics*, Chapter 17 (ed. W. Flugge), 1st edn, McGraw-Hill, New York, pp. 17.1- 17.6.

Lanczos, C. (1956) *Applied Analysis*, Prentice-Hall, Englewood Cliffs, NJ. Republished by Dover, Mineola, New York.

Mark, W.D. (1978) Analysis of the vibratory excitation of gear systems: basic theory. *Journal of the Acoustical Society of America*, **63**, 1409–1430.

Mark, W.D. (1979) Analysis of the vibratory excitation of gear systems. II: tooth error representations, approximations, and application. *Journal of the Acoustical Society of America*, **66**, 1758–1787.

Mark, W.D. (1983) Analytical reconstruction of the running surfaces of gear teeth using standard profile and lead measurements. *ASME Journal of Mechanisms, Transmissions, and Automation in Design*, **105**, 725–735.

Mark, W.D. (1987) Use of the generalized transmission error in the equations of motion of gear systems. *ASME Journal of Mechanisms, Transmissions, and Automation Design*, **109**, 283–291.

Mark, W.D. (1989) The generalized transmission error of parallel– axis gears. *ASME Journal of Mechanisms, Transmissions, and Automation in Design*, **111**, 414–423.

Mark, W.D. (1992b) Elements of gear noise prediction, in *Noise and Vibration Control Engineering: Principles and Applications*, Chapter 21, (eds L.L. Beranek and I.L. Ver), John Wiley & Sons, Inc., New York, pp. 735–770.

Mark, W.D., Lee, H., Patrick, R., and Coker, J.D. (2010) A simple frequency-domain algorithm for early detection of damaged gear teeth. *Mechanical Systems and Signal Processing*, **24**, 2807–2823.

Mark, W.D. and Reagor, C.P. (2007) Static-transmission-error vibratory-excitation contributions from plastically deformed gear teeth caused by tooth bending-fatigue damage. *Mechanical Systems and Signal Processing*, **21**, 885–905.

Mark, W.D., Reagor, C.P., and McPherson, D.R. (2007) Assessing the role of plastic deformation in gear-health monitoring by precision measurement of failed gears. *Mechanical Systems and Signal Processing*, **21**, 177–192.

Papoulis, A. (1968) *Systems and Transforms with Applications in Optics*, McGraw-Hill, New York.

Papoulis, A. (1977) *Signal Analysis*, McGraw-Hill, New York.

Press, W.H., Teukolsky, S.A., Vetterling, W.T., and Flannery, B.P. (1999) *Numerical Recipes in Fortran 77: The Art of Scientific Computing*, 2nd edn, Vol. **1** of Fortran Numerical Recipes, Cambridge University Press, Cambridge.

Remmers, E.P. (1972) *Analytical Gear Tooth Profile Design*, ASME Technical Paper 72-PTG-47, American Society of Mechanical Engineers, New York.

Tolstov, G.P. (1962) *Fourier Series*, Prentice-Hall, Englewood Cliffs, NJ. Republished by Dover, Mineola, NY.

Welker, W.E. (1996) *A method for accurate finite-element computation of the contact deformation of gear teeth*. MS thesis in Mechanical Engineering, The Pennsylvania State University, University Park, Pennsylvania.

Woodward, P.M. (1964) *Probability and Information Theory, with Application to Radar*, 2nd edn, Pergamon, Oxford.

8

Discussion and Summary of Computational Algorithms

A major purpose of this book is description of an efficient systematic method for measuring generic parallel-axis helical (or spur) gears using present-day dedicated computer-numerically-controlled (CNC) gear-metrology equipment, and development from such measurements useful metrics of gear performance and manufacturing diagnostic information. In this final chapter, some of the main results are summarized, in particular, methods of gear measurement and computation of various performance metrics. This summary includes descriptions of:

- systematic method of tooth-working-surface measurement
- working-surface-deviation representation by normalized Legendre polynomials
- simple metrics, Equation (8.9), assessing coincidence of base-cylinder axis and gear-measurement axis
- representation of working-surface-deviations of individual teeth, mean working-surface-deviation, and differences of individual working-surfaces from the mean working-surface
- simple metrics, Equations (8.10–8.12), assessing working-surface variability
- rotational-harmonic spectra of working-surface-deviations
- working-surface-deviation contributions causing user-identified rotational-harmonic spectrum contributions
- explanation of rotational-harmonic contributions to kinematic-transmission-error spectra
- computation of normalized mesh stiffness and its reciprocal
- computation of mesh-attenuation functions
- sign conventions to distinguish right-hand versus left-hand helices and driver versus follower gears
- computation of kinematic transmission error as a function of roll distance
- computation of rms kinematic-transmission-error spectrum.

Performance-Based Gear Metrology: Kinematic-Transmission-Error Computation and Diagnosis,
First Edition. William D. Mark.
© 2013 John Wiley & Sons, Ltd. Published 2013 by John Wiley & Sons, Ltd.

8.1 Tooth-Working-Surface Measurements

Present day dedicated gear-metrology equipment can carry out line-scanning measurements in the axial direction and in the radial direction on tooth-working-surfaces. As described in Section 3.6, one set of measurements, either profile (radial) or lead (axial), is to be chosen as the primary set. Measurement data is obtainable effectively continuously *along* the line-scanning measurements, but must be obtained by an interpolation procedure *across* the line-scanning measurements of the chosen primary set, either profile or lead.

Two methods have been described for obtaining the required number of line-scanning measurements in the primary set. If no "ghost-tone" harmonics are believed to exist, the method described in Appendix 3.A based on capturing all working-surface-deviations with amplitudes larger than the *rms* surface roughness is likely the most suitable method. If ghost tones are believed to exist, or are possible, the method described in Appendix 3.B is the more suitable, where the largest rotational-harmonic-number n of a possible ghost tone must be specified. The methods described in Appendices 3.A and 3.B each suggest how to determine whether profile measurements or lead measurements should be chosen as the primary set in order to minimize the measurement effort and duration. Methods to determine the required number of primary line-scanning measurements are explained in these appendices. It is suggested that a secondary set of perhaps five line-scanning measurements of the non-primary set also be used, primarily to capture surface-roughness characteristics in the non-primary direction. In order to predict rotational-harmonic amplitudes of working-surface-deviations, and of resultant transmission-error amplitudes, all teeth must be measured.

In making either lead measurements or profile measurements, the direction of measurement probe displacement should be that defined by the intersection of the plane of contact and transverse plane. Because the plane of the paper in the lower sketch in Figure 2.6 is the transverse plane, this desired direction of probe displacement coincides with the direction of the plane of contact in the lower sketch there. For spur gear measurement, this direction is normal to the tooth-working-surfaces. For a helical gear, this direction is not normal to the working-surfaces, as can be seen from the upper sketch in Figure 2.6, but is offset from the normal direction by an angle equal to the base-cylinder helix angle ψ_b. Therefore, if the probe measurement direction is taken normal to the tooth of a helical gear, that measurement value should be divided by $cos\ \psi_b$ to obtain its component in the direction defined by the intersection of the plane of contact and transverse plane.

It is assumed that scanning lead and profile measurements are absolute measurements measuring deviations from the equispacd perfect involute working-surfaces described in Chapter 2. Thus, the *consistency* of such non-smoothed scanning measurements can be checked by computing the *rms* differences between scanning lead and profile measurements made on the same tooth at their intersections. A computed *rms* difference between non-smoothed measurements at their intersections significantly larger than $\sqrt{2}\sigma$, where σ is *rms* surface roughness, would be an indication of measurement inconsistency. Successful implementation of the methods described in this book requires measurement consistency.

Analysis Region and Measurement Locations

A rectangular *analysis region* of axial facewidth F and radial tooth depth D must be chosen by the analyst. This rectangular analysis region is the analyst's best estimate of the actual tooth contact region that the subject gear would experience when meshing with a mating gear. Consequently, the chosen analysis region of axial width F and radial depth D would be a rectangular region likely inscribed within the full region illustrated in Figure 3.4. The analysis facewidth F is measured by the axial coordinate y, and the analysis tooth-depth D is measured by roll distance $R_b \epsilon$ using Equation (3.7), where subscripts t and r designate the tip and root limits of D, and φ is transverse pressure angle.

In Equation (3.7), roll-angle ϵ is "measured" in radians. To obtain radian measure from degree measure, multiply degree measure by $2\pi/360$. Base-circle radius R_b is related to pitch-circle radius R_p and transverse pressure angle φ by Equation (2.3).

In choosing the axial dimension F of the analysis region, the analysis region should be kept inside of any axial end chamfers that might be present. In choosing the upper roll angle ϵ_t near the tooth tip, roll angle ϵ_t should be chosen "below" any tip chamfer that might be present. The roll-angle value ϵ_r near the tooth root should not be taken smaller than the value at the Start of Active Profile (SAP) (Drago, 1988, pp. 59, 60) and probably should be chosen at a slightly larger value. This root roll-angle value cannot be negative. If a mating gear to the subject gear is known, its design can be used to estimate the SAP on the subject measured gear, and also possibly, the appropriate axial endpoints of the analysis region on the subject gear.

The axial positions y_α of line-scanning profile measurements must be located at the zeros of a Legendre polynomial of degree equal to the number of line-scanning profile measurements to be taken. Because Legendre polynomials, Equation (3.8), are defined over the interval $-1 \leq \xi \leq 1$, and their zero locations are computed for this interval, whereas, our axial analysis region is $(-F/2) \leq y \leq (F/2)$, to convert the Legendre polynomial zero locations to our profile axial measurement locations, the Legendre zero locations must be multiplied by $F/2$. In a completely analogous manner, the radial positions z_β of line-scanning lead measurements must be converted from the locations of Legendre polynomial zeros by multiplying these zero locations by $D/2$.

Estimating Required Number of Primary Line-Scanning Measurements

At this juncture, the method outlined in either Appendix 3.A or 3.B for estimating the number of line-scanning measurements of the primary measurement set, either profile or lead measurements, must be chosen.

The surface-roughness-criteria method described in Appendix 3.A requires three lead measurements to be made on a typical tooth at Legendre zeros z_β and three profile measurements made at Legendre zeros y_α. The locations of these zeros defined on the interval $-1 \leq \xi \leq 1$ are (Hildebrand, 1974, p. 391), 0 and $\pm\sqrt{15}/5 = \pm0.774596669$. The Legendre expansion coefficients required in the expressions (3.A.3) and (3.A.4) are to be computed by Equations (3.32) and (3.41), respectively. Also, required in expressions (3.A.3) and (3.A.4) are integrations of the squared working-surface-deviations on the left-hand sides of these expressions. Unsmoothed metrology measurements should be used in carrying out all line-scanning measurements, and because such measurements are digitized and include surface roughness, the best results should be obtained using the trapezoidal rule of integration (Lanczos, 1956, pp. 380, 381). The *same* sample points and (trapezoidal) integration rule must be used in the integrations in Equations (3.32) and (3.A.3); and likewise for the integrations in Equations (3.41) and (3.A.4). Once all required integrations are carried out and the quantities described by the expressions (3.A.3) and (3.A.4) are tabulated or displayed as functions of K and L', respectively, for each of the six line-scanning measurement locations, the procedure for choosing the primary measurement set, either profile measurements or lead measurements, and the required number of those measurements in the primary set, is as described in Appendix 3.A.

The ghost-tone rotational-harmonic-number criterion for choosing the primary measurement set and required number of line-scanning measurements in that set is described in Appendix 3.B. This method requires the analyst to specify the rotational-harmonic-number n of the largest ghost-tone harmonic of potential interest. Once this rotational-harmonic-number n is chosen, it is required to compute the axial Q_a and transverse Q_t contact ratios, which are uniquely determined by the above-described rectangular analysis region.

The axial contact ratio is *defined* by Equation (2.10). Using Equation (2.11), it is readily computed by

$$Q_a = \frac{F \tan \psi_b}{\Delta} = \frac{F \cos \phi \tan \psi}{\Delta} \tag{8.1}$$

where Equation (2.8b) was used in obtaining the second expression. F is the axial facewidth of the *analysis region* illustrated in Figures 2.6 and 3.4, Δ is the base pitch

$$\Delta = 2\pi R_b / N, \tag{8.2}$$

ψ_b is base-cylinder helix angle, φ is transverse pressure angle, and ψ is pitch-cylinder helix angle.

The transverse contact ratio is *defined* by Equation (2.9), where L is the "roll-distance" length in the transverse plane of the *analysis* contact region illustrated in the upper portion of Figure 2.6. Using the notation of Equations (3.4–3.7), ϵ_t is the involute roll angle of the upper location of the analysis region near the tooth tip and ϵ_r is the involute roll angle of the lower location of the analysis region near the tooth root. Thus, using radian measure of roll angles,

$$L = R_b \left(\epsilon_t - \epsilon_r \right), \tag{8.3}$$

and from Equation (2.9),

$$Q_t \triangleq \frac{L}{\Delta} = \frac{R_b \left(\epsilon_t - \epsilon_r \right)}{\Delta}. \tag{8.4}$$

But, from Equation (8.2), $\Delta = 2\pi R_b / N$, and therefore, from Equation (8.4),

$$Q_t = \frac{R_b \left(\epsilon_t - \epsilon_r \right)}{2\pi R_b / N} = \frac{\epsilon_t - \epsilon_r}{2\pi / N} \tag{8.5}$$

where roll angles are "measured" in radians. Equivalently (Lynwander, 1983, p.42), then,

$$Q_t = \frac{\text{roll angle span in degrees}}{360/N}, \tag{8.6}$$

where this roll angle span is the span of the analysis region "measured" in degrees.

According to the discussion accompanying Equations (3.B.3) and (3.B.6), if $Q_t > Q_a$, there are more radial undulation cycles D/λ_ℓ than axial undulation cycles F/λ_k, and therefore in this case, choice of profile scanning measurements as the primary set would be chosen to minimize measurement effort, thereby requiring interpolation *across* these profile measurements in an axial direction where there are fewer undulation cycles. Conversely, if $Q_a > Q_t$ lead scanning measurements would be chosen as the primary set, requiring interpolation in a radial direction *across* these lead measurements.

The above-described numbers of undulation cycles, D/λ_ℓ and F/λ_k, computed by Equations (3.B.6) and (3.B.3), respectively, will almost never be integers. Let the number of undulation cycles in the transverse (radial) direction be denoted by m_t and the number in the axial direction by m_a. Then, from Equations (3.B.6) and (3.B.3), respectively,

$$m_t = \frac{n}{N}Q_t \qquad m_a = \frac{n}{N}Q_a, \tag{8.7a,b}$$

where n is the rotational-harmonic number of the ghost tone of interest. Then for either of Equation (8.7a) or (8.7b), denote by m the number of undulation cycles, m_t or m_a. Denote by n' the required number of line-scanning measurements to accurately interpolate *across* the primary set of line-scanning measurements, either profile measurements or lead measurements. Then, from Equation (3.B.7), the required number of line-scanning measurements made at the locations of the Legendre polynomial zeros is

$$n' = m\pi + k, \tag{8.8}$$

where m is taken as m_t, Equation (8.7a), for radial-direction interpolation across n' line-scanning lead measurements located radially at the normalized Legendre polynomial zeros z_β, and m is taken as m_a, Equation (8.7b), for axial-direction interpolation across n' line-scanning profile measurements located axially at the normalized Legendre polynomial zeros y_α. For any value of m, the value of k determined by m, is to be found from Table 3.B.1. To determine the value of k from Table 3.B.1, the value of m_t or m_a should be rounded *up* to the nearest integer m for use in the table. To determine the integer number of required line-scanning measurements, the number n' computed by Equation (8.8) is to be rounded to the *closest* integer.

As described above, computed values of the zeros of Legendre polynomials are based on their interval of definition of $-1 \leq \xi \leq 1$. Therefore, the locations of the n' Legendre zeros defined over this interval must be multiplied by $D/2$ for the case where line-scanning lead measurements are taken as the primary measurement set. Similarly, the locations of the n' Legendre zeros, defined over $-1 \leq \xi \leq 1$, must be multiplied by $F/2$ for the case where line-scanning profile measurements are taken as the primary set.

The same rules apply to the Legendre zeros of the secondary set of perhaps five line-scanning measurements. Algorithms for computing the values of the zeros of Legendre polynomials are available, for example, Press *et al.* (1999, pp. 144–151). Tables of the zeros of the Legendre polynomials also are available for all polynomial orders from 1 to 64 (Stroud and Secrest, 1966).

8.2 Computation of Two-Dimensional Legendre Expansion Coefficients

For each of the methods described in Appendices 3.A and 3.B, the choice of which measurement set is primary, either scanning profile or scanning lead, is made as the set that has the larger number of working-surface significant-amplitude fluctuations in the direction *along* the scanning measurements. This choice minimizes the number of primary line-scanning measurements that must be interpolated *across* in order to achieve an adequate representation of the working-surface-deviations over the entire rectangular analysis region.

Although we conceptually envision the required representation *across* the primary line-scanning measurements as obtained by interpolation, the manner in which it is computed, as explained in Section 3.6, almost always will provide a considerably more accurate representation of working-surface-deviations than *direct* interpolations across the line-scanning measurements would provide. Consider the case described first in Section 3.6 where lead-scanning measurements are chosen as the primary set. For each of the n lead-scanning measurements located at radial positions z_β, $\beta = 1, 2, \ldots, n$, the Legendre expansion coefficients $c_{j,k\cdot}(z_\beta)$ are computed by Equation (3.32) for all coefficients $k = 0, 1, 2, \ldots$ up to $k = M$, limited only by the density of the discrete line-scanning samples. Hence, these coefficients can be computed with great accuracy. The two-dimensional Legendre expansion coefficients $c_{j,k\ell}$, represented exactly by Equation (3.31), are computed using the approximate integration representation $c'_{j,k\ell}$, Equation (3.33), which is an integration across the line-scanning lead measurements, and is evaluated by Gaussian quadrature, Equation (3.37), since the "data-point values" $c_{j,k\cdot}(z_\beta)\psi_{z\ell}(z_\beta)$ are located at the normalized zeros z_β, $\beta = 1, 2, \ldots, n$, of a Legendre polynomial of degree n. We normally can expect the behavior of the low-order coefficients $c_{j,k\cdot}(z)$, Equation (3.30), $k = 0, 1, 2, \ldots$, to vary smoothly with z, of which the $c_{j,k\cdot}(z_\beta)$, $\beta = 1, 2, \ldots, n$ of Equation (3.32) are samples. In contrast, we normally would expect raw line-scanning lead measurements to vary considerably less smoothly with z. But we know (Cheney, 1982, pp. 109, 110; Lanczos, 1956, pp. 396–400) that the Gaussian quadrature result, Equation (3.37), yields the accuracy achievable by exact integration of a polynomial representation of $c_{j,k\cdot}(z)$ of order $2n - 1$. Hence, we can expect, especially the lower-order coefficients $c'_{j,k\ell}$, $k = 0, 1, 2, \ldots$ to be computed very accurately by Equation (3.37).

It is true that the Gaussian quadrature formula also interpolates, exactly, the coefficients $c_{j,k\cdot}(z_\beta)$ across the locations of the line-scanning lead measurements, and from an interpolation viewpoint the choice of the Legendre zeros, z_β, $\beta = 1, 2, \ldots, n$ is optimum (Hildebrand, 1974, pp. 466–468). But this interpretation obscures interpreting the coefficients $c'_{j,k\ell}$, derived in Chapter 3, as the two-dimensional Legendre expansion coefficients, Equation (3.20). For example, if the continuous sine wave in the lower Figure 3.B.2b were not present, one would be hard pressed to envision a *smooth* interpolation through the sample points shown there, yet as exhibited by the upper Figure 3.B.2a, the Legendre polynomial expansion of the sinusoid shown in the upper figure, whose expansion coefficients were evaluated only from the discrete samples shown in the lower Figure 3.B.2b by Gaussian quadrature, reconstructed, virtually exactly, the sinusoid that the sample points were taken from. Hence, representation by Legendre polynomial expansion, rather than interpolation, is the preferred interpretation.

The case described second in Section 3.6, where profile-scanning measurements are chosen as the primary set, is completely analogous to the above-described case where lead-scanning measurements are chosen as primary. In the case of primary profile-scanning measurements, at each of the axial normalized Legendre zeros y_α $\alpha = 1, 2, \ldots, m$ where the m profile measurements are located, the Legendre coefficients $c_{j,\cdot\ell}(y_\alpha)$, Equation (3.41), are computed for $\ell = 0, 1, 2, \ldots, N'$ up to a value of N' limited only by the density of measurement sample points of the profile measurements. The two-dimensional Legendre expansion coefficients $c_{j,k\ell}$, Equation (3.40), represented approximately by Equation (3.42), are computed by the Gaussian quadrature formula, Equation (3.43).

Legendre expansion coefficients of the non-primary line-scanning measurements are computed in exactly the same manner, as described above.

Metrics Assessing Coincidence of Base-Cylinder Axis and Gear-Measurement Axis

In principle, the base-cylinder axis is the axis of rotation of a gear when the involute teeth are cut. If the axis of rotation of the gear during its measurement differs from this base-cylinder axis, apparent working-surface errors will be measured, caused by the non-coincidence of base-cylinder axis and axis-of-rotation in gear measurement. If these two axes are parallel but do not coincide when a gear is measured, an apparent once-per-revolution sinusoidal accumulated tooth-spacing (index) error will be present in the measurements. This index error for a tooth j is described by the Legendre coefficients $k = 0, \ell = 0$ in Table 3.1, and by Equation (3.26), where, as shown by Equation (3.24), the Legendre term $c'_{j,00}$ describes the average value over the working-surface of tooth j of its position. In addition, if the two axes are parallel, but do not coincide, an apparent once-per-revolution sinusoidal variation in linear profile deviation also will be present in the measurements. This behavior is described by the Legendre term $k = 0$, $\ell = 1$ in Table 3.1. Hence in this case, a once-per-revolution sinusoidal variation of the Legendre term $c_{j,01}$ is to be expected. *For a helical gear,* when the two axes are offset, but parallel, an apparent *small* once-per-revolution variation in linear lead deviation also will be present, which is the Legendre term $k = 1, \ell = 0$ in Table 3.1.

Additional apparent working-surface errors will be present if the above-described two axes are out of parallel. In this case, there will exist an apparent linear profile error $\ell = 1$ that varies linearly axially. This behavior is described by the Legendre term $k = 1$,

$\ell=1$ in Table 3.1. The resultant Legendre term $c_{j,11}$ $(k=1, \ell=1)$, will have a once-per-revolution sinusoidal variation. Moreover, in this non-parallel axis case, there also will be a stronger apparent once-per-revolution sinusoidal variation in the linear lead term $k=1, \ell=0$ of Table 3.1, which is described by the Legendre coefficient $c_{j,10}$.

Thus, the first four pairs of Legendre terms listed in Table 3.1 are useful for diagnosing the coincidence of the base-cylinder axis used in tooth generation and the axis of rotation used in gear measurement. For any $k\ell$ pair, plots of these expansion coefficients $c_{j,k\ell}$ as a function of tooth number $j=0,1,2, \ldots, N-1$, can be generated. In particular, a plot of $c_{j,00}$ for $j=0,1,2, \ldots, N-1$, representing accumulated tooth-spacing (index) errors, will provide an accurate representation of such errors, because, as indicated by Equations (3.24–3.26), $c_{j,00}$ represents the deviation of the *average* working-surface location, as opposed to the deviation obtained by a *point* measurement.

It follows directly from Equation (3.23) and the orthogonal property of the two-dimensional normalized Legendre expansion functions $\psi_{yk}(y)\psi_{k\ell}(z)$ that the absolute value $|c_{j,k\ell}|$ of each individual expansion coefficient is the *rms* value of the contribution of that term $k\ell$ in Equation (3.18) to the deviation $n_{Cj}(y,z)$ on the tooth number j, where this *rms* value of the term $k\ell$ is obtained from an average formed over the rectangular analysis region $(-F/2) \le y \le (F/2), (-D/2) \le z \le (D/2)$.

As described above, the once-per-revolution sinusoidal contributions of the first four $k\ell$ pairs of the expansion coefficients $c_{j,k\ell}$ listed in Table 3.1 are useful in assessing the coincidence of the base-cylinder axis used in cutting the teeth and the gear-measurement axis of rotation. It follows directly from Equations (4.26) to (4.30) and the accompanying discussion that the *rms* value of the once-per-revolution sinusoidal contribution of the term $c_{j,k\ell} \psi_{yk}(y)\psi_{k\ell}(z)$ in Equation (3.18), for a generic $k\ell$ pair, is given by

$$rms\ rotational\ fundamental\ contribution = \sqrt{2}\,|B_{k\ell}(1)|, \qquad (8.9)$$

where $B_{k\ell}(n)$ is computed from the sequence of coefficients $c_{j,k\ell}, j=0,1, \ldots, N-1$ by Equation (4.21a).

8.3 Regeneration of Working-Surface-Deviations

For some applications of this technology, only the primary set of line-scanning measurements made on every tooth is required. However, if an approximation of surface roughness features in both axial and radial directions is desired, then a secondary set of line-scanning measurements made on every tooth in the non-primary direction also is required.

When both primary and secondary sets of line-scanning measurements are made, there exists rectangular regions in the k, ℓ plane, Figure 3.9, where the Legendre coefficients $c_{j,k\ell}$ from both measurement sets are obtained. Figure 3.10 illustrates the suggested region in the k, ℓ plane where the coefficients $c_{j,k\ell}$ from the primary set of measurements and those from the secondary set of measurements should be retained.

Those from the primary set might be retained on the diagonal locations separating the two regions shown in Figure 3.10.

As illustrated, for example, by Figures 3.12 and 3.13, the Legendre coefficients $c_{j,k\ell}$, obtained as described above and in Chapter 3, can be used to re-generate the working-surface-deviations of a generic tooth j by utilizing Equation (3.18) and the available expansion coefficients illustrated in Figure 3.10. The number of coefficients M or N' obtained in the Legendre expansions *along* the line-scanning measurements is limited only by the densities of the discrete measurement points available along the line-scanning measurements. As is illustrated by Figures 3.12 and 3.13, the resultant re-generations of the working-surface-deviations provide a description of the deviations at *all* locations within the analysis region $(-F/2) \leq y \leq (F/2)$, $(-D/2) \leq z \leq (D/2)$.

Normally, a gear designer will prescribe a desired modification, from the involute, of the working-surfaces, where every tooth ideally should have the same modification. By displaying the deviations of the individual working-surfaces, as described above, the designer can obtain a visual assessment of the tooth-to-tooth variability in the working-surfaces.

One possible goal in prescribing such working-surface modifications is to minimize the tooth-meshing harmonic contributions to the transmission error. The working-surface-deviation contribution to the tooth-meshing harmonics from a single gear of a meshing pair is the contribution provided by the *average* working-surface-deviation, where this average deviation surface is obtained by forming the average of the deviation surfaces over all $j = 0, 1, 2, \ldots, N-1$ working-surfaces on the gear. Because the Legendre expansion functions $\psi_{yk}(y)\psi_{z\ell}(z)$, used in a particular application, are the same for every tooth j, it follows by forming an average over $j = 0, 1, \ldots, N-1$ of Equation (3.18) that the average working-surface-deviation, Equation (3.45), can be computed from the average value, Equation (3.46), of the individual Legendre expansion coefficients, as indicated by Equation (3.47). Figure 3.14 illustrates one such average deviation surface (with little or no intentional modification). This procedure gives the designer a quantitative capability to compare the actually achieved modification of the working-surfaces with that specified by his or her design. Moreover, by subtracting, for each $k\ell$ pair, the mean coefficient $\bar{c}'_{k\ell}$, Equation (3.46), from the individual coefficient $c'_{j,k\ell}$ for any tooth j, thereby yielding the coefficient $c'_{j,k\ell} - \bar{c}'_{k\ell}$, Equation (3.18) then can be used to generate the deviations of the working-surface of any tooth j from the mean working-surface for that gear.

Useful Metrics of Working-Surface Variability

In most applications, it is desirable to modify, from a perfect involute surface, every tooth-working-surface on a gear in exactly the same way. The normalizations of the Legendre expansion functions described by Equations (3.13) and (3.14) allow the expansion coefficients to be directly interpreted, as illustrated by Equation (3.23) and explained below.

The deviation of the working-surface of a generic tooth j from the mean working-surface $\bar{\eta}_C(y, z)$, Equation (4.15), is $\epsilon_{Cj}(y, z)$, Equation (4.16). The *rms* value of these deviations, obtained as the square-root of the mean-square deviation from the mean working-surface-deviation, obtained by an average over the analysis area FD and over all N teeth, is a useful metric of manufacturing accuracy. This mean-square deviation is described as a function of the expansion coefficients by Equation (4.C.5). It follows directly from Equations (4.C.5) and (4.C.6) that the above-described *rms* variation of the N working-surfaces from the mean working-surface is

$$\text{rms variation of } \epsilon_{Cj}(y,z) = \left\{ \sum_{k=0}^{\infty} \sum_{\ell=0}^{\infty} \left[\left(\frac{1}{N} \sum_{j=0}^{N-1} c_{j,k\ell}^2 \right) - (\bar{c}_{k\ell})^2 \right] \right\}^{1/2} \qquad (8.10)$$

where, from Equation (4.C.1a),

$$\bar{c}_{k\ell} \triangleq \frac{1}{N} \sum_{j=0}^{N-1} c_{j,k\ell} \,. \qquad (8.11)$$

Accumulated tooth-spacing errors, $k=0$, $\ell=0$, Table 3.1, affect gear performance very differently from all other types of errors. Therefore, it is sensible to *exclude* the term $k=0$, $\ell=0$ in assessing the *rms* variation of $\epsilon_{Cj}(y,z)$ described by Equation (8.10).

Equation 8.10 is easily evaluated from a tabulation of the expansion coefficients $c_{j,k\ell}$. Because an *rms* error is a linear length metric, it is easy to interpret. The degree to which Equation (8.10) will include surface roughness is dependent on the number of line-scanning working-surface measurements included in the secondary set of measurements.

A metric comparable to that of Equation (8.10) can be computed to obtain the *rms* deviation of the mean working-surface, $\bar{\eta}_C(y,z)$, Equation (4.15), from a perfect involute surface. This metric is the square root of the square of $\bar{\eta}_C(y,z)$ averaged over the analysis area *FD*. It is obtained directly from the square-root of Equation (4.C.10a),

$$\text{rms deviation of } \bar{\eta}_C(y,z) = \left\{ \sum_{\substack{k=0 \\ except \\ k=0 \ \ell=0}}^{\infty} \sum_{\ell=0}^{\infty} (\bar{c}_{k\ell})^2 \right\}^{1/2} \qquad (8.12)$$

where $(\bar{c}_{k\ell})$ is given by Equation (8.11) and the single term $k=0$, $\ell=0$ is to be excluded in Equation (8.12).

For intentionally modified working-surfaces, Equation (8.12) provides an *rms* measure of the actual mean modification that was achieved. However, Equation (8.10) provides an *rms* metric of the tooth-to-tooth variability in the actual working-surfaces. If that variability is a small fraction of the intentional modification, then most of the teeth would exhibit approximately the same modification. Hence, the ratio of Equations (8.10) to (8.12), *excluding in both equations the single term $k=0$, $\ell=0$*, is an easily computed metric describing the degree to which all working-surfaces have approximately the same modification.

8.4 Rotational-Harmonic Decomposition of Working-Surface-Deviations

It was shown by Equation (4.26) that the mean-square working-surface-deviations of all teeth on a gear can be decomposed into a rotational-harmonic spectrum which is periodic in rotational-harmonic number n, the period being the number N of teeth, which is the rotational-harmonic spacing between tooth-meshing-harmonic locations. The resultant one-sided mean-square rotational-harmonic spectrum $G_n(n)$ is defined by Equations (4.30a, b). It is computed from the squared absolute values of the discrete Fourier transforms, Equation (4.21a), of the Legendre expansion coefficients $c_{j,k\ell}$ of the working-surface-deviations, where these squared absolute values $|B_{k\ell}(n)|^2$ are computed by Equation (4.B.3). Because linear length measures are easier to interpret than their squared values, the square-root $[G_n(n)]^{1/2}$ of the rotational-harmonic spectrum of the working-surface-deviations of the teeth on the pinion in Figure 3.11 are shown in Figure 4.1. All rotational-harmonic contributions in Figure 4.1 except those at integer multiples of $N = 38$, the number of teeth, are caused by tooth-to-tooth working-surface variations.

Working-Surface-Deviations Causing Specific Rotational Harmonics

The working-surface error pattern on a specified tooth j, caused by a specific rotational harmonic contribution n of the working-surface-deviations, other than a tooth-meshing harmonic $n = pN, p = 0, 1, 2, \ldots$, can be computed by Equation (4.42). Because it should be possible for a user of this technology to relate such an error pattern on one or several teeth to the kinematic properties of gear cutting, Equation (4.42) provides a powerful diagnostic tool for identifying the manufacturing source of any undesirable rotational harmonic, such as a "ghost tone." Figures 4.2, 4.3, and 4.5, illustrate computation of working-surface-deviation patterns that are the cause of rotational harmonics $n = 1, 19$, and 20 of the *rms* spectrum shown in Figure 4.1.

Equation (4.46a), which is equivalent to Equation (3.47), computes the working-surface-deviation causing the contributions to the tooth-meshing harmonics of the tooth-deviation spectrum. The rotational-harmonic spectrum $G_n(n)$, Equation (4.30), describes the mean-square rotational-harmonic contributions of working-surface-deviations before the attenuation of these deviations, resulting from the meshing action with a mating gear, takes place.

8.5 Explanation of Attenuation Caused by Gear Meshing Action

The material in Chapters 3 and 4 has dealt with geometric working-surface-deviations, and their rotational-harmonic spectra, without consideration of the attenuating effects on these deviations brought about by the meshing action with a mating gear. These attenuating effects on the working-surface-deviations generally reduce the transmission-error contributions arising from the working-surface-deviations. Because the "exact" derivation of these attenuating effects, found in Chapter 7, is long and detailed, it was decided to provide a physical explanation of these effects earlier in Chapter 5. This physical explanation is carried out with the aid of Equations (5.1)–(5.13) leading to the relatively easily understood explanation following Equation (5.13).

Because the attenuating effects on working-surface-deviations provided by gear-meshing action is well modeled as a linear system, and the working-surface-deviations were expressed as the superposition, Equation (3.18), of two-dimensional normalized Legendre polynomials, expressed in frequency domain by Equation (4.22a), the Fourier series coefficients of the resultant transmission-error contributions were explained to be given by Equation (5.16), where the attenuation from gear-meshing action on a generic two-dimensional Legendre term $k\ell$ is characterized by the "mesh-attenuation function" $\phi_{k\ell}(n/N)$, as can be seen by a comparison of Equations (4.22a) and (5.16). (The derivation of Equation (5.16) is carried out in Chapter 7, leading to Equation (7.95).) The resultant "one-sided" mean-square rotational-harmonic spectrum of the working-surface-deviation contributions to the transmission error is given by Equations (5.18a, b).

8.6 Diagnosing and Understanding Manufacturing-Deviation Contributions to Transmission-Error Spectra

The product $B_{k\ell}(n)\phi_{k\ell}(n/N)$ of the individual Legendre terms $k\ell$ in the expression, Equation (5.16), for the rotational harmonic Fourier series coefficients, and in the mean-square spectra, Equation (5.18b), enables one to potentially diagnose and understand the transmission-error rotational harmonic spectrum contributions arising from the various two-dimensional Legendre terms $k\ell$, Equation (3.18), of working-surface-deviations. To facilitate this capability, the approximate mesh-attenuation function, Equation (6.3), derived in Equation (7.129), is used in Chapter 6 to assess the low-order rotational harmonic and so-called "sideband" contributions arising from accumulated tooth-spacing errors, $k = 0$, $\ell = 0$, and higher-order Legendre contributions k and ℓ not both zero. These results are summarized in the italicized explanations following Equation (6.49). Log-log plots of the mesh-attenuation function approximation, Equation (6.3), for various $k\ell$ pairs, are shown in Figure 6.3, which complement the analytical approximations to the mesh attenuation functions used to describe the attenuating effects on the tooth-deviation spectrum $B_{k\ell}(n)$ provided by the meshing action with a mating gear.

It can be seen by a comparison of Equations (4.30b) and (5.18b) that it is the aggregate of all terms, $k = 0, 1, 2, \ldots; \ell = 0, 1, 2, \ldots$ in the kinematic transmission-error spectrum $G_\zeta(n)$, Equation (5.18b), that determines the attenuation of the individual rotational harmonics n of the mean-square tooth-deviation spectrum $G_\eta(n)$, Equation (4.30b). Comparison of the *rms* spectra in Figures 5.1a and 5.1b shows strong attenuation of all rotational harmonics in Figure 5.1a except those at rotational harmonics $n = 1$, $n = 2$, and $n = 20$. As described above, rotational harmonics $n = 1$ and $n = 2$ are caused by accumulated tooth-spacing errors which experience negligible attenuation, as illustrated by Figure 6.3a. The remaining harmonic $n = 20$ that experienced almost no attenuation was seen in Figure 4.5 to be caused by a relatively long wavelength sinusoidal-like undulation error. Harmonic $n = 20$ generally is called a ghost tone or phantom tone.

As seen by this comparison of Figures 5.1a and 5.1b, and the several illustrations in Chapter 6, the Legendre polynomial representation of working-surface-deviations is fully capable of representing sinusoidal-like undulation deviations (and any others), but in order to *understand* how such lack of attenuation takes place, it was necessary to utilize two-dimensional sinusoids, as illustrated by Figure 6.4. Equations (6.64a, b) and (6.65a, b), and the accompanying discussion, describe the working-surface-deviation requirements for no attenuation of working-surface undulation errors to take place from gear-meshing action. Such undulation errors with amplitudes significantly smaller than a micron can lead to unacceptable ghost-tone amplitudes.

8.7 Computation of Mesh-Attenuated Kinematic-Transmission-Error Contributions

Within the hypothesis that tooth-pair contact of meshing gear pairs takes place over the rectangular contact region $(-F/2) \leq y \leq (F/2)$ and $(-D/2) \leq z \leq (D/2)$ illustrated in Figures 2.6 and 3.2, an "exact" method for computing the transmission-error contributions from the working-surface-deviations of a generic parallel-axis helical gear was derived in Chapter 7. For any parallel-axis helical (or spur) gear characterized by its base-cylinder radius R_b, base-helix angle ψ_b, and number of teeth N, it follows from Equations (8.1–8.4) that the axial Q_a and transverse Q_t contact ratios uniquely describe this contact region.

It was necessary to use complex analysis to obtain the Fourier series representation, Equations (7.174a, b) and (7.177) of this transmission-error contribution. But in order to facilitate software implementation of the means to compute this Fourier series representation, and the mean-square rotational harmonic spectrum of these transmission-error contributions, the results of this complex analysis were reduced to computations involving only real quantities in Chapter 7. Outlined below is the sequence of computations required to obtain this Fourier series representation and rotational harmonic spectrum.

Computation of Normalized Mesh Stiffness

As pointed out at the beginning of Section 7.5, evaluation of the mesh-attenuation function, Equation (7.94), requires evaluation of the Fourier-series coefficients of the reciprocal normalized mesh stiffness $\bar{K}_M/K_M(s)$. These coefficients are obtained from the Fourier-series coefficients of the normalized mesh stiffness $K_M(s)/\bar{K}_M$, Equations (7.113), (7.150a, b), and (7.151a, b). For the general case where the tooth-pair stiffness per unit length of line of contact $K_{TC}(y,z)$ is *not* assumed to be constant, the complex Fourier-series coefficients of $K_M(s)/\bar{K}_M$ are given by Equations (7.113) and (7.110). The real Fourier-series coefficients A_p and B_p, Equations (7.151a,b), of $K_M(s)/\bar{K}_M$ are given by Equations (7.168a, b) in terms of $a_{k\ell}$, Equation (7.110), with $n = p$, the tooth-meshing harmonic number.

In the practically important case where the tooth-pair stiffness per unit length of line of contact is assumed to be constant, the Fourier-series coefficients, Equations (7.150a, b) of the normalized mesh stiffness $K_M(s)/\bar{K}_M$ are real, that is, $B_p = 0$, and are given by Equations (7.117a, b) and (7.169a, b), which are functions only of the axial Q_a and transverse Q_t contact ratios, and of tooth-meshing harmonic number $p = n$. In particular, these Fourier-series coefficients are independent of the assumed constant value of the tooth-pair stiffness per unit length of line of contact, as pointed out earlier in Equation (5.13).

Computation of Reciprocal Normalized Mesh Stiffness

Two methods were described in Section 7.7 for computing the Fourier-series coefficients of the reciprocal normalized mesh stiffness $\bar{K}_M/K_M(s)$ from the Fourier-series coefficients of $K_M(s)/\bar{K}_M$. The power-series method utilizing Equations (7.120a, b) was used to develop the approximation, Equation (7.129), of the mesh-attenuation functions, and of Equations (7.137–7.144), but was not developed in complete detail. However, especially for the case of Equations (7.117a, b) of constant tooth-pair stiffness per unit length of line of contact, this power-series method described by Equations (7.117a, b) and (7.139–7.141) should be useful when employed using software capable of carrying out the discrete convolutions of Equations (7.140) and (7.141). See Equations (7.C.22a, b), and for the general complex case, Equations (7.194–7.196a, b). As pointed out in connection with Equations (7.145) and (7.146), the power-series method has the advantage of allowing the user to choose the number of terms required to limit the truncation error in its use.

Because software is readily available for solving simultaneous linear algebraic equations, the linear equation method of Section 7.7 is likely the easier of the two methods to implement for accurate computation of the Fourier-series coefficients of the reciprocal normalized mesh stiffness. For the general case of non-constant tooth-pair stiffness per unit length of line of contact, these simultaneous equations are described by Equations (7.159) and (7.160), where the coefficients A_p and B_p, with $p = m, n + m$, and $n - m$, are the Fourier-series coefficients, Equations (7.151a, b), (7.162a, b), and (7.163a, b) of the normalized mesh stiffness $K_M(s)/\bar{K}_M$, described above. The solution $a_n, b_n, n = 0, 1, 2, \ldots$ of the simultaneous Equations (7.159) and (7.160) are the real Fourier-series coefficients, Equations (7.149a, b), $n = p$, of the reciprocal normalized mesh stiffness.

The set of simultaneous Equations (7.159) and (7.160), $m = 0, 1, 2, \ldots, L'$ are "exact" only as $L' \to \infty$. But harmonic numbers n of the coefficients and of the solutions a_n and b_n are tooth-meshing harmonics. We seldom are interested in frequencies beyond about the 10th tooth-meshing harmonic, and usually less. Consequently, because the coefficients A_n and B_n in Equations (7.159) and (7.160), which are A_p and B_p, $p = n$, the normalized mesh-stiffness Fourier-series coefficients, Equations (7.151a, b), decay rapidly as p and n increase, we can expect good accuracy to be achieved in the solutions of Equations (7.159) and (7.160) for finite large values of L', say L' equal to, perhaps, 10 times the largest tooth-meshing harmonic in the rotational harmonic range of interest. Solving Equations (7.159) and (7.160), for two or more such large values of L', and comparing the solutions a_n and b_n for values of n within the harmonic number range of interest, will provide a good idea of the achieved accuracy.

In the practically important case where the tooth-pair stiffness per unit length of line of contact is assumed to be constant, the set of general simultaneous equations, Equations (7.159) and (7.160), reduces to the single set, Equation (7.172a), which is written out in detail in Equation (7.172b), where the above-mentioned decaying behavior of the coefficients is discussed in more detail. In this practically important case, the solutions b_n of the set, Equations (7.159) and (7.160), are zero, $b_n = 0$, Equation (7.171). The solutions a_p of the set, Equations (7.172a,b), yield the real Fourier-series coefficients, Equation (7.149a), of the normalized reciprocal mesh stiffness. In this practically important case, the coefficients A_n of the simultaneous Equations (7.172a,b) are given by Equations (7.169a,b).

Computation of Mesh-Attenuation Functions

The complex form of the mesh-attenuation functions, $\phi_{k\ell}(n/N)$, is given by Equation (7.189) in terms of the Fourier-series coefficients, α'_p, Equations (7.148), (7.149a,b), and (7.190), of the reciprocal normalized mesh stiffness and the functions $\hat{\psi}'_{k\ell}\left(\frac{n}{N} - p\right)$ given by Equations (7.191) and (7.192a,b). The real and imaginary parts of the mesh-attenuation functions are described by Equation (7.182), where the real part is an even function of n/N, and the imaginary part is an odd function of n/N. These even and odd contributions are given by Equations (7.198a,b), respectively, in terms of the above-mentioned even and odd parts, Equation (7.190), of the Fourier-series coefficients of the reciprocal normalized mesh stiffness and the even and odd parts, Equations (7.192a,b), of the functions $\hat{\psi}'_{k\ell}(g)$. These $\hat{\psi}'_{k\ell}(g)$ functions are to be evaluated from the spherical Bessel functions $j_k(\pi Q_a g)$ and $j_\ell(\pi Q_t g)$ found in Equations (7.192a,b), for example, Press *et al.* (1999, p. 245). Evaluation of the Fourier-series coefficients, $\left(\alpha'_p\right)_e = a_p$ and $\left(\alpha'_p\right)_o = b_p$, of the reciprocal normalized mesh stiffness was discussed above.

In the practically important case where the tooth-pair stiffness per unit length of line of contact is assumed to be constant, the coefficients A_n of the set of simultaneous Equations (7.172a,b) are given by Equations (7.169a,b), which are functions only of tooth-meshing harmonic number n and the axial Q_a and transverse Q_t contact ratios. Consequently, the solutions a_n of this set of Equations (7.172a,b) also are functions only of these same quantities, n, Q_a, and Q_t. These solutions a_n are the even Fourier-series coefficients $(\alpha_p')_e$, $n = p$, of the reciprocal normalized mesh stiffness, Equation (7.149a). Moreover, the quantities $\hat{\psi}'_{k\ell}(g)_e$ and $\hat{\psi}'_{k\ell}(g)_o$, Equations (7.192a,b), are functions only of k, ℓ, g, and Q_a and Q_t. It then follows from Equations (7.198a,b) that the mesh-attenuation functions, Equations (7.182) and (7.198a,b), for each Legendre $k\ell$ pair, are functions only of n/N, and the axial Q_a and transverse Q_t contact ratios which describe the rectangular analysis region. The ratio n/N is rotational-harmonic number divided by the number of teeth (where $(n/N) = 1$ at the tooth-meshing fundamental location and $(n/N) = 2$ at twice tooth-meshing fundamental, and so on). Furthermore, in this practically important case $\left(\alpha'_p\right)_o = b_p = 0$; hence, only the first line in each of Equations (7.198a,b) is non-zero.

Required Algebraic Signs to Distinguish Driving Gear versus Following Gear and Right-Hand Helix versus Left-Hand Helix

Apart from the discussion in Appendix 7.B and Table 7.1, the means to distinguish driving gear versus following gear and right-hand helix versus left-hand helix of parallel-axis external helical gears has not been addressed. *In each application of this technology, the correct choice among these four possibilities must be made in order to obtain an accurate computation of the transmission-error contribution from working-surface-deviations, as explained below.*

Correct use of the sign convention of working-surface measurements is absolutely essential. That sign convention is defined in Figures 3.2 and 3.4. As one faces the working-surface of a tooth, as in Figures 3.2 and 3.4, radial-coordinate z increases from root to tip, and axial-coordinate y increases from left to right. If tooth measurements are not stored using this convention they must be corrected to be consistent with this convention. Moreover, increasing tooth numbers $j = 0, 1, 2, \ldots$ of the tooth measurements must coincide with the direction of interaction of the teeth with a mating gear, which is the direction of increasing x in Equation (3.1).

We explain below a very simple method to distinguish the above-mentioned four different gear configurations in the required computations. For each of the four configurations given in Table 7.B.1, we first solve for roll-distance coordinate s, yielding

$$RHD \qquad s = y \tan \psi_b + z / \sin \phi \qquad \qquad (8.13\text{a})$$

$$LHD \qquad s = -y \tan \psi_b + z / \sin \phi \qquad \qquad (8.13\text{b})$$

$$RHF \qquad s = y \tan \psi_b - z / \sin \phi \qquad \qquad (8.13\text{c})$$

$$LHF \qquad s = -y \tan \psi_b - z / \sin \phi \qquad \qquad (8.13\text{d})$$

where RH and LH refer to right-hand and left-hand helices, and D and F refer to driver and follower. To distinguish the four gear configurations, we shall utilize the fact that a change in the sign of the independent variable in carrying out a Fourier transform changes the sign of the frequency variable, for example, Poularikas (1999, p. 3), as is easily shown.

The key Fourier transform with respect to roll-distance s of the kinematic transmission error contribution is described by Equations (7.41) and (7.42). Using the transformation, Equations (7.18–7.21a, b), this Fourier transform with respect to s was changed to Fourier transforms with respect to y and z by Equations (7.61–7.63a, b), where g is the transform variable of s, Equation (7.42). From Equations (7.61–7.63a, b), we note that the coefficient of the y-axis transform variable is $\tan \psi_b$ and that of the z-axis transform variable is $\csc \varphi = 1 / \sin \varphi$, just as in Equations (8.13a–d). The result of Equations (7.61–7.63a, b) is applicable throughout all analyses in Chapter 7, for both working-surface-deviations and tooth-pair stiffness representations.

All such quantities are represented in Chapter 7 by the two-dimensional normalized Legendre polynomials, Equations (3.13) and (3.14). As illustrated by Equations (7.90a, b) and (7.91) for working-surface-deviations, and Equation (7.113) for local tooth-pair stiffnesses, Fourier transforms with respect to y and z become spherical

Table 8.1 Signs for Q_a and Q_t

	Sign of Q_a	Sign of Q_t
RHD	+	+
LHD	−	+
RHF	+	−
LHF	−	−

Bessel functions $j_k(\ldots)$ and $j_\ell(\ldots)$, respectively. In every case, in the final formulas, for example, Equations (7.113), (7.168a,b), and (7.192a,b), the coefficient of the frequency variable g, p, or n, in the axial y transform is proportional to Q_a, and in the radial z transform is proportional to Q_t. *Therefore, as in Equations (7.61–7.63a,b), the algebraic signs given by Equations (8.13a–8.13d) can be incorporated into all final computations by assigning the algebraic signs to Q_a and Q_t that are given in Table 8.1.*

In the case of an internal helical gear, the "hand of the helix" is the same as that of the mating pinion (Drago, 1988, p.484).

Computation of Kinematic Transmission Error as a Function of Roll Distance

The kinematic transmission error caused by working-surface-deviations on the subject gear is given by Equation (7.177), where x is roll-distance $R_b\theta$, Equation (3.1), of the subject gear. The real Fourier-series coefficients α_{ne} and α_{no} in Equation (7.177) are to be computed by Equations (7.186) and (7.187), respectively, by using Equations (7.184) and (7.185), where the even and odd contributions of the mesh-attenuation functions were obtained by Equations (7.198a,b), respectively. The even and odd contributions, $B_{k\ell}(n)_e$ and $B_{k\ell}(n)_o$ of the DFT of the Legendre expansion coefficients $c_{j,k\ell}$ are given by Equations (7.180) and (7.181), where these expansion coefficients are given by Equation (7.86), or (7.89) for the case of assumed constant local tooth-pair stiffness. Hence, these DFTs characterize the working-surface-deviation rotational harmonic contributions, and the mesh-attenuation functions characterize the computed attenuation of the working-surface-deviations that would be caused by the meshing action of the subject gear with a mating gear.

Computation of One-Sided rms Kinematic-Transmission-Error Spectrum

Equation (7.199) yields the one-sided rms kinematic-transmission-error rotational harmonic spectrum caused by working-surface-deviations on the subject gear. It is computed from the Fourier series coefficients given by Equations (7.186) and (7.187).

References

Cheney, E.W. (1982) *Introduction to Approximation Theory*, 2nd edn, Chelsea Publishing Company, New York.

Drago, R.J. (1988) *Fundamentals of Gear Design*, Butterworths, Boston, MA.

Hildebrand, F.B. (1974) *Introduction to Numerical Analysis*, 2nd edn, McGraw-Hill, New York. Republished by Dover, New York.

Lanczos, C. (1956) *Applied Analysis*, Prentice-Hall, Englewood Cliffs, NJ. Republished by Dover, New York.

Lynwander, P. (1983) *Gear Drive Systems: Design and Application*, Marcel Dekker, New York.

Poularikas, A.D. (1999) *The Handbook of Formulas and Tables for Signal Processing*, CRC Press, Boca Raton, FL.

Press, W.H., Teukolsky, S.A., Vetterling, W.T., and Flannery, B.P. (1999) *Numerical Recipes in Fortran 77: The Art of Scientific Computing*, 2nd edn, Vol. 1 of Fortran Numerical Recipes, Cambridge University Press, Cambridge.

Stroud, A.H. and Secrest, D. (1966) *Gaussian Quadrature Formulas*, Prentice-Hall, Englewood Cliffs, NJ.

Subject Index

accumulated tooth-spacing error, 42–44, 121

accuracy
  computational, 7–9, 45–49, 159, 200–203, 209, 210
  measurement, 15–19, 45, 46, 158–161
addendum cylinder, 27
aliasing, 78, 129, 144, 159
analysis region, 12, 34, 243
average deviation surface, 57, 58, 156
averaging
  statistical, 19, 161
axial contact ratio, 28, 29, 62, 101, 114, 119, 135, 199, 244, 245, 257
axial coordinate, 34–36
axial pitch, 28, 29
axial working-surface location, 34–36

base circle, 23
  circumference, 25
  pitch, 25
  radius, 25
base cylinder, 21, 23, 25, 35
  axis, 38, 241, 247, 248
  helix angle, 24
  radius, 25
base plane, 23
basic rack, 220–222
Bessel functions, 116, 185
  first kind, 116, 185
  spherical, 116, 117, 119, 124, 185, 207

binomial coefficients, 200, 201
binomial theorem, 200

Cauchy-Schwartz inequality, 102
circular pitch, 26
complete set, 6, 181, 183
complex frequency response function, 109
complex variable, xv, 70
computer numerically controlled (CNC), 1, 6, 45
contact lines, 5, 28, 33
contact ratio, 27, 114
  axial, 12, 27, 29, 114, 244
  face, 27
  profile, 27
  role in attenuation, 12, 101, 125, 132, 135, 136
  signs for, 256, 257
  total, 115
  transverse, 12, 27, 63, 115, 244, 245
convolution
  discrete, 201, 202, 215, 228, 236–238
  two-dimensional, 233, 235
convolution theorem, 177, 180, 201, 227, 228
  for Fourier series, 228
  for Fourier transform, 177, 180, 227
coordinate
  axial, 34, 36
  radial, 34–36
  tooth contact line, 171, 172, 220–225
  tooth working surface, 34–36, 220–225

Performance-Based Gear Metrology: Kinematic-Transmission-Error Computation and Diagnosis,
First Edition. William D. Mark.
© 2013 John Wiley & Sons, Ltd. Published 2013 by John Wiley & Sons, Ltd.

Figure Index

Bold case indicates Figure numbers.

Performance-Based Gear Metrology: Kinematic-Transmission-Error Computation and Diagnosis,
First Edition. William D. Mark.
© 2013 John Wiley & Sons, Ltd. Published 2013 by John Wiley & Sons, Ltd.

Table Index

Bold case indicates Table numbers.

Performance-Based Gear Metrology: Kinematic-Transmission-Error Computation and Diagnosis,
First Edition. William D. Mark.
© 2013 John Wiley & Sons, Ltd. Published 2013 by John Wiley & Sons, Ltd.